VOLUME FIVE HUNDRED AND SEVENTEEN

# METHODS IN
# ENZYMOLOGY

Natural Product Biosynthesis by
Microorganisms and Plants, Part C

# METHODS IN ENZYMOLOGY

*Editors-in-Chief*

**JOHN N. ABELSON and MELVIN I. SIMON**
*Division of Biology*
*California Institute of Technology*
*Pasadena, California*

*Founding Editors*

**SIDNEY P. COLOWICK and NATHAN O. KAPLAN**

VOLUME FIVE HUNDRED AND SEVENTEEN

# METHODS IN
# ENZYMOLOGY

Natural Product Biosynthesis by
Microorganisms and Plants, Part C

Edited by

**DAVID A. HOPWOOD**
*Department of Molecular Microbiology*
*John Innes Centre*
*Norwich, UK*

AMSTERDAM • BOSTON • HEIDELBERG • LONDON
NEW YORK • OXFORD • PARIS • SAN DIEGO
SAN FRANCISCO • SINGAPORE • SYDNEY • TOKYO
Academic Press is an imprint of Elsevier

Academic Press is an imprint of Elsevier
525 B Street, Suite 1900, San Diego, CA 92101-4495, USA
225 Wyman Street, Waltham, MA 02451, USA
The Boulevard, Langford Lane, Kidlington, Oxford, OX51GB, UK
32, Jamestown Road, London NW1 7BY, UK
Radarweg 29, PO Box 211, 1000 AE Amsterdam, The Netherlands

First edition 2012

For information on all Academic Press publications
visit our website at store.elsevier.com

ISBN: 978-0-12-404634-4
ISSN: 0076-6879

Printed and bound in United States of America
12   13   14   15   16      11   10   9   8   7   6   5   4   3   2   1

Working together to grow
libraries in developing countries

www.elsevier.com | www.bookaid.org | www.sabre.org

ELSEVIER        BOOK AID        Sabre Foundation
                International

# CONTENTS

## 7. Genomic Approaches for Interrogating the Biochemistry of Medicinal Plant Species     139

Elsa Góngora-Castillo, Greg Fedewa, Yunsoo Yeo, Joe Chappell, Dean DellaPenna, and C. Robin Buell

## 8. Phylogenetic Approaches to Natural Product Structure Prediction     161

Nadine Ziemert and Paul R. Jensen

# Section 3
# Heterologous Expression of Pathways

## 9. Using a Virus-Derived System to Manipulate Plant Natural Product Biosynthetic Pathways     185

Frank Sainsbury, Pooja Saxena, Katrin Geisler, Anne Osbourn, and George P. Lomonossoff

## Section 4
## Waking Up Silent Genes

# CONTRIBUTORS

**Bertrand Aigle**
Génétique et Microbiologie, UMR UL-INRA 1128, IFR110 EFABA, Université de Lorraine, Vandœuvre-lès-Nancy, France

**Mervyn J. Bibb**
Department of Molecular Microbiology, John Innes Centre, Norwich, United Kingdom

**Elizabeth Bradshaw**
Department of Molecular Microbiology, John Innes Centre, Norwich, United Kingdom

**Sean F. Brady**
Laboratory of Genetically Encoded Small Molecules, Howard Hughes Medical Institute, The Rockefeller University, New York, USA

**Axel A. Brakhage**
Department of Molecular and Applied Microbiology, Leibniz Institute for Natural Product Research and Infection Biology – Hans Knöll Institute, and Institute of Microbiology, Friedrich Schiller University, Jena, Germany

**C. Robin Buell**
Department of Plant Biology, Michigan State University, East Lansing, Michigan, USA

**Joe Chappell**
Department of Plant and Soil Sciences, University of Kentucky, Lexington, Kentucky, USA

**Jon Clardy**
Department of Biological Chemistry and Molecular Pharmacology, Harvard Medical School, Boston, Massachusetts, USA

**Christophe Corre**
Department of Chemistry, University of Warwick, Coventry, United Kingdom

**Dean DellaPenna**
Department of Biochemistry & Molecular Biology, Michigan State University, East Lansing, Michigan, USA

**Elke Dittmann**
Department of Microbiology, Institute of Biochemistry and Biology, University of Potsdam, Golm, Germany

**Greg Fedewa**
Department of Biochemistry & Molecular Biology, Michigan State University, East Lansing, Michigan, USA

**Ben Field**
Laboratoire de Génétique et de Biophysique des Plantes, Unité Mixte de Recherche 7265 Centre National de la Recherche Scientifique-Commissariat à l'Energie Atomique et aux Energies Alternatives-Aix-Marseille Université, Marseille, France

**Elsa Góngora-Castillo**
Department of Plant Biology, Michigan State University, East Lansing, Michigan, USA

**Katrin Geisler**
Department of Metabolic Biology, John Innes Centre, Norwich, United Kingdom

**Juan Pablo Gomez-Escribano**
Department of Molecular Microbiology, John Innes Centre, Norwich, United Kingdom

**Rebecca J.M. Goss**
School of Chemistry, University of East Anglia, Norwich, United Kingdom

**Sabine Grüschow**
School of Chemistry, University of East Anglia, Norwich, United Kingdom

**Matthew I. Hutchings**
School of Biological Sciences, University of East Anglia, Norwich Research Park, Norwich, United Kingdom

**Paul R. Jensen**
Center for Marine Biotechnology and Biomedicine, Scripps Institution of Oceanography, University of California San Diego, La Jolla, California, USA

**Cheng-Lin Jiang**
Yunnan Institute of Microbiology, Yunnan University, Kunming, Yunnan, China

**Khomaizon A. K. Pahirulzaman**
School of Biological Sciences, University of Bristol, Bristol, United Kingdom

**Dimitris Kallifidas**
Laboratory of Genetically Encoded Small Molecules, Howard Hughes Medical Institute, The Rockefeller University, New York, USA

**Nancy P. Keller**
Department of Medical Microbiology and Immunology, University of Wisconsin, Madison, Wisconsin, USA

**Iris Keren**
Antimicrobial Discovery Center and Department of Biology, Northeastern University, Boston, Massachusetts, USA

**Roberto Kolter**
Department of Microbiology and Immunobiology, Harvard Medical School, Boston, Massachusetts, USA

**Colin M. Lazarus**
School of Biological Sciences, University of Bristol, Bristol, United Kingdom

**Kim Lewis**
Antimicrobial Discovery Center and Department of Biology, Northeastern University, Boston, Massachusetts, USA

**Fang Yun Lim**
Department of Medical Microbiology and Immunology, University of Wisconsin, Madison, Wisconsin, USA

**George P. Lomonossoff**
Department of Biological Chemistry, John Innes Centre, Norwich, United Kingdom

**Gillian MacNevin**
Department of Biological Sciences, University of Calgary, Calgary, Alberta, Canada

**Michael McArthur**
Department of Molecular Microbiology, John Innes Centre, Norwich, United Kingdom

**Jane M. Moore**
Department of Molecular Microbiology, John Innes Centre, Norwich, United Kingdom

**Lawrence R. Mulcahy**
Antimicrobial Discovery Center and Department of Biology, Northeastern University, Boston, Massachusetts, USA

**Hans-Wilhelm Nützmann**
Department of Molecular and Applied Microbiology, Leibniz Institute for Natural Product Research and Infection Biology – Hans Knöll Institute, and Institute of Microbiology, Friedrich Schiller University, Jena, Germany

**Trinh-Don Nguyen**
Department of Biological Sciences, University of Calgary, Calgary, Alberta, Canada

**Anne Osbourn**
Department of Metabolic Biology, John Innes Centre, Norwich Research Park, Norwich, United Kingdom

**Kalliopi K. Papadopoulou**
Department of Biochemistry & Biotechnology, University of Thessaly, Larissa, Greece

**Xiaoquan Qi**
Key Laboratory of Plant Molecular Physiology, Institute of Botany, Chinese Academy of Sciences, Beijing, PR China

**Dae-Kyun Ro**
Department of Biological Sciences, University of Calgary, Calgary, Alberta, Canada

**Frank Sainsbury**
Département de Phytologie, Pavillon des Services, Université Laval, Québec, QC, Canada

**James F. Sanchez**
Department of Pharmacology and Pharmaceutical Sciences, University of Southern California, School of Pharmacy, Los Angeles, California, USA

**Pooja Saxena**
Department of Biological Chemistry, John Innes Centre, Norwich, United Kingdom

**Volker Schroeckh**
Department of Molecular and Applied Microbiology, Leibniz Institute for Natural Product Research and Infection Biology – Hans Knöll Institute, Jena, Germany

**Ryan F. Seipke**
School of Biological Sciences, University of East Anglia, Norwich Research Park, Norwich, United Kingdom

**Mohammad R. Seyedsayamdost**
Department of Biological Chemistry and Molecular Pharmacology, Harvard Medical School, Boston, Massachusetts, USA

**Zengyi Shao**
Department of Chemical and Biomolecular Engineering, University of Illinois at Urbana-Champaign, Urbana, Illinois, USA

**John D. Sidda**
Department of Chemistry, University of Warwick, Coventry, United Kingdom

**Matthew F. Traxler**
Department of Microbiology and Immunobiology, Harvard Medical School, Boston, Massachusetts, USA

**Hans von Döhren**
Technische Universität Berlin, Fakultät II, Institut für Chemie, Sekretariat OE 2, Franklinstraße, Berlin, Germany

**Clay C.C. Wang**
Department of Pharmacology and Pharmaceutical Sciences, and Department of Chemistry, University of Southern California, School of Pharmacy, College of Letters, Arts, and Sciences, Los Angeles, California, USA

**Eva Wegel**
Department of Biochemistry, University of Oxford, South Parks Road, Oxford, United Kingdom

**Martin Welker**
Technische Universität Berlin, FG Umweltmikrobiologie, Sekr. BH 6-1, Ernst-Reuter-Platz, Berlin, Germany

**Katherine Williams**
School of Biological Sciences, University of Bristol, Bristol, United Kingdom

**Li-Hua Xu**
Yunnan Institute of Microbiology, Yunnan University, Kunming, Yunnan, China

**Yunsoo Yeo**
Department of Plant and Soil Sciences, University of Kentucky, Lexington, Kentucky, USA

**Huimin Zhao**
Department of Chemical and Biomolecular Engineering; Institute for Genomic Biology, and Departments of Chemistry, Biochemistry, and Bioengineering, University of Illinois at Urbana-Champaign, Urbana, Illinois, USA

**Li-Xing Zhao**
Yunnan Institute of Microbiology, Yunnan University, Kunming, Yunnan, China

**Nadine Ziemert**
Center for Marine Biotechnology and Biomedicine, Scripps Institution of Oceanography, University of California San Diego, La Jolla, California, USA

# PREFACE

In 2009, I edited two volumes of *Methods in Enzymology* (Volumes 458 and 459) entitled "Complex enzymes in microbial natural product biosynthesis." The project was motivated by two main factors. The first was the development, over the previous few years, of a novel toolbox of practical techniques for the study of natural product biosynthesis, involving a fusion of chemistry, genetics, enzymology, and structural studies, thereby bringing within reach an understanding of the "programming" of complex, multifunctional enzyme systems that had not been attainable previously and opening the possibility of creating "unnatural natural products" by genetic engineering. The second was the increasing need for novel bioactive natural products, especially antibiotics and anticancer drugs, and the new possibilities for addressing this need by carrying out "chemistry through genetics" and by studying the gamut of potential natural products revealed by the sequencing of microbial genomes. Three years later, these driving forces are still very much alive, hence the motivation to extend the project.

As well as including overview articles, the two 2009 volumes covered many of the hotspots in peptide and polyketide research, plus aminocoumarin compounds and some aspects of carbohydrate-based natural products. Therefore, the main emphasis this time is on chemical classes that were not included in the previous volumes, notably terpenoids and alkaloids, as well as further coverage of peptides and inclusion of Type III polyketides, which did not make it into the previous volumes. Interesting tailoring reactions, which often give natural products their biological activity by adding functional groups to the carbon skeletons assembled by complex enzyme systems, are also included.

Less obvious, in a series dedicated to enzymology, is the inclusion of sections dealing with the isolation and study of novel classes of organisms and of organisms from novel habitats. Other chapters describe heterologous pathway expression and methods for waking up sleeping gene clusters. The reasoning is that getting hold of the enzymes is an essential prerequisite for their study. Apart from its intrinsic scientific interest, this is a growth area in relation to natural product discovery and is revealing an Aladdin's cave of novel metabolism, much of it unexpressed under the conditions employed in traditional natural product screening campaigns.

In contrast to the previous two-volume set, which was focused on microorganisms, this time I have widened the coverage to include plants. Plants have long been known to produce an enormous number of important natural products, but study of their enzymology at the detailed molecular biological level lagged behind that of microbial products, largely for technical reasons. With the rise of plant genomics, made possible by the recent development of next-generation sequencing technologies and the analysis of transcriptomes by RNAseq, which together bring the large genomes of higher plants within reach, this deficiency has been redressed and there are now many examples of penetrating analysis of plant metabolites. Where appropriate, chapters on microbial systems—both bacterial and fungal—are grouped with chapters on plant metabolites so that interesting comparisons can be made. In a further difference from last time, the main criterion for inclusion in the new volumes is good biochemical and/or genetic understanding of a biosynthetic pathway, combined with interesting chemistry and/or unusual producing organisms. Thus, the choice is not confined to "complex enzymes" *per se*.

Especially in the descriptions of methods to study plant systems, there is overlap between a few of the chapters, even if they deal with the analysis of different classes of metabolites. I should like to regard this as a strength rather than a defect in editing! These technologies are still developing, so having more than one set of protocols to explore may be helpful to those who wish to extend the techniques to classes of compounds that are not explicitly covered in these volumes.

Volume A opens with a major section on terpenoids. Members of this huge class of natural products are derived from five-carbon isoprene units, ranging in number from hemiterpenes with one unit, monoterpenes with two, sesquiterpenes with three, diterpenes with four, tetraterpenes with eight, and polyterpenes with many. At one time thought to be rare or even absent from prokaryotes, they are now known to be important in bacteria as well as in the eukaryotic fungi. Coverage of the terpenoids in Volume A reflects this wide distribution and importance, with eight chapters devoted to various aspects of the study of terpenoid compounds. Several of the chapters introduce novel, cutting-edge technology. Appropriately for *Methods in Enzymology*, Chapter 1 describes novel enzymology: the steady-state kinetic characterization of plant sesquiterpene synthases by gas chromatography–mass spectrometry (GC–MS). Likewise, Chapter 4 represents a fine example of enzymology, defining and describing the specialized polyterpenoid synthase that makes natural rubber. Other chapters in this section deal with the genetic engineering of terpenoids and their expression in heterologous hosts, while providing a mine of information on the biosynthesis of the individual compounds, ranging from fungal mycotoxins to bacterial menaquinones.

The next group of chapters in Volume A is devoted to the alkaloids and glucosinolates. Alkaloids represent perhaps the longest known group of plant natural products. They are highly diverse in their structures and biological activities but are united by the presence of a basic nitrogen atom at some position in the molecule. Chapters 9 and 10 describe techniques for the discovery and analysis of monoterpene-derived indole alkaloids, which include the crucial antitumor Vinca alkaloids as well as several other classes of molecules with important medicinal uses, while Chapter 11 deals with the L-tyrosine-derived benzylisoquinoline alkaloids from opium poppy and related species, again compounds of extreme pharmacological interest, including morphine. Chapter 12 introduces the ergot alkaloids, classically associated with plants but actually made by the fungi that parasitize them, notably species of *Claviceps*, but now known to be made by a wider range of fungi. This section ends with a chapter on the amino acid-derived glucosinolates of plants, notable as beneficial dietary components found in brassicas, and their heterologous expression in *Nicotiana* and in yeast.

As mentioned above, polyketide synthases dominated the 2009 volumes in this series, but did not include the Type III systems, best known for the biosynthesis of anthocyanin pigments in higher plants but responsible for a wide range of other important compounds and now well established also as the producers of metabolites of microorganisms, both bacterial and fungal. They differ from the Type I and II synthases in consisting of small homodimeric proteins rather than large multifunctional enzymes with a multitude of separate active sites, making them in some ways easier to study biochemically but more cryptic in their programming. Chapters 14–16 describe the analysis and manipulation of these important enzyme systems.

Peptide-derived natural products are perhaps best known as being derived by nonribosomal assembly-line mechanisms very distinct from those depending on the ribosome. Several examples were covered in the 2009 volumes, along with a single class of compounds—the lantibiotics—resulting from ribosomal biosynthesis followed by extensive posttranslational modifications. Volume B opens with four further examples of ribosomally derived metabolites: the thiopeptide antibiotics produced by *Streptomyces* and *Bacillus* species, microviridin made by cyanobacteria, the plant cyclotides, and the cyclic peptide toxins of mushrooms, including the infamous amatoxins. This section ends with a special activity of a novel nonribosomal peptide synthetase, the Pictet-Spengler mechanism involved in the biosynthesis of tetrahydroisoquinoline antitumor antibiotics.

The next section of Volume B contains three chapters describing very diverse enzymology. The P–C bonds in phosphonate and phosphinate natural products endow them with a high level of stability and the ability to mimic phosphate esters and carboxylates, so their biosynthesis is particularly intriguing. The radical SAM enzymes carry out remarkable chemical transformations by releasing an active radical via the cleavage of S-adenosyl-L-methionine; the second chapter in this section describes novel methods for their purification and characterization. The third chapter describes methods for probing the biosynthesis of novel high-carbon sugar nucleosides containing up to 11 contiguous carbons.

Very often, natural product biosynthesis proceeds by the assembly of a core backbone—perhaps a polyketide, peptide, or terpene—followed by reactions that add functional groups that endow the molecules with their specific biological activities. Volume B continues with a section containing nine chapters devoted to such important tailoring reactions. It begins with one of the most famous classes of tailoring enzymes, the heme-dependent cytochromes P450, followed by the less well known nonheme iron-dependent enzymes. Then come two chapters on the halogenating enzymes of microorganisms and plants, first those that introduce fluorine and then chlorinating and brominating enzymes. Next comes prenylation, here represented by fungal enzymes of the dimethylallyltryptophan superfamily. This chapter relates to the biosynthesis of ergot alkaloids and so could have been placed in the alkaloid section of Volume A but is included in the tailoring enzyme section of Volume B because of the widespread importance of prenylation in determining the biological activity of molecules. Acylation is another crucial tailoring step in conferring biological activity on natural products; the section includes a chapter on one of the most important classes of acylating enzymes of plants, the serine carboxypeptidase-like acyltransferases. The actinomycete-derived enediynes are some of the most remarkable natural products both structurally and for their extreme cytotoxicity. Two chapters in this section derive from aspects of their biosynthesis, but they are included for the much wider applicability of the resulting enzymology. Chapter 15 deals with 4-methylideneimidazole-5-one (MIO)-containing aminomutases that catalyze β-amino acid formation, and Chapter 16 deals with tailoring enzymes acting on carrier-protein-tethered substrates, an approach that promises to open new vistas in the engineering of designer natural products. The section ends with a chapter on glycosylation. This topic received substantial coverage in the 2009 volumes but has recently been refined by the development of high-throughput colorimetric assays for nucleotide sugar formation and glycosyl transfer.

Volume C is devoted to methods for the discovery of novel secondary metabolite-synthesizing gene clusters and their analysis. First comes a group of five chapters describing novel sources of natural products or novel methods for their discovery. This is a growth area in the field of natural product research and could have included a much larger number of studies. Three of the selected chapters deal with the isolation of endophytic microorganisms from the tissues of traditional Chinese medicinal plants, with methods for handling cyanobacteria in relation to natural product discovery, and with the fascinating ecosystem represented by the nests of leaf-cutting ants that farm fungi as a food source and depend on beneficial microbes—mostly actinomycetes in the known examples—to protect their fungus gardens from parasitic fungi. The other two chapters in this section cover small molecule-mediated interactions within and between microbial colonies, an aspect of microbial ecology that is revealing interesting new metabolites in increasing numbers.

The next section containing three chapters represents another growth area in natural product research, namely, computational and bench-level approaches to the analysis of the gene clusters that contain the sets of genes encoding natural product biosynthetic genes. It has been a long-established paradigm that such gene sets are clustered together on bacterial genomes, with spectacular examples first discovered in the actinomycetes and later extended to myxobacteria and other groups of differentiating bacteria. Cotranscription of operons of clustered genes for primary metabolism in bacteria was an early discovery of the golden age of bacterial genetics, providing one driving force for clustering, though rarely do the secondary metabolic gene clusters represent a single operon. It came as a mild surprise that clustering is also the rule in the eukaryotic fungi, in which operons are not found. Recently, clustering has also been found to be a feature of some but not all secondary metabolic pathways in higher plants, though the clusters are very different from those of microorganisms in containing long stretches of untranslated DNA between the protein-encoding genes.

A powerful approach to the functional analysis of gene clusters is to express them in nonnative hosts, where they may be reassembled and/or engineered, sometimes to make unnatural products. A group of six chapters deal with various systems for such analysis, ranging from the use of virus vectors for heterologous expression in plants to systems for heterologous expression in streptomycetes, filamentous fungi, and, especially, yeast.

The final section of Volume C reflects the crucial discovery that the genomes at least of microorganisms contain far more clusters of genes potentially encoding natural product biosynthesis than are expressed under a given set of conditions. The challenge is to find generally applicable methods to wake up such "sleeping" gene sets and so give access to a range of potentially valuable compounds that would otherwise go unexplored. Two chapters in this section describe diverse approaches in filamentous fungi and two in *Streptomyces* species. The section ends with a chapter on the intriguing and important problem of "persisters" that represent a sub-population of a bacterial pathogen in which the whole cell is "sleeping." They evade killing by antibiotics but may be outwitted by judicious treatments, some of which block the normal wake-up process.

It goes without saying that the value of any edited work depends on the renown of the invited authors and their willingness to write chapters. I was particularly impressed by the enthusiastic response of nearly all my invitees and the quality of the submitted manuscripts. Indeed, such was the enthusiasm of the authors that an original two-volume project expanded to the final three-volume version. My grateful thanks go to all of you as well as to the small army of coauthors who were recruited to the task. I am very grateful also to colleagues who made suggestions for the content of these volumes, especially Greg Challis, Wilfred van der Donk, Sarah O'Connor, Paul O'Maille, Ben Shen, and Anne Osbourn. I also thank Shaun Gamble of Elsevier, who never failed to offer timely advice and reassurance during the entire project.

DAVID A. HOPWOOD

# METHODS IN ENZYMOLOGY

# Metabolites from Novel Sources

# CHAPTER ONE

# Methods for the Study of Endophytic Microorganisms from Traditional Chinese Medicine Plants

## Li-Xing Zhao, Li-Hua Xu, Cheng-Lin Jiang[1]

Yunnan Institute of Microbiology, Yunnan University, Kunming, Yunnan, China
[1]Corresponding author: e-mail address: chenglinj@yahoo.com

## Contents

## Abstract

Plant endophytes are very numerous and widely distributed in nature, their relationships being described as a balanced symbiotic continuum ranging from mutualism through commensalism to parasitism during a long period of coevolution. Traditional Chinese medicines have played a very important role in disease treatment in China and other Asian countries. Investigations show that these medicinal plants harbor endophytes with different kinds of ecological functions, and some of them have potential to produce bioactive small-molecule compounds. This chapter will focus on the selective isolation methods, the diversity of some endophytes (actinobacteria and fungi) isolated from Traditional Chinese Medicine (TCM) plants, and the bioactive compounds from selected endophytic actinobacteria reported in the past 3 years.

# 1. INTRODUCTION

Natural products are naturally derived metabolites and/or by-products from microorganisms, plants, or animals (Baker, Mocek, & Garr, 2000) and continue to be an incredible resource for drug discovery (Cragg, Grothaus, & Newman, 2009). Medicinal plants and soil microbes remain the most popular sources of natural products for pharmaceutical research and development. However, it is becoming very difficult to find novel microbial metabolites after more than 60 years focusing researches on soil microbes, particularly members of the genus *Streptomyces*, from which many antibiotics and other bioactive secondary metabolites with unique pharmacophores have been discovered (Bérdy, 2005; Bull, 2003). To avoid rediscovery of known compounds from microbes, many approaches are employed to obtain high-quality isolates and novel microbes. Some studies have focused on poorly investigated extreme biological habitats (Bull, 2011). Plants supply another kind of extreme biosphere for microbes—endophytes.

The term endophyte, first introduced by De Bary in 1866 (Rodrigues, 1996), was used to define the organisms occurring within plant tissues. Then, scientists widened the definition. Now, the inclusive and widely accepted definition of endophytes was given by Bacon and White (2000): "microbes that colonize living, internal tissues of plants without causing any immediate, overt negative effects," which includes virtually any organism residing inside a plant host.

Although the first discovery of endophytes happened in 1904, they did not attract attention till the finding of the world's first billion-dollar anticancer drug, paclitaxel (Taxol) from *Pestalotiopsis microspora*, a fungus that colonizes the Himalayan yew tree *Taxus wallichiana* (Haheshwari, 2006). It is believed that each plant on earth is host to one or more endophytes (Strobel & Daisy, 2003). During the long period of coevolution, endophyte and host plant gradually established mutually beneficial relationships. The host plant can supply nutriment and habitation for the survival of its endophytes, while the endophytes would produce bioactive substances that help the host plants to resist external biotic and abiotic stresses, benefiting host growth in return (Rodriguez, White, Arnold, & Redman, 2009; Silvia, Sturdikova, & Muckova, 2007). Till now, research results from selected plants show that endophytes exist widely in their tissues, and the diversity is also rich in a single plant (Benhizia et al., 2004; Hallmann, Quadt-Hallmann, & Mahaffee, 1997; Rosenblueth & Martinez-Romero, 2006; Sturz, Christie, Matheson, & Nowak, 1997).

In the past 20 years, investigations on bioactive metabolites from endophytes were developed and many compounds with antimicrobial, insecticidal, cytotoxic, and anticancer activities were discovered from endophytes. These compounds were structurally classified as alkaloids, lactones, phenols, quinines, terpenoids, steroids, and lignans (Zhang, Song, & Tan, 2006). Therefore, endophytes represent a diverse potential source of new products for use in medicine and biotechnology (Gunatilaka, 2006; Hostettmann & Marston, 2007; Kusari, Zuhlke, & Spiteller, 2009; Pezzuto, 1997; Stierle, Strobel, & Stierle, 1993; Strobel, 2002, 2006; Strobel & Daisy, 2003; Strobel, Daisy, Castillo, & Harper, 2004; Tan & Zou, 2001; Zhang et al., 2006).

Traditional Chinese medicine (TCM) originated in ancient China thousands of years ago and has played a very important role in Chinese medicine. Medicinal plants, minerals, and animals are used to treat different kinds of diseases by TCM practitioners. There are roughly 13,000 medicines used in China and over 100,000 medicinal recipes recorded in the ancient literature (Chen & Yu, 1999). Today, TCM is considered part of complementary and alternative medicine (Barnes, Bloom, & Nahin, 2008). Medicinal and pharmaceutical researchers pay more and more attention to TCM and Chinese herbs. The Chinese are by far the largest users of traditional medicines with over 5000 plants and plant products in their pharmacopeia. Some TCM herbs or plants were intensively investigated for their active components and pharmaceutical actions. There are millions of natural products in medicinal plants with a series of pharmacological functions (Niero & Cechinel, 2008; Zhou et al., 2007). Even one species can possess a mass of various compounds with different biological activities. Some single compounds produced by TCM herbs and plants have been widely used in the clinic, such as aspirin, paclitaxel (taxol), reserpine, colchicines, camptothecin, etc. Yet, little is known about the endophytic organisms behind TCMs.

Natural products produced by endophytes usually differ from those produced by the plant itself, although there are examples of endophytes producing the same kinds of defensive compounds as their host (Kusari et al., 2009; Stierle et al., 1993). Recent researches confirm that endophytes also are a significant reservoir of novel bioactive secondary metabolites (Strobel et al., 2004; Tan & Zou, 2001) with promising medicinal or agricultural applications (Aly, Debbab, Kjier, & Proksch, 2010; Deshmukh & Verekar, 2012; Wang, Zhang, Lin, Hu, & Zhang, 2011; Zhang et al., 2006). It is believed that each plant on earth is host to one or more endophytes (Strobel & Daisy, 2003). Considering the wide use of TCM plants and the limited investigation of their endophytes, it is highly likely that novel natural products will be found from TCM endophytes. The first step is to obtain high-quality cultural endophytic microbial strains, which are crucial for further investigation on their bioactive nature products.

Medicinal plants can be chosen according to some reasonable hypotheses: (i) medicinal plants in the Chinese pharmacopeia could be the best choices; (ii) plants with specific uses or applications of interest described in the Chinese literature can be selected for study; (iii) both plants and some microorganisms can produce the same or similar compounds—for example, the plant *Maytenus hookeri* and certain *Streptomyces* strains can produce ansamycin or it analogs; (iv) medicinal plants from unique biological niches should be considered for study.

This chapter describes methods used in Yunnan Institute of Microbiology for selective isolation of endophytes from TCM plants.

# 2. EXPERIMENTAL METHODS

## 2.1. Medicinal plant selection and sampling

The sampling regime should be designed with the intention of isolating as many endophytic species as possible from TCM plants. Plant selection should be based on their economic importance in the market and their perennial nature. Rare and valuable plants are taken into consideration also. As endophytes show some degree of tissue preference (Huang, Cai, Hyde, Croke, & Sun, 2008; Kumaeresan & Suryanarayanan, 2002; Pang, Vrijmoed, Goh, Plaingam, & Jones, 2008), different tissues of medicinal plants should be screened to obtain more endophytic strains for further investigations.

1. Select living, healthy, and authenticated medicinal plants.
2. Cut branches or side roots into 20-cm segments for woody plants. Collect old and clean leaves from branches directly. Uproot whole plants for Chinese herbs.
3. Record sample site and time, and the names of the plants.

## 2.2. Pretreatment and surface-sterilization

This is a very important step for the isolation of endophytic organisms. Pretreatment should be done 4–10 h after sampling to avoid contamination by outside microorganisms. Effective surface sterilization ensures that isolates are true endophytes. Some groups prefer to set up their own surface-sterilization protocol. Here, we describe the general procedure used in our laboratory.

### 2.2.1 Pretreatment

1. Cut off about 5 cm from the two ends of the selected 20-cm branches or side roots.
2. Seal the two ends using wax immediately to exclude exogenous microorganisms.
3. Keep and seal leaves and whole grasses in clean plastic bags after removing soil.
4. Store plant samples at 4 °C for isolation after 1 or 2 days, or at −80 °C for preservation for up to 1 month.

### 2.2.2 Surface sterilization

1. Wash samples with running tap water to remove adhering dirt or soil particles.
2. Clean with distilled water using a 160-W ultrasonic cleaner for 1 min.
3. Immerse samples in 0.01% (v/v) tween-20 for 1 min.
4. Rinse thrice with sterile water.
5. Soak for 3 min in 5% aqueous NaOCl.
6. Rinse thrice with sterile water.
7. Immerse in 2.5% aqueous sodium thiosulfate solution for 10 min.
8. Rinse thrice with sterile water.
9. Soak in 75% ethanol for 30 s (leaves) or 1–5 min (root and branch samples).
10. Rinse thrice with sterile water.
11. Dry with sterile paper towels or overnight in sterile Petri dishes.
12. Check the effectiveness of the surface-sterilization procedure by two methods. (1) Plate 0.1 mL of the final sterile water rinse on Luria–Bertani (LB) agar (see Section 2.3), International *Streptomyces* project medium 2 (ISP-2) (yeast extract 4 g, malt extract 10 g, D-glucose 4 g, agar 15–20 g, distilled water 1000 mL, pH 7.2), or potato dextrose agar (PDA) plates (see Section 2.3). (2) Roll the sterilized sample surface several times on LB, ISP-2, or PDA plates. Incubate at 28 and 37 °C for 2 days; any microbial growth of microorganisms indicates the effectiveness of the protocol.

## 2.3. Isolation of endophytes

**1.** Prepare isolation media as follows

Media for isolation of endophytic bacteria and actinobacteria

**a.** Luria–Bertani agar (M1): tryptone 10 g, yeast extract 5 g, NaCl 10 g, agar 15 g, distilled water 1 L, pH 7.2.

**b.** Nutrient agar (M2): peptone 5 g, beef extract 3 g, NaCl 5 g, agar 15 g, distilled water 1 L, pH 7.0.

**c.** M9 minimal medium (M3): $Na_2HPO_4$ 10 g, $KH_2PO_4$ 3 g, NaCl 0.6 g, $NH_4Cl$ 20 g, glucose 5 g, agar 20 g, distilled water 1 L, pH 7.2.

**d.** Gause's synthetic agar (M4): $KNO_3$ 10 g, soluble starch 20 g, $K_2HPO_4$ 0.5 g, $MgSO_4 \cdot 7H_2O$ 0.5 g, NaCl 0.5 g, $FeSO_4$ 0.01 g, agar 15 g, distilled water 1 L, pH 7.2–7.4.

**e.** TWYE agar (El-Shatoury, Abdulla, El-Karaaly, El-Kazzaz, & Dewedar, 2006) (M5): yeast extract 0.25 g, $K_2HPO_4$ 0.5 g, agar 15 g, tap water 1 L, pH 7.2.

**f.** HV agar (Hayakawa & Nonomura, 1987) (M6): humic acid 1 g, KCl 1.7 g, $Na_2HPO_4$ 0.5 g, $MgSO_4$ 0.5 g, $CaCO_3$ 0.02 g, $FeSO_4$ 0.01 g, $V_B$ stock solution (0.00005% each of thiamine–HCl, riboflavin, niacin, pyridoxine–HCl, inositol, Ca–pantothenate, $p$-aminobenzoic acid, and 0.000025% of biotin) 1 mL, 10 g agar, 1000 mL distilled water, pH 7.2.

**g.** Glycerol–asparagine agar (M7): asparagine 1 g, glycerol 10 g, $K_2HPO_4$ 1 g, trace salt solution ($FeSO_4$ 0.1 g, $MnCl_2$ 0.1 g, $ZnSO_4$ 0.1 g, 100 mL distilled water) 1 mL, 10 g agar, 1 L distilled water, pH 7.2.

**h.** Chitin medium (Lingappa & Lockwood, 1961) (M8): colloidal chitin 2.5 g, $K_2HPO_4$ 0.7 g, $KH_2PO_4$ 0.3 g, $MgSO_4$ 0.5 g, $FeSO_4$ 0.01 g, $ZnSO_4$ 0.01 g, agar 20 g, distilled water 1 L, pH 7.0.

**i.** Inhibitors used for selective isolation of endophytic actinomycete strains: cycloheximide 50 mg/L, nalidixic acid 25 mg/L, nystatin 50 mg/L, $K_2Cr_2O_7$ 25 mg/L. These inhibitors can be used singly or in combination when the aim is to isolate actinomycetes. They should be sterilized appropriately with diethyl ether before use.

Media for isolation of endophytic fungi

**a.** Potato dextrose agar (PDA) (M9): broth of fresh potato 200 g, dextrose 20 g, distilled water 1 L.

**b.** Malt extract agar (M10): malt extract 30 g, mycological peptone 5 g, agar 15 g, distilled water 1 L, pH 5.4.

    **c.** Water agar (M11): agar 15 g, distilled water 1 L, pH 5.4.

    **d.** Inhibitors used for isolation of endophytic fungal strains: 200 U/mL penicillin G, 200 μg/mL streptomycin sulfate for suppressing the growth of bacteria.

**2.** Dissect or grind pretreated samples into small pieces or pulverize in a small autoclaved grinder.

**3.** Isolate endophytes.

    **a.** Method one: homogeneously disperse 0.1 g small pieces or powder on 90-mm Petri dishes containing fresh isolation media. Incubate at $28 \pm 1\ °C$ for bacteria, $25 \pm 1\ °C$ for fungi.

    **b.** Method two: add 1 g ground sample and 9 mL sterile water to a sterile homogenizer. The mixed solution is maintained at $28\ °C$ with occasional stirring. Immediately after incubation for 30 min, 10-fold dilutions of up to $10^{-3}$ are prepared in sterile water, and 1 mL aliquots are inoculated on 90-mm Petri dishes containing fresh isolation media. Incubate at $28 \pm 1\ °C$ for bacteria, $25 \pm 1\ °C$ for fungi.

**4.** Monitor the growth of endophytic clones during the incubation period.

## 2.4. Purification and maintenance

**1.** Prepare media for purification and maintenance.

**2.** TWYE and 1/2 ISP-2 agar are used for purification of actinobacteria, LB agar for bacteria, and PDA for fungi. TSA (tryptone 15 g, soytone (enzymatic digest of soybean meal) 5 g, sodium chloride 5 g, agar 15 g) and modified ISP-2 (YIM 38 medium, 10 g malt extract, 4 g yeast extract, 4 g glucose, vitamin mixture (0.5 mg each of thiamine–HCl, riboflavin, niacin, pyridoxine–HCl, inositol, calcium pantothenate and *p*-aminobenzoic acid and 0.25 mg biotin), 20 g agar; pH 7.2) (Jiang et al., 2007) are employed to maintain actinobacteria.

**3.** Pick colonies from isolation plates after incubation for 3 days to 12 weeks.

**4.** Streak colonies on plates containing purification media.

**5.** Visually and microscopically examine the purity of isolates. When necessary, mixed endophytic cultures can be purified by diluting cultures grown on selective media to separate different microbes.

**6.** Grow purified single colonies on selected media, PDA for fungi, LB agar for bacteria, TSA, and modified ISP-2 for actinobacteria.

**7.** Maintain the grown strains on slants of PDA, LB agar, TSA, or modified ISP-2 according to the kinds of endophytes and the growth conditions.

**8.** Store mature strains (spores or mycelium) in 20% glycerol at $-70\ ^{\circ}C$ and in sealed lyophilized milk tubes at $4\ ^{\circ}C$.

## 2.5. Taxonomic analysis

A general polyphasic taxonomy approach for bacterial systematics was reviewed by Vandamme et al. (1996) and Gillis, Vandamme, De Vos, Swings, and Kersters (2001). Taxonomic analysis of endophytic bacteria will not be described here, please refer to methods in the literature. Potential new strains will be selected by experiments and DNA analysis, and identification will be based on the experimental results of morphological and cultural characteristics, physiological, and biochemical tests, chemotaxonomic analysis, DNA $G+C$ content, phylogenetic analysis of 16S rRNA gene, and DNA–DNA hybridization.

Fungal taxonomy is traditionally based on comparative morphological features, and the development of sexual or asexual reproductive structures, which depend on medium and cultural conditions. There are difficulties in disciminating similar endophytes and because some isolates fail to sporulate in culture. Therefore, analysis of endophytes should be carried out on a comprehensive phylogenetic classification, as described by Hibbett et al. (2007) and Göker, García-Blázquez, Voglmayr, Tellería, and Martín (2009).

## 3. NOVEL BIOACTIVE COMPOUNDS FROM ENDOPHYTIC ACTINOBACTERIA

There is mounting evidence that many bioactive compounds isolated from various endophytic fungi. Review articles about bioactive compounds from endophytes were published during the period of 2000–2010, and mainly focused on the compounds from endophytic fungi (Deshmukh & Verekar, 2012; Gunatilaka, 2006; Guo, Wang, Sun, & Tang, 2008; Ryan, Germaine, Franks, Ryan, & Dowling, 2008; Staniek, Woerdenbag, & Kayser, 2008; Verma, Kharwar, & Strobel, 2009; Wang, Zhang, et al., 2011; Zhang et al., 2006). Bioactive compounds from endophytic actinobacteria and their biotechnological potential applications were reviewed in detail by Qin et al. (2011). In this chapter, we just describe the novel compounds (Fig. 1.1) from endophytic actinobacteria from 2009 to the writing time of 2011. Presented in Table 1.1 are novel natural products, their endophytic actinomycete producers, host plants, reported biological activities, and the literature references.

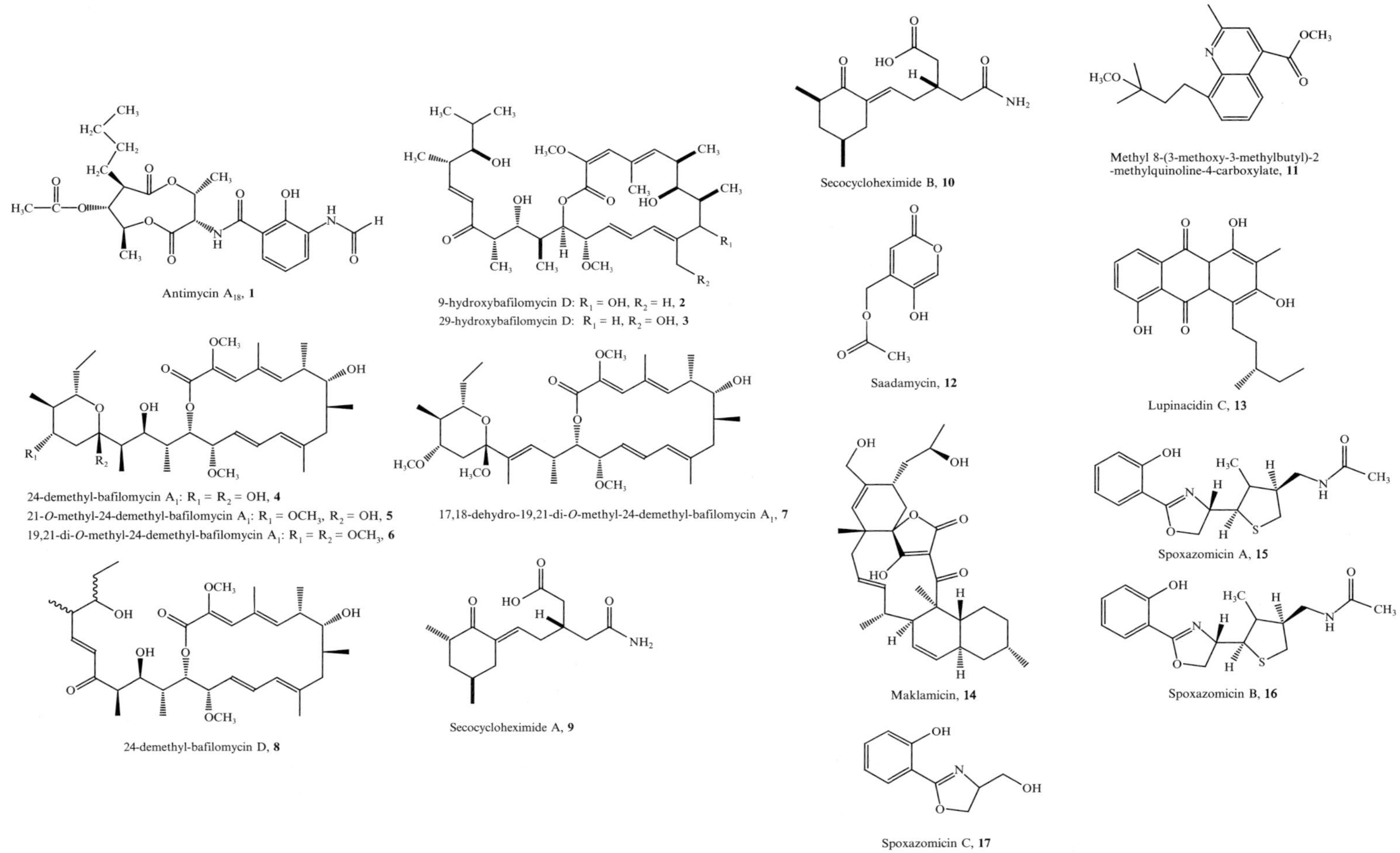

**Figure 1.1** Structures of novel natural products reported in past 3 years.

**Table 1.1** Novel natural products of endophytic actinobacteria reported in past 3 years

| Natural products | Microbial strain | Plant host | Biological activity | References |
|---|---|---|---|---|
| Lupinacidin C | *Micromonospora lupine* Lupac 08 | *Lupinus angustifolius* | Antitumor | Igarashi, Yanase, et al. (2011) |
| Spoxazomicins A–C | *Streptosporangium oxazolinicum* K07-0460 | Orchid plant | Antitrypanosomal | Inahashi et al. (2011) |
| Antimycin A18 | *Streptomyces albidoflavus* I07A-01824 | *Bruguiera gymnorrhiza* | Antifungal | Yan et al., 2010 |
| Saadamycin | *Streptomyces* sp. Hedaya48 | *Aplysina fistularis* | Antifungal | El-Gendy and El-Bondkly (2010) |
| Methyl-8-(3-methoxy-3-methylbutyl)-2-methylquinoline-4-carboxylate | *Streptomyces* sp. neau50 | Soybean | Antitumor | Wang et al. (2010) |
| 24-Demethyl-bafilomycin A1 21-O-Methyl-24-demethyl-bafilomycin A1 19,21-Di-O-methyl-24-demethyl-bafilomycin A1 17,18-Dehydro-19,21-di-O-methyl-24-demethyl-bafilomycin A1 24-Demethyl-bafilomycin D | *Streptomyces* sp. CS | *Maytenus hookeri* | Antitumor | Li, Lu, and Shen (2010) |
| Maklamicin | *Micromonospora* sp. GMKU326 | *Abrus pulchellus* Wall. Ex Thwaites | Antibaterial | Igarashi, Ogura, et al. (2011) |

**Table 1.1** Novel natural products of endophytic actinobacteria reported in past 3 years—cont'd

| Natural products | Microbial strain | Plant host | Biological activity | References |
|---|---|---|---|---|
| | | subsp. *pulchellus* | | |
| 9-Hydroxybafilomycin D 29-Hydroxybafilomycin D | *Streptomyces* sp. YIM56209 | *Drymaria cordata* | Antitumor | Yu et al. (2011) |
| Secocycloheximide A Secocycloheximide B | *Streptomyces* sp. YIM56132 *Streptomyces* sp. YIM56141 | *Carex baccaus* *Fagopyrum cymosum* | Antitumor | Huang et al. (2011) |

Although it is becoming difficult to discover novel compounds, actinomycetes, especially *Streptomyces*, are still considered attractive sources of bioactive compounds. Antimycin $A_{18}$ (**1**, Fig. 1.1), the first naturally occurring antimycin with an acetoxy group at C-8, was isolated from edophytice *Streptomyces albidoflavus* I07A-01824, which was isolated from the leaf of *Bruguiera gymnrrhiza*. The compound showed better antifungal activity to plant pathogenic fungi—*Colletotrichum lindemuthianum*, *Botrytis cinerea*, *Alternaria solani*—than blasticidin S, a commercialized fungicide (Yan et al., 2010). New bafilomycin derivatives (**2–8**, Fig. 1.1) were isolated from the culture broth of two endophytic *Streptomyces* spp., which were isolated from the rainforest plants *Drymaria cordata* and *Maytenus hookei* in Xishuangbanna, Yunnan (Li et al., 2010; Yu et al., 2011). All these bafilomycins showed cytotoxic against tumor cell lines, and **2** was able to reduce PRL-mediated signals to ERK1/2 (Yu et al., 2011). Endophtyitc *Streptomyces* spp. YIM56132 from *Carex baccaus* and YIM56141 from *Fagopyrum cymosum* showed inhibition of eukaryotic protein synthesis and were cytotoxic in a bicistronic mRNA translation screen. Extraction of the culture broth of these strains afforded two new cycloheximide congeners—secocycloheximide A (**9**, Fig. 1.1) and B (**10**, Fig. 1.1), which contain carboxylic acid and primary amide functionalities instead of an intact glutarimide. Compared to other cycloheximide congeners, translation inhibition assays showed that the presence of a hydroxyl group at C-8 significantly improves inhibitory activity. This discovery could, therefore, serve as a scaffold for the chemical synthesis and biosynthesis of more potent modified eukaryotic protein synthesis inhibitors (Huang, Yu, et al., 2011). A new quinoline derivative, methyl 8-(3-methoxy-3-methylbutyl)-2-methylquinoline-4-carboxylate (**11**, Fig. 1.1), produced by the endophytic *Streptomyces* sp. neau50, showed cytotoxicity against human lung adenocarcinoma cell line A549 with an $IC_{50}$ value of 29.3 µg/mL (Wang, Gong, et al., 2011). A novel antimycotic, saadamycin (12, Fig. 1.1), active against dermatophytes, and other clinical fungi was isolated from endophytic *Streptomyces* sp. Hedaya48 (El-Gendy & El-Bondkly, 2010).

Other endophytic actinomycete genera such as *Micromonospora* and *Streptosporangium* also can produce novel compounds. Lupinacidin C (**13**, Fig. 1.1), an anthraquinone derivative, was isolated from the endophytic actinomycete *Micromonospora lupine* Lupac 08 (Igarashi, Yanase, et al., 2011), an endophytic strain from *Lupinus angustifolius*. The molecule exhibited the most potent inhibitory effects among the congeners on the invasion of murine colon carcinoma cells into the reconstituted basement membrane. Another compound, maklamicin (**14**, Fig. 1.1), produced by the endophytic *Micromonospora* sp. GMKU326, showed strong to modest antimicrobial activity against Gram-positive bacteria (Igarashi, Ogura, et al., 2011). A new endophtyic actinomycete species, *Streptosporangium oxazolinicum* K07-0460$^{T}$, afforded three novel antitrypanosomal alkaloids, named spoxazomicins A–C (**15-17**, Fig. 1.1). Spoxazomicin A showed potent and selective antitrypanosomal activity with an $IC_{50}$ value of 0.11 µg/mL *in vitro* without cytotoxicity against MRC-5 cells ($IC_{50} = 27.8$ µg/mL) (Igarashi et al., 2011c).

## 4. CONCLUSION

The growth requirements of microorganisms cover a very wide range of conditions. No single isolation medium and condition is suitable for the growth of all microorganisms. Enormous numbers of microorganisms exist in soil, plant tissues, seawater, and other environments (Andrews & Harris, 2000; Rosselló-Mora & Amann, 2001; Rusch et al., 2007; Torsvik & Øvreås, 2002; Venter et al., 2004; Yooseph et al., 2007). The established isolation may be effective for fast-growing, ubiquitous species, while rare species with limited competitive strength and more specialized requirements still remain undiscovered (Sun, Guo, & Hyde, 2011; Unterseher & Schnittler, 2009). Therefore, some special media and sample pretreatment conditions should be considered to recover any endophytes present in plant tissues. For example, to recover more endophytic actinobacteria, more than eight media will be employed for plant samples. Other reported methods can also be used to recover new endophytes, such as the sodium dodecyl sulfate-yeast extract method (Hayakawa & Nonomura, 1989), the moist incubation and desiccation method (Matukawa, Nakagawa, Limura, & Hayakawa, 2007), the membrane filter method (Nagai, Khan, Tamura, Takagi, & Shin-ya, 2011), lectin-modified microengines (Campuzano et al., 2012), and other methods (Khan et al., 2011). In medium design, nutrient media containing extracts of the plant host, combined with controlled growing conditions, may be helpful to recover, grow, and sporulate the endophyte (Moricca & Ragazzi, 2011).

Selective media have been used to isolate pathogenic microorganisms in diagnostic medicine and to isolate commercially useful microorganisms from environmental samples. Selective isolation methods were also successfully used to obtain pure cultures of previously unculturable environmental micro-organisms (Kaeberlein, Lewis, & Epstein, 2002; Zengler et al., 2002). Most selective media have been developed in empirical studies, while Kawanishi et al. (2011) designed highly specific media by a selective medium–design algorithm restricted by two constraints (SMART) to specifically grow targeted bacteria of *Burkholderia glumae*, *Acidovorax avenae*, *Pectobacterium carotovorum*, *Ralstonia solanacearum*, and *Xanthomonas campestris*. Selective isolation of rare endophytes is very important, especially for actinobacteria, as they are a potential source for novel antibiotics (Tiwari & Gupta, 2011). Using such selective isolation procedures, a total of 228 isolates from *Artemisia annua* representing at least 19 different genera of actinobacteria were obtained and several of them should be novel taxa (Huang, Li, et al., 2011; Li, Zhao, Huang, Qin, et al., 2011; Li, Zhao, Huang, Zhu, et al., 2011; Li, Zhao, Zhu, et al., 2011; Zhao, Li, Huang, Zhu, Zhao, et al., 2011; Zhao, Li, Zhu, Klenk, et al., 2011; Zhao, Li, Huang, Zhu, Park, et al., 2011; Zhao, Zhu, et al., 2011; Zhao et al., 2010; Zhao, Li, Zhu, Wei, et al., 2011). Another 312 pure culture isolates were obtained from different parts of the Chinese medicinal plant *Maytenus austroyunnanensis*, and were classified in 21 genera. Some genera, such as *Jiangella, Polymorphospora, and Cellulosimicrobium*, were first isolated from the endophytic environment. Also, novel endophytic species were discovered from the plant (Qin, Chen, et al., 2009; Qin, Li, Zhang, et al., 2009; Qin, Zhao, Klenk, et al. (2009); Qin et al., 2009e, 2010, 2011).

Even in the surface-sterilization step, improvements are possible for special aims. Sodium thiosulfate solution can suppress the detrimental effects of residual NaOCl on the plant material surface (Qin, Li, Chen, et al., 2009). To suppress the growth of endophytic fungi, surface-sterilized plant samples can be immersed in 10% $NaHCO_3$ solution for 10 min when isolating endophytic actinobacteria (Tan et al., 2004). Plant tissues differ from each other in species, age, and components. Therefore, the surface-sterilization procedure should be optimized according to the plant tissue, targeted endophytic microorganisms, and specific requirements.

Pure culture of microorganisms is still very important in the field of microbiology and other related subjects, although culture-independent techniques display their potential power. Culture-independent methods should give some valuable information and guide for pure culture from environmental samples. Strategies for culture of "unculturable" bacteria from soil and aquatic environments were reviewed by Vartoukian et al. (2010). The approaches should be used to obtain more endophytic clones. TCM plants produce many bioactive small molecules. Investigations on TCM endophytes and their bioactive secondary metabolites have been developed over the past two decades, especially for endophytic actinomycetes. It is very important to isolate more endophytes by using various isolation methods for obtaining new bioactive compounds (Takgi & Shin-ya, 2011) and effective usage of TCM medicinal plants.

## ACKNOWLEDGMENT

This work was support in part by National Natural Science Foundation of China 2162028 (for Zhao, L. X.) and U0932601(for Xu, L. H.), and a grant SRF for ROCS (for Zhao, L. X.), Chinese Ministry of Education.

## REFERENCES

Aly, A. H., Debbab, A., Kjier, J., & Proksch, P. (2010). Fungal endophytes from higher plants: A prolific source of phytochemicals and other bioactive natural products. *Fungal Diversity, 41*, 1–6.

Andrews, J. H., & Harris, R. F. (2000). The ecology and biogeography of microorganisms on plant surfaces. *Annual Review of Phytopathology, 38*, 145–180.

Bacon, C. W., & White, J. F. (2000). *Microbial endophytes*. New York, USA: Marcel Dekker Inc.

Baker, D., Mocek, U., & Garr, C. (2000). Natural products vs. combinatories: A case study. In S. K. Wrigley, M. A. Hayes, R. Thomas, E. J. T. Chrystal & N. Nicholson (Eds.), *Biodiversity: New leads for pharmaceutical and agrochemical industries* (pp. 66–72). Cambridge, United Kingdom: The Royal Society of Chemistry.

Barnes, P. M., Bloom, B., & Nahin, R. (2008). Complementary and alternative medicine use among adults and children: United States, 2007. *CDC National Health Statistics Report #12 2008*

Benhizia, Y., Benhizia, H., Benguedouar, A., Muresu, R., Giacomini, A., & Squartini, A. (2004). Gamma proteobacteria can nodulate legumes of the genus Hedysarum. *Systematic and Applied Microbiology*, *27*, 462–468.

Bérdy, J. (2005). Bioactive microbial metabolites—A personal view. *The Journal of Antibiotics*, *58*, 1–26.

Bull, A. T. (2003). *Microbial diversity and bioprospecting*. Washington: ASM Press.

Bull, A. T. (2011). Actinobacteria of the extremobiosphere. In K. Horikoshi, G. Antranikian, A. T. Bull, F. Robb & K. O. Steter (Eds.), *Extremophiles handbook 2*, (pp. 1204–1240). Tokyo: Springer.

Campuzano, S., Orozco, J., Kagan, D., Guix, M., Gao, W., Sattayasamitsathit, S., et al. (2012). Bacterial Isolation by lectin-modified microengines. *Nano Letters*, *12*, 396–401.

Chen, Y., & Yu, B. (1999). Certain progress of clinical research on Chinese integrative medicine. *Chinese Medical Journal*, *112*, 934–937.

Cragg, G. M., Grothaus, P. G., & Newman, D. J. (2009). Impact of natural products on developing new anti-cancer agents. *Chemical Reviews*, *109*, 3012–3043.

Deshmukh, S. K., & Verekar, S. A. (2012). Fungal endophytes: A potential source of antifungal compounds. *Frontiers in Bioscience (Elite Edition)*, *4*, 2045–2070.

El-Gendy, M. M. A., & El-Bondkly, A. M. A. (2010). Production and genetic improvement of a novel antimycotic agent, saadamycin, against Dermatophytes and other clinical fungi from endophytic Streptomyces sp. Hedaya48. *Journal of Industrial Microbiology & Biotechnology*, *37*, 831–841.

El-Shatoury, S., Abdulla, H., El-Karaaly, O., El-Kazzaz, W., & Dewedar, A. (2006). Bioactivities of endophytic actinomycetes from selected medicinal plants in the world heritage site of saint Katherine, Egypt. *International Journal of Botany*, *2*, 307–312.

Gillis, M., Vandamme, P., De Vos, P., Swings, J., & Kersters, K. (2001). Polyphasic taxonomy. In D. R. Boone & R. W. Castenholz (Eds.), *Bergey's Manual of Systematic Bacteriology*, Vol. 1, (pp. 43–48). Springer New York.

Göker, M., García-Blázquez, G., Voglmayr, H., Tellería, M. T., & Martín, M. P. (2009). Molecular taxonomy of phytopathogenic fungi: A case study in *Peronospora*. *PLoS One*, *4*, e6319.

Gunatilaka, A. A. L. (2006). Natural products from plant-associated microorganisms: Distribution, structural diversity, bioactivity, and their occurrence. *Journal of Natural Products*, *69*, 509–526.

Guo, B., Wang, Y., Sun, X., & Tang, K. (2008). Bioactive natural products from endophytes: A review. *Applied Biochemistry and Microbiology*, *44*, 153–158.

Haheshwari, R. (2006). What is an endophytic fungus? *Current Science*, *90*, 1039.

Hallmann, J., Quadt-Hallmann, A., & Mahaffee, W. F. (1997). Bacterial endophytes in agricultural crops. *Canadian Journal of Microbiology*, *3*, 895–914.

Hayakawa, M. T., & Nonomura, H. (1987). Humic acid-vitamin agar, a new method for the selective isolation of soil actinomycetes. *Journal of Fermentation and Bioengineering*, *65*, 501–509.

Hayakawa, M., & Nonomura, H. (1989). A new method for the intensive isolation of actinomycetes from soil. *Actinomycetologica*, *3*, 95–104.

Hibbett, D. S., Binder, M., Bischoff, J. F., Blackwell, M., Cannon, P. F., Eriksson, O. E., et al. (2007). A higher-level phylogenetic classification of the Fungi. *Mycological Research*, *111*, 509–547.

Hostettmann, K., & Marston, A. (2007). The search for new drugs from higher plants. *Chimia*, *61*, 322–326.

Huang, W. Y., Cai, Y. Z., Hyde, K. D., Croke, H., & Sun, M. (2008). Biodiversity of endophytic fungi associated with 29 traditional Chinese medicinal plants. *Fungal Diversity*, *33*, 61–75.

Huang, H. Y., Li, J., Zhao, G. Z., Zhu, W. Y., Yang, L. L., Tang, H. Y., et al. (2011). *Sphingomonas ednophytica* sp. nov., a novel bacterium isolated from *Artemisia annua* L. *International Journal of Systematic and Evolutionary Microbiology*, doi:10.1099/ijs.0.031484-0.

Huang, S. X., Yu, Z., Robert, F., Zhao, L. X., Jiang, Y., Duan, Y. W., et al. (2011). Cycloheximide and congeners as inhibitors of eukaryotic protein synthesis from endophytic actinomycetes *Streptomyces* sps. YIM56132 and YIM56141. *The Journal of Antibiotics*, *64*, 163–166.

Igarashi, Y., Ogura, H., Furihata, K., Oku, N., Indananda, C., & Thanmchaipenet, A. (2011). Maklamicin, an antibacterial polyketide from an endophytic *Micromonospora* sp. *Journal of Natural Products*, *74*, 670–674.

Igarashi, Y., Yanase, S., Sugimoto, K., Enomoto, M., Miyanaga, S., Trujillo, M. E., et al. (2011). Lupinacidin C, an Inhibitor of tumor cell invasion from *Micromonospora lupine*. *Journal of Natural Products*, *74*, 862–865.

Inahashi, Y., Iwatsuki, M., Ishiyama, A., Namatame, M., Nishihara-Tsukashima, A., Matsumoto, A., et al. (2011). Spoxazomicins A–C, novel antitrypanosomal alkaloids produced by an endophytic actinomycete, *Streptosporangium oxazolinicum* K07-0460T. *The Journal of Antibiotics*, *64*, 303–307.

Jiang, Y., Tang, S.-K., Wiese, J., Xu, L.-H., Imhoff, J. F., & Jiang, C.-L. (2007). *Streptomyces hainanensis* sp. nov., a novel member of the genus *Streptomyces*. *International Journal of Systematic and Evolutionary Microbiology*, *57*, 2694–2698.

Kaeberlein, T., Lewis, K., & Epstein, S. S. (2002). Isolating "uncultivable" microorganisms in pure culture in a simulated natural environment. *Science*, *296*, 1127–1129.

Kawanishi, T., Shiraishi, T., Okano, Y., Sugawara, K., Hashimoto, M., Maejima, K., et al. (2011). New detection system of bacteria using highly selective media designed by SMART: Selective medium-design algorithm restricted by two constraints. *PLoS One*, *6*, e16512.

Khan, S. T., Komaki, H., Motohashi, K., Kozone, I., Mukai, A., Takagi, M., et al. (2011). *Streptomyces* associated with a marine sponge *Haliclona* sp.; biosynthetic genes from secondary metabolites and products. *Environmental Microbiology*, *13*, 729–731.

Kumaeresan, V., & Suryanarayanan, T. S. (2002). Endophyte assemblages in young, mature, and senescent leaves of *Rhizophora apiculata*: Evidence for the role of endophytes in mangrove litter degradation. *Fungal Diversity*, *9*, 81–89.

Kusari, S., Zuhlke, S., & Spiteller, M. (2009). An endophytic fungus from *Camptotheca acuminata* that produces camptothecin and analogs. *Journal of Natural Products*, *72*, 2–7.

Li, J., Lu, C., & Shen, Y. (2010). Macrolides of the bafilomycin family produced by *Streptomyces* sp. CS. *The Journal of Antibiotics*, *63*, 595–599.

Li, J., Zhao, G. Z., Huang, H. Y., Qin, S., Zhu, W. Y., Zhao, L. X., et al. (2011). Isolation and characterization of culturable endophytic actinobacteria associated with *Artemisia annua* L. *Antonie Van Leeuwenhoek*, doi:10.1007/s10482-011-9661-3.

Li, J., Zhao, G. Z., Huang, H. Y., Zhu, W. Y., Lee, J. C., Xu, L. H., et al. (2011). *Nonomurea endophytica* sp. nov., an endophytic actinomycete isolated from *Artemisia annua* L. *International Journal of Systematic and Evolutionary Microbiology*, *61*, 757–761.

Li, J., Zhao, G. Z., Zhu, W. Y., Huang, H. Y., Xu, L. H., Zhang, S., et al. (2011). *Phytomonospora endophytica* gen. nov., sp. nov., isolated from the roots of *Artemisia annua* L. *International Journal of Systematic and Evolutionary Microbiology*, *61*, 2967–2973.

Lingappa, Y., & Lockwood, J. A. (1961). Chitin medium for isolation, growth and maintenance of actinomycetes. *Nature*, *189*, 158–159.

Matukawa, E., Nakagawa, Y., Limura, Y., & Hayakawa, M. (2007). A new enrichment method for selective isolation of *Streptomyces* from root surfaces of herbaceous plants. *Actinomycetologica*, *21*, 66–69.

Moricca, S., & Ragazzi, A. (2011). The holomorph *Apiognomonia quercina/ Discula quercina* as a pathogen/endophyte in Oak. In A. M. Pirttilä & A. G. Frank (Eds.), *Endophytes of forest trees, biology and application* (pp. 47–66). Springer Netherlands.

Nagai, A., Khan, S. T., Tamura, T., Takagi, M., & Shin-ya, K. (2011). *Streptomyces aoemiensis* sp. nov., a novel species of *Streptomyces* isolated from a soil sample using the membrane filter method. *International Journal of Systematic and Evolutionary Microbiology*, *61*, 947–950.

Niero, R., & Cechinel, V. (2008). Therapeutic potential and chemical composition of plants from the genus Rubus: A mini review of the last 10 years. *Natural Product Communications*, *3*, 437–444.

Pang, K. L., Vrijmoed, L. P., Goh, T. K., Plaingam, N., & Jones, E. B. G. (2008). Fungal endophytes associated with *Kandelia candel* (Rhizophoraceae) in Mai Po Nature Reserve, Hong Kong. *Botanica Marina*, *51*, 171–178.

Pezzuto, J. M. (1997). Plant-derived anticancer agents. *Biochemical Pharmacology*, *53*, 121–133.

Qin, S., Chen, H. H., Klenk, H. P., Zhao, G. Z., Li, J., Xu, L. H., et al. (2009). *Glycomyces scopariae* sp. nov. and *Glycomyces mayteni* sp. nov., isolated from medicinal plants in China. *International Journal of Systematic and Evolutionary Microbiology*, *59*, 1023–1027.

Qin, S., Jiang, J. H., Klenk, H. P., Zhu, W. Y., Zhao, G. Z., Zhao, L. X., et al. (2011). *Promicromonospora xylanilytica* sp. nov., an endophytic actinomycete isolated from surface-sterilized leaves of the medicinal plant *Maytenus austroyunnanensis*. *International Journal of Systematic and Evolutionary Microbiology*, *62*, 84–89. doi:10.1099/ijs.0.032185-0.

Qin, S., Li, J., Chen, H. H., Zhao, G. Z., Zhu, W. Y., Jiang, C. L., et al. (2009). Isolation, diversity, and antimicrobial activity of rare actinobacteria from medicinal plants of tropical rain forests in Xishuangbanna, China. *Applied and Environmental Microbiology*, *75*, 6176–6186.

Qin, S., Li, J., Zhang, Y. Q., Zhu, W. Y., Zhao, G. Z., Xu, L. H., et al. (2009). *Plantactinospora mayteni* gen. nov., sp. nov., a member of the family Micromonosporaceae. *International Journal of Systematic and Evolutionary Microbiology*, *59*, 2527–2533.

Qin, S., Zhao, G. Z., Klenk, H. P., Li, J., Zhu, W. Y., Xu, L. H., et al. (2009). *Nonomuraea antimicrobica* sp. nov., an endophytic actinomycete isolated from a leaf of *Maytenus austroyunnanensis*. *International Journal of Systematic and Evolutionary Microbiology*, *59*, 2747–2751.

Qin, S., Zhao, G. Z., Li, J., Zhu, W. Y., Xu, L. H., & Li, W. J. (2009a). *Actinomadura flavalba* sp. nov., an endophytic actinomycete isolated from leaves of *Maytenus austroyunnanensis*. *International Journal of Systematic and Evolutionary Microbiology*, *59*, 2453–2457.

Qin, S., Zhao, G. Z., Li, J., Zhu, W. Y., Xu, L. H., & Li, W. J. (2009b). *Jiangella alba* sp. nov., an endophytic actinomycete isolated from the stem of *Maytenus austroyunnanensis*. *International Journal of Systematic and Evolutionary Microbiology*, *59*, 2161–2165.

Qin, S., Zhu, W. Y., Jiang, J. H., Klenk, H. P., Li, J., Zhao, G. Z., et al. (2010). *Pseudonocardia tropica* sp. nov., an endophytic actinomycete isolated from the stem of *Maytenus austroyunnanensis*. *International Journal of Systematic and Evolutionary Microbiology*, *60*, 2524–2528.

Rodrigues, K. F. (1996). Fungal endophytes of palms. In S. C. Redlin & L. M. Carris (Eds.), *Endophytic fungi in grass and woody plants: Systematics, ecology, and evolution* (pp. 31–65). St. Paul, Minnesota, USA: APS Press.

Rodriguez, R. J., White, J. F., Arnold, A. E., & Redman, R. S. (2009). Fungal endophytes: Diversity and functional roles. *The New Phytologist*, *182*, 314–330.

Rosenblueth, M., & Martinez-Romero, E. (2006). Bacterial endophytes and their interactions with hosts. *Molecular Plant–Microbe Interactions*, *19*, 827–837.

Rosselló-Mora, R., & Amann, R. (2001). The species concept for prokaryotes. *FEMS Microbiology Reviews*, *25*, 39–67.

Rusch, D. B., Halpern, A. L., Sutton, G., Heidelberg, K. B., Williamson, S., Yooseph, S., et al. (2007). The Sorcerer II global ocean sampling expedition: Northwest Atlantic through eastern tropical Pacific. *PLoS Biology*, *5*, e77.

Ryan, R. P., Germaine, K., Franks, A., Ryan, D. J., & Dowling, D. N. (2008). Bacterial endophytes: Recent development and applications. *FEMS Microbiology Letters, 278*, 1–9.

Silvia, F., Sturdikova, M., & Muckova, M. (2007). Bioactive secondary metabolites produced by microorganisms associated with plants. *Biologia, 62*, 251–257.

Staniek, A., Woerdenbag, H. J., & Kayser, O. (2008). Endophytes: Exploiting biodiversity for the improvement of natural product-based drug discovery. *Journal of Plant Interactions, 3*, 75–93.

Stierle, S., Strobel, G. A., & Stierle, D. (1993). Taxol and taxane production by Taxomyces andreanae, an endophytic fungus of Pacific yew. *Science, 260*, 214–216.

Strobel, G. A. (2002). Rainforest endophytes and bioactive products. *Critical Reviews in Biotechnology, 22*, 315–333.

Strobel, G. (2006). Harnessing endophytes for industrial microbiology. *Current Opinion in Microbiology, 9*, 240–244.

Strobel, G., & Daisy, B. (2003). Bioprospecting for microbial endophytes and their natural products. *Microbiology and Molecular Biology Reviews, 67*, 491–502.

Strobel, G., Daisy, B., Castillo, U., & Harper, J. (2004). Natural products from endophytic microorganisms. *Journal of Natural Products, 67*, 257–268.

Sturz, A. V., Christie, B. R., Matheson, B. G., & Nowak, J. (1997). Biodiversity of endophytic bacteria which colonize red clover nodules, roots, stems and foliage and their influence oh host growth. *Biology and Fertility of Soils, 25*, 13–19.

Sun, X., Guo, L.-D., & Hyde, K. D. (2011). Community composition of endophytic fungi in *Aver truncatum* and their role in decomposition. *Fungal Diversity, 47*, 85–95.

Takgi, M., & Shin-ya, K. (2011). New species of actinomycetes do not always produce new compounds with high frequency. *The Journal of Antibiotics, 64*, 699–701.

Tan, H. M., Cao, L. X., He, Z. F., Su, G. J., Lin, B., & Zhou, S. N. (2004). Isolation of endophytic actinomycetes from different cultivars of tomato and their activities against *Ralstonia solanacearum in vitro*. *World Journal of Microbiology and Biotechnology, 22*, 1275–1280.

Tan, R. X., & Zou, W. X. (2001). Endophytes: A rich source of function metabolites. *Natural Product Reports, 18*, 448–459.

Tiwari, K., & Gupta, R. K. (2012). Rare actinomycetes: A potential storehouse for novel antibiotics. *Critical Reviews in Biotechnology, 32*, 108–132.

Torsvik, V., & Øvreås, L. (2002). Microbial diversity and function in soil: From genes to ecosystems. *Current Opinion in Microbiology, 5*, 240–245.

Unterseher, M., & Schnittler, M. (2009). Dilution-to-extinction cultivation of leaf-inhabiting endophytic fungi in beech (*Fagus sylvatica* L.) different cultivation techniques influence fungal biodiversity assessment. *Mycological Research, 113*, 645–654.

Vandamme, P., Pot, B., Gillis, M., Vos, P. D., Kersters, K., & Swings, J. (1996). Polyphasic taxonomy, a consensus approach to bacterial systematics. *Microbiology and Molecular Biology Reviews, 60*, 407–438.

Vartoukian, S. R., Palmer, R. M., & Wade, W. G. (2010). Strategies for culture of 'unculturable' bacteria. *FEMS Microbiology Letters, 309*, 1–7.

Venter, J. C., Remington, K., Heidelberg, J. F., Halpern, A. L., Rusch, D., Eisen, J. A., et al. (2004). Environmental genome shotgun sequencing of the Sargasso Sea. *Science, 304*, 66–74.

Verma, V. C., Kharwar, R. N., & Strobel, G. A. (2009). Chemical and functional diversity of natural products from plant associated endophytic fungi. *Natural Product Communications, 4*, 1511–1532.

Wang, X. J., Gong, D. L., Wang, J. D., Zhang, J., Liu, C. X., & Xiang, W. S. (2011). A new quninoline derivative with cytotoxic activity from *Streptomyces* sp. neau 50. *Bioorganic & Medicinal Chemistry Letters, 21*, 2313–2315.

Wang, L. W., Zhang, Y. L., Lin, F. C., Hu, Y. Z., & Zhang, C. L. (2011). Natural products with antitumor activity from endophytic fungi. *Mini Reviews in Medicinal Chemistry, 11,* 1056–1074.

Yan, L. L., Han, N. N., Zhang, Y. Q., Yu, L. Y., Chen, J., Wei, Y. Z., et al. (2010). Antimycin A18 produced by an endophytic *Streptomyces albidoflavus* isolated from a mangrove plant. *The Journal of Antibiotics, 63,* 259–261.

Yooseph, S., Sutton, G., Rusch, D. B., Halpern, A. L., Williamson, S. J., Remington, K., et al. (2007). The Sorcerer II global ocean sampling expedition: Expanding the universe of protein families. *PLoS Biology, 5,* e16.

Yu, Z., Zhao, L. X., Jiang, C. L., Duan, Y., Wong, L., Carver, K. C., et al. (2011). Bafilomycins produced by an endophytic actinomycete *Streptomyces* sp. YIM56209. *The Journal of Antibiotics, 64,* 159–162.

Zengler, K., Toledo, G., Rappe, M., Elkins, J., Mathur, E. J., Short, J. M., et al. (2002). Cultivating the uncultured. *Proceedings of the National Academy of Sciences of the United States of America, 99,* 15681–15686.

Zhang, H. W., Song, Y. C., & Tan, R. X. (2006). Biology and chemistry of endophytes. *Natural Product Reports, 23,* 753–771.

Zhao, G. Z., Li, J., Huang, H. Y., Zhu, W. Y., Park, D. J., Kim, C. J., et al. (2011a). *Pseudonocardia kunmingensis* sp. nov., an actinobacterium isolated from surface-sterilized roots of *Artemisia annua* L. *International Journal of Systematic and Evolutionary Microbiology, 61,* 2292–2297.

Zhao, G. Z., Li, J., Huang, H. Y., Zhu, W. Y., Zhao, L. X., Tang, S. K., et al. (2011b). *Pseudonocardia artemisiae* sp. nov., isolated from surface-sterilized *Artemisia annua* L. *International Journal of Systematic and Evolutionary Microbiology, 61,* 1061–1065.

Zhao, G. Z., Li, J., Qin, S., Huang, H. Y., Zhu, W. Y., Xu, L. H., et al. (2010). *Streptomyces artemisiae* sp. nov., isolated from surface-sterilized tissue of *Artemisia annua* L. *International Journal of Systematic and Evolutionary Microbiology, 60,* 27–32.

Zhao, G. Z., Li, J., Zhu, W. Y., Klenk, H. P., Xu, L. H., & Li, W. J. (2011). *Nocardia artemisiae* sp. nov., an endophytic actinobacterium isolated from a surface-sterilized stem of *Artemisia annua* L. *International Journal of Systematic and Evolutionary Microbiology, 61,* 2933–2937.

Zhao, G. Z., Li, J., Zhu, W. Y., Wei, D. Q., Zhang, J. L., Xu, L. H., et al. (2011). *Pseudonocardia xishanensis* sp. nov., a novel endophytic actinomycete isolated from the roots of *Artemisia annua* L. *International Journal of Systematic and Evolutionary Microbiology,* doi:10.1099/ijs.0.037028-0.

Zhao, G. Z., Zhu, W. Y., Li, J., Xie, Q., Xu, L. H., & Li, W. J. (2011). *Pseudonocardia serianimatus* sp. nov., a novel actinomycete isolated from the surface-sterilized leaves of *Artemisia annua* L. *Antonie Van Leeuwenhoek, 100,* 521–528.

Zhou, X. W., Lin, J., Yin, Y. Z., Zhao, J. Y., Sun, X. F., & Tang, K. X. (2007). Ganodermataceae: Natural products and their related pharmacological functions. *The American Journal of Chinese Medicine, 35,* 559–574.

CHAPTER TWO

# Cyanobacteria as a Source of Natural Products

## Martin Welker*, Elke Dittmann[†,1], Hans von Döhren[‡]

*Technische Universität Berlin, FG Umweltmikrobiologie, Sekr. BH 6-1, Ernst-Reuter-Platz, Berlin, Germany
[†]Department of Microbiology, Institute of Biochemistry and Biology, University of Potsdam, Golm, Germany
[‡]Technische Universität Berlin, Fakultät II, Institut für Chemie, Sekretariat OE 2, Franklinstraße, Berlin, Germany
[1]Corresponding author: e-mail address: editt@uni-potsdam.de

## Contents

## Abstract

Cyanobacteria or blue-green algae from various environments have been recognized as sources of a variety of bioactive metabolites. Strategies of strain isolation from aquatic habitats, and cultivation and harvesting for metabolite production are described. Strategies for screening of compounds are discussed, including their direct MALDI-TOF mass spectrometric detection in whole cells. Genetic approaches including genomic mining, mutagenesis including transcriptional activation, heterologous expression, and in vitro reconstitution of pathways are presented.

*Methods in Enzymology*, Volume 517
ISSN 0076-6879
http://dx.doi.org/10.1016/B978-0-12-404634-4.00002-4

23

# 1. INTRODUCTION

Cyanobacteria are increasingly recognized as a prolific source of bioactive natural products. These oxygenic phototrophic bacteria proliferate in very varied environments, including marine and freshwater as well as terrestrial habitats (Rippka, 1988). They may flourish at the surface of lakes or the ocean or form mats in benthic environments. Cyanobacteria are also frequently found in symbiotic associations with invertebrates, plants, or fungi. Natural products from cyanobacteria were originally discovered through bioactivity-guided screening programs and revealed a truly fascinating variety of structures and inhibitory activities (Patterson, Larsen, & Moore, 1994).

The two most intensively studied habitats are tropical marine environments and freshwater habitats. Notably, compounds isolated from these different environments show distinct metabolite profiles, with respect to both structural properties and bioactivity features. Strains from tropical marine habitats such as *Lyngbya majuscula* or *Moorea producta* frequently produce cytotoxic compounds, some of them having high potential for drug development. The structures of these metabolites, such as curacin A, jamaicamide, barbamide, hectochlorin, lyngbyatoxin, and aplysiatoxin, possess a polyketide backbone, but often contain amino acid constituents (Jones, Gu, Sorrels, Sherman, & Gerwick, 2009). On the other hand, bloom-forming freshwater cyanobacteria such as *Microcystis*, *Planktothrix*, and *Anabaena* produce a variety of peptide backbones, eventually having polyketide side chains. Examples include microcystin, aeruginosin, anabaenopeptin, cyanopeptolin, microginin, and microviridin showing activity against various proteases or protein phosphatases (Welker & von Döhren, 2006). Cyanobacterial strains from terrestrial sources were only incidentally screened but revealed yet different types of compounds, such as cryptophycin, nostopeptolide and nostocyclopeptide, scytonemin, and mycosporic acid (Kehr, Gatte Picchi, & Dittmann, 2011). Although certain types of cyanobacterial compounds, including microcystin and nostopeptolide, have been detected in different habitats the majority of them seem to be specific for a particular environmental niche, pointing toward distinct ecological and physiological roles for these compounds. There are, however, exceptions to this rule. Patellamide and related cyanobactins, for example, show an almost ubiquitous distribution combined with a range of bioactivities, although the function of these compounds cannot yet be anticipated (Donia, Ravel, & Schmidt, 2008).

Here, we give an overview of methodological developments for the analytical detection, gene assignment, and structural elucidation of cyanobacterial secondary metabolites. A special focus is given to sampling strategies in freshwater lakes, the advancement of mass spectrometric techniques, and genetic tools for genomic mining of cyanobacterial secondary metabolites.

## 2. CYANOBACTERIA AS STARTING MATERIAL FOR NATURAL PRODUCTS

Despite the recognized diversity and bioactivity of cyanobacterial natural products (Chlipala, Mo, & Orjala, 2011; Sielaff, Christiansen, & Schwecke, 2006; Welker & von Döhren, 2006), the number of individual compounds for which a clinical application is considered is rather limited. Reasons for this imbalance may be the difficulty of obtaining pure isolates, the low growth rates and biomass yields exhibited by most cyanobacterial isolates, and the comparatively low cell quota of most natural products in wild-type isolates. These reasons apply in principle to the vast majority of new natural products (Beutler, 2012; Molinari, 2009)—probably only rather more to cyanobacteria. Nevertheless, recent developments in analytics, molecular engineering, and cultivation techniques allow some optimism that limitations encountered in the past may be overcome.

These limitations can be demonstrated by the example of microcystins, probably the best studied cyanobacterial natural products (Dittmann & Wiegand, 2006; van Apeldoorn, van Egmond, Speijers, & Bakker, 2007; Zurawell, Chen, Burke, & Prepas, 2005). The hepatotoxic microcystins are cyclic peptides commonly produced by freshwater, bloom-forming species through a hybrid NRPS/PKS (nonribosomal synthetase/polyketide synthase) biosynthesis occurring in some 90 structural variants (Tillett et al., 2000; Welker, 2008). Microcystins have a major economic impact owing to the potential contamination of drinking water when this is produced from surface water sources (Haddix, Hughley, & Lechevallier, 2007; Westrick, Szlag, Southwell, & Sinclair, 2010). The surveillance of microcystin concentration has been included in guidelines in a number of countries (WHO, 2006) and toxicological studies are ongoing (Zegura, Straser, & Filipic, 2011). Therefore, microcystins are compounds of commercial interest, on the one hand, as reference material for chemical analysis and, on the other hand, as material for studies to solidify the toxicological characteristics of the compounds. Yet, only a few out of more than 90 known structural variants are commercially available and only for a price of some 3000 €/mg—about 50,000 times the price of gold (as of February 2012). If it was easy to produce pure microcystins from essentially costless field collections of biomass or cultivated cell material, considerable profits could be made. Yet, this is apparently not the case and what applies for microcystins applies even more for other cyanobacterial compounds for which no market has yet been established. What applies to microcystins applies the more to other cyanobacterial compounds for which as of now only an indication of bioactivity against selected enzymes is available.

Newly described cyanobacterial compounds are isolated from either laboratory strains or field collections, the former generally applying mostly to freshwater cyanobacteria while the latter mostly to marine species. The reason for this is partly historical because in freshwater ecology a long tradition exists to study interactions between organisms under laboratory conditions using strains, including cyanobacteria. Therefore, numerous freshwater cyanobacterial strains were available before they were recognized as prolific sources of natural products.

## 2.1. Isolation of cyanobacterial strains

Pure cyanobacterial isolates, so-called axenic strains, are only obtained after—generally numerous—laborious purification steps (Bolch & Blackburn, 1996; Wolk, 1988). Contaminants are often caused by heterotrophic bacteria that thrive in close association with cyanobacterial cells, for example, within the mucilaginous sheaths surrounding trichomes. Also small picoplanktonic phototrophs can be observed in cyanobacterial cultures, potentially overgrowing the original isolate. Axenic cultures are readily available only from a few recognized culture collections such as the Pasteur Culture Collection (PCC, France) and the National Institute of Environmental Studies (NIES, Japan) and have been the source of new natural products (Ishida, Matsuda, & Murakami, 1998; Murakami et al., 1995; Shin, Murakami, Matsuda, Ishida, & Yamaguchi, 1995). Non-axenic cyanobacterial strains of many species have been studied for natural products, generally assuming that the natural product eventually isolated is indeed produced by the cyanobacterium and not by accompanying heterotrophs. Although this might be a valid assumption in the majority of cases, bias due to production by heterotrophic bacteria has to be considered.

## 2.2. Mass cultivation of cyanobacterial strains

A number of cyanobacterial species and strains are cultivated traditionally for high biomass yields with success. "Spirulina" (*Arthrospira* sp.), for example, was historically harvested in Central America and Africa for human consumption and in our days is cultivated in open, swimming-pool-sized raceway ponds for the production of nutritional complements or *Nostoc flagelliforme* (fat choy) that is traditionally cultivated in Asia for human consumption. In recent years, a new aspect of cyanobacteria has received increased attention, the production of biofuels (Brennan & Owende, 2010; Parmar, Singh, Pandey, Gnansounou, & Madamwar, 2011). As a prerequisite, mass cultivation of cyanobacteria has been established for full-scale production. Though only a few strains are at present cultivated in an industrial scale, the success underlines the expectation that upscaling might not be a principle limitation but can be overcome by technological know-how and selection of suitable strains.

Cultivation of a particular producer strain in larger volumes is required to yield enough biomass to extract the target compound in amounts allowing structure elucidation and bioassays. What is considered as mass cultivation in most laboratories, however, will scarcely lead to biomass yields as it would be expected from mass cultivation of heterotrophic bacteria or fungi. This is due to the fact that biomass yields per volume are generally an order of magnitude lower for cyanobacteria than for heterotrophs. The reason for this is in part the obligate phototrophic metabolism. With increasing cell densities the availability of the primary energy source, the light penetrating the culture, is decreasing (Huisman, 1999). In particular, in larger vessels required for mass cultivation, further growth can become light limited. This, in consequence, increases the risk of a collapse of the culture either through lysis (van Hannen et al., 1999) or, in the case of *Nostocales*, through the massive formation of resting stages, so-called akinetes (Adams & Duggan, 1999). Therefore, the mass cultivation of cyanobacteria generally requires optimization for individual strains, especially when an upscaling is considered.

### 2.2.1 Protocol for mass cultivation of cyanobacteria in a laboratory scale

Different cyanobacterial species and strains exhibit markedly different behavior in culture. Hence, the protocol should be understood more as a guideline that requires adaptation to selected strains. Light can be provided by white fluorescent tubes. Light intensity requirements differ largely between individual species and strains.

1. When multiple producer strains are available select one that has been stably cultivated in a particular medium for several months in batch culture. A selection of recipes for common media for cyanobacteria is available (Rippka, 1988).

2. Scale up the culture volume stepwise with the inoculum volume at least 20% of the final volume from small culture flasks to up to 2 L cultures.

3. In volumes exceeding 500 mL aeration with compressed, filtered air is required. In-line air filters are available from Sartorius, for example. Aeration is achieved through a system of glass or plastic pipes connected to the source of compressed air by silicon tubing.

4. Prepare larger volumes of medium from deionized water and stock solutions directly in the final culture vessels. Polycarbonate bottles of up to 20 L volume have been found practical (e.g., Nalgene). If possible, glassware should be avoided to reduce the risk of accidents.

5. Sterilize directly in the medium in the large-volume vessel in an appropriate autoclave. Run a program with at least 45 min at 121 °C. Consider extended cooling times, inside the autoclave as well as outside of it.

6. When cooled to room temperature, add trace elements and carbonates according to the medium recipe.

7. Connect the aeration tubes to the compressed air supply system and adjust aeration to a constant bubbling.

8. Add the inoculum to the vessel. For axenic cultures this needs to be done under a flow cabinet. For non-axenic cultures this can be done on a standard laboratory bench.

9. Place the culture vessel under the light source. In the initial incubation phase, light intensity should be lowered to some 20–40 µE at the outer wall of the vessel closest to the light source.

10. The incubation may take several weeks or even months to reach a density that makes a cell harvest efficient. Incubation duration is always a trade-off between higher biomass yields and increasing risk of culture collapse. Thus, the culture must be checked at regular intervals for viability and the light intensity and aeration adjusted. Perform quantitative analysis of target compound concentrations.

11. Harvest cells with a flow-through centrifuge or in aliquots in a large centrifuge. Collect pellets and freeze them in appropriate vessels until processed.

## 2.3. Harvesting cyanobacteria from natural habitats

One possibility to overcome the problem of biomass production to isolate natural products is to use material directly harvested from natural sources—as is done in many marine studies. A prerequisite for this approach is the abundance of strains or species of interest in large quantities and the ability to harvest the cells efficiently. Species growing as single cells or colonies consisting of only a few cells cannot be harvested efficiently from field populations. Especially planktonic, freshwater cyanobacterial blooms of the genera *Microcystis*, *Planktothrix*, *Aphanizomenon*, and *Anabaena* have been exploited in this respect. During bloom conditions, these genera tend to form dense surface blooms that can be "skimmed" from natural waters, easily yielding biomass in the range of kilograms of dry weight that can be processed in the laboratory. Similarly, marine cyanobacteria that grow in macroscopic tufts in shallow marine environments (e.g., *L. majuscula*, the "fireweed," new classification: *M. producta*) can be harvested in larger quantities (Luesch, Williams, Yoshida, Moore, & Paul, 2002). It should be noted that the diversity of natural product structures in natural assemblages can be orders of magnitude higher than in clonal isolates (Forchert, Neumann, & Papendorf, 2001; Welker, Christiansen, & von Döhren, 2004; Welker, Marsalek, Sejnohova, & von Döhren, 2006), making the purification of target compounds potentially very tedious. Further, repeated supply with the same material may be impossible owing to the dynamic changes of the chemotype composition in natural blooms. Eventually, harvesting may be restricted for conservational reasons, for example, in reef habitats.

### 2.3.1 Protocol: Harvesting cyanobacterial cells from field habitats

Note since no two sites where harvesting is intended are the same, nor are the conditions at one site the same at different dates, the protocol necessarily gives general recommendations rather than detailed guidance that could be followed literally. Species growing in single cells or colonies consisting of only a few cells cannot be harvested efficiently from field populations.

1. Obtain permissions for sampling. Especially for sensitive areas, this must be planned well ahead of the sampling campaign.
2. In the laboratory prepare anything required for the first sample processing like filtering devices, place in a freezer, homogenizers, etc.
3. Prepare sampling devices such as nets, sieves, gloves, rakes, plastic funnels, etc. For SCUBA diving, the general and local safety regulations must be followed.

**4.** Prepare cooling devices such as a cooling box and appropriate sample containers such as wide-mouth plastic bottles or polyethylene zipper-bags in sufficient numbers to contain the collected material. Prepare labels when multiple separate samples are envisaged.

**5.** Collection on site depends primarily on the targeted type of cyanobacteria.

    **a.** Benthic cyanobacteria (e.g., *Lyngbya* spp.) can be harvested by man-ually collecting macroscopic assemblages.

    **b.** Planktonic cyanobacteria forming surface scums (e.g., *Microcystis* and *Aphanizomenon* spp.) can be collected from the water surface with a bucket or directly in wide-mouth bottles at high cell densities. When a further concentration is desired this can be achieved with a plankton net or in a larger scale, by passing the water containing the cyano-bacteria through a cloth.

    **c.** Planktonic cyanobacteria homogeneously distributed in the water column (e.g., *Planktothrix* and *Cylindrospermopsis* spp.) can be col-lected with a plankton net. The mesh size should allow the efficient retention of cyanobacteria, generally not exceeding 30 μm. Alterna-tively, water in large quantities can be passed through a cloth laid out on the ground with appropriate support.

**6.** Further processing depends on the planned procedure.

    **a.** When freezing is intended for storage, aliquot the cyanobacterial sam-ples in portions allowing efficient lyophilization at a later time, that is, thickness of frozen samples not exceeding 3 cm.

    **b.** When the material is to be used fresh, keep the sample cool and start extraction as soon as possible. Freeze excess material for later use.

Given that a larger amount of biomass has been harvested the next step is the extraction of compounds of interest. It should be mentioned that the natural products isolated from cyanobacteria so far are present in only relatively low concentrations in the starting material. The yields are often far below 1% of dry biomass, for example, Micropeptins 478 A and B from *Microcystis* at 0.006% (w/w) (Ishida, Matsuda, Murakami, & Yamaguchi, 1997), Schizopeptin 791 from *Schizotrix* sp. at 0.003% (w/w) (Reshef & Carmeli, 2001), and taveuniamide A from a *Lyngbya* and *Schizothrix* consortium at 0.02% (w/w) (Williamson, Singh, & Gerwick, 2004).

Even with an increase in purification efficiency by an order of magni-tude, the yields would still be poor. One strategy to overcome this could be the heterologous production of cyanobacterial natural products (see below).

# 3. SCREENING FOR NEW COMPOUNDS AND PRODUCERS

The discovery of new natural products generally follows one of two principle strategies. With the first, a new compound has been found in a sample during a screening and is then isolated and structurally characterized, and eventually tested for bioactivity in an array of tests. Alternatively, the isolation starts directly with bioactivity assays on crude extracts that further guide the entire fractionation and purification process. For both tracks, advantages and disadvantages can be outlined.

## 3.1. Screening for new compounds

An advantage of the compound-guided approach is that the chemical novelty of a newly identified compound can be estimated based on preliminary analytical data (see below). The major disadvantage, however, is the risk that after all purification efforts no pronounced bioactivity is evident. In this case, the isolated compound may be useful when included in a compound library that can be used to test multiple compounds in multiple test systems in a single experiment (Bindseil et al., 2001). Examples for a compound-guided isolation are georgamide (Wan & Erickson, 2001), prenylagaramides (Murakami, Itou, Ishida, & Shin, 1999), or microcyclamides 7806A and 7806B (Ziemert et al., 2008).

For the bioactivity-guided approach, the advantage is in turn the certainty that the compound eventually purified shows bioactivity in the chosen test system(s). The disadvantage, on the other hand, is the risk that the compound purified with considerable efforts proves to be an already known compound. To balance the risks associated with both strategies, in practice often combinations of the two strategies are applied by screening bioactive fractions for known compounds, for example. Examples for the bioactivity-guided approach are much more numerous, and even when not stated explicitly, a screening of bioactive fractions for known compounds is generally done at the same time.

As examples of the principal strategies and of combinations thereof, four recent (randomly selected) papers on new cyanobacterial peptides will be summarized.

Gademann, Portmann, Blom, Zeder, and Jüttner (2010) reported a new variant of the cyanopeptolin class, cyanopeptolin 1020, that shows toxicity to the freshwater crustacean *Thamnocephalus platyurus*. Fractions of a crude extract of a *Microcystis* strain (UV-006) were subjected to the bioassay and in eight out of nine active fractions microcystin congeners could be identified based on absorption and mass spectra. In the remaining active fraction, a new peptide was identified that was then purified from raw material by HPLC and characterized by NMR and Marfey's analysis.

Taniguchi et al. (2010) isolated and characterized the new compound palmyramide A together with the known peptides curacin D and malyngamide C from a field collection of a cyanobacterium/red alga consortium originating from the Palmyra atoll in the Pacific Ocean. After subjecting the dried crude extract to silica gel vacuum liquid chromatography, the resulting fractions were subjected to sodium channel-blocking activity assays. From the active fraction, palmyramide A was isolated together with malyngamide C by chromatographic methods supported by mass spectrometry. MALDI imaging eventually confirmed that the cyanobacterium is indeed the producer of palmyramide A.

Okumura, Philmus, Portmann, and Hemscheidt (2009) isolated new variants of the cyanopeptolin- and anabaenopeptin-type peptide classes after they realized an increase in peak area in HPLC chromatograms of a *Planktothrix agardhii* (CYA 126/8) to which homotyrosine was fed. The initial rationale for the experiments was to study some details of the biosynthesis of microcystins. However, once the peptides caught the authors' attention these were isolated and characterized by standard methodology.

These examples highlight the diverse "starting point" of the discovery of new natural products, ranging from the directed search for compounds showing activity in a particular biotest to pure serendipity. Equally, the starting material is diverse, from exotic field collections to axenic laboratory strains isolated decades ago. Further, the examples underline the need for detailed information on already published structures to avoid the laborious purification of compounds already known and respective techniques to reliably identify them.

### 3.1.1 Protocol: Screening by whole-cell mass spectrometry of freeze-dried cyanobacterial biomass for the presence of microcystins

This protocol is applicable in principle to any other class of compounds that have been shown to be detectable with matrix-assisted laser desorption/ionization time-of-flight mass spectrometry (MALDI-TOF MS).

1. Transfer a small amount of material to reaction tube equivalent to 5–50 μL; the amount to be transferred is mostly dependent of the texture of the dry material.

**1b**. *Alternative*: fresh cells are harvested from live cultures by centrifugation or filtration, including a washing step with pure water to remove excess salts from the medium → 5b.

2. Add a mixture of acetonitrile:ethanol:water (1:1:1) acidified with 0.05% trifluoro-acetic acid (room temperature); the volume needs to be adjusted to fully suspend the cyanobacterial material.

3. Sonicate the suspension in a sonication bath for 1 min at room temperature.

4. Spin down larger particles; the supernatant has not to be clear but free of larger particles to avoid contamination of the instrument.

5. Transfer 1 μL to a MALDI target and let evaporate the solvents to near dryness; preferably in replicates to have spots available for further analyses (postsource-decay, PSD).

    **5b**. *Alternative*: transfer fresh cells equivalent to a biovolume of 0.1–0.5 μL to a spot on the MALDI target and allow to dry nearly. Especially in strains with strong mucilage, a complete drying should be avoided to allow efficient extraction.

6. Add 1 μL of DHB (2,5-dihydroxy benzoic acid) matrix solution consisting of 10 mg DHB dissolved in the solution given in step 2. Alternative matrix compounds are a-cyano-4-hydroxycinnamic acid and sinapic acid, the use of which may result in an improved response of target compounds in mass spectra.

7. Allow solvents to evaporate at room temperature and the matrix to crystallize; when no crystal formation is observed, reduce the volume of extract transferred to the target.

8. Prepare an appropriate calibrant mix to one or more spots on the same target as the samples and allow to dry.

9. Introduce loaded target into MALDI-TOF MS instrument and calibrate.

10. Analyze sample in automode or manually with appropriate settings for $DHB^2$ matrix in reflector mode.

11. Analyze resulting mass spectra for expected $m/z$ values of known microcystin variants; consider $m/z$ values of $M + H^+$, $M + Na^+$, $MK^+$.

12. Optional: recalibrate the spectrum by using chlorophyll-a derivatives as internal standards (pheophytin-a: $m/z$, pheophorbide-a $m/z$).

13. Analyze possible microcystin peaks (known variants or new ones) for isotopic distribution (most MALDI softwares allow the computing of an expected isotopic distribution based on a molecular formula).

14. Perform a fragment analysis by PSD fragmentation and collision-induced dissociation.

**15.** Analyze PSD spectrum for signature fragments and compare sample fragment spectrum to reference fragment spectra of other microcystins; for possible new variants, compare fragment spectrum to theoretical fragment spectra. Signature fragments can be found in Fastner et al. (1999) and Welker et al. (2004).

## 3.2. Characterization of (new?) natural products

The first steps to characterize new natural products, bioactive fractions, or new producing organisms generally include analytical techniques that allow moderate- to high-throughput screening. In this light, NMR, for example, is not a very useful technique for the initial steps, as it requires considerable amounts of more or less pure compounds that could turn out to be already known. Considering the immense structural diversity—and hence broad variability in physicochemical properties—it is not surprising that essentially any analytical method suitable to detect small- to medium-sized organic molecules has been applied for cyanobacterial compound characterization, from various chromatographic methods (TLC, HPLC, SPE) and mass spectrometric methods (MALDI-TOF, HRFAB, ESI-TOF) to classical methods like refraction analysis or spectrophotometry. It is beyond the scope of this chapter to discuss all methods that have been applied in one or the other study, so we will concentrate on a few that are universally applicable.

A screening for cyanobacterial natural products can have two principal targets: first, the discovery of new compounds, and second, the discovery of a new producer of an already well-known compound.

To detect entirely new compounds, the methods should be only moderately specific, that is, it should be possible to detect molecules over a broad range of physicochemical properties. Further, the time per sample should be short but the information obtained accurate and complete enough to allow the identification of positive samples (i.e., with new compounds or known ones of interest) as well as negative sample (i.e., without compounds of interest). These requirements are met, for example, by MALDI-TOF MS, which allows the simultaneous detection of a wide diversity of compounds over a defined mass range. For peptides and many polyketides, for example, a mass range from 400 to 2500 Da can be scanned with high accuracy directly from samples after a minimal sample preparation. In the mass spectra, peak masses and isotope patterns can then be compared to characteristics of known compounds and from intense peaks fragment spectra can be obtained directly from the same sample by PSD fragmentation (Li, Garden,

Romanova, & Sweedler, 1999; Spengler, 1997). By comparison to fragment spectra of known compounds and *in silico* fragmentation a new compound can often be characterized and the (flat) structure partially elucidated (Czarnecki, Lippert, Henning, & Welker, 2006; Welker et al., 2006). This applies especially for new structural variants of known compounds, for example, new congeners of a known peptide type. With a MALDI-TOF MS approach, a large number of new structural variants of known compound classes could be identified and characterized as well as entirely new structures. A similar rapid screening can also be performed with LC–MS and LC–MS/MS (Rohrlack et al., 2008). While requiring more preparation steps and time compared to MALDI-TOF MS, LC–MS has the advantage of being more sensitive, especially with respect to minor components that can be suppressed in MALDI-TOF mass spectra. The closer the compound isolation process comes to the point where a compound is considered as pure, the more refined and accurate mass spectral analyses are generally applied, eventually often high-resolution FAB, ESI-TOF, or ESI-Q to establish the molecular formula. Pure compounds are normally subjected to NMR studies and amino acid analysis (in the case of peptides) to elucidate the flat structure and Marfey's analyses to reveal chirality in absolute configurations.

## 3.3. Finding new sources of known compounds

When the discovery of a new producer, for example, one that produces a compound of interest in higher cell quota, is the target of the screening, preferably methods are applied that, on the one hand, are sensitive to allow the quantification and, on the other hand, specific for an unambiguous identification. To this end, various chromatographic techniques are available, often coupled to mass spectrometry, such as LC–MS, HPLC, and GS–MS. However, owing to the fact that only a very limited number of cyanobacterial compounds are of commercial and pharmaceutical interest, extended screenings for producers of a particular compound have rarely been carried out. The best known example, again, is the production of toxic compounds such as microcystins, saxitoxins, and cylindrospermopsins by a broad variety of cyanobacterial taxa as has been studied intensively with accurate and quantitative screening methods such as HPLC-PDA, LC–MS, HILIC, ELISA, and others (e.g., Antal et al., 2011 Ballot, Fastner, & Wiedner, 2010; Bigham, Hoyer, & Canfield, 2009; Liu & Scott, 2011). For a reliable detection, the methods require optimization specific for the target compound, thus being unfit for the detection or recognition of unknown compounds in a sample. For example, antibodies for ELISA are selected to specifically interact with the target molecule or parts of it while being inert to all other compounds in a sample. Less specific is HPLC and, as the example above shows, a close examination of chromatographic data can lead to the isolation of new compounds.

Although specific screenings for known compounds other than cyanobacterial toxins are not widely performed to date, the experiences made in this field will surely support the search for new sources, once a cyanobacterial natural product has been considered as an interesting pharmaceutical candidate.

## 4. GENOMIC MINING OF CYANOBACTERIAL NATURAL PRODUCTS

The rapidly rising availability of cyanobacterial genome sequences has inspired the discovery of novel cyanobacterial natural products and the assignment of biosynthetic pathways through genomic mining strategies. Genomes can provide the clue for the elucidation of biosynthetic pathways in two possible directions: (1) structures of interest are known, but biosynthetic pathways have not yet been assigned and (2) genome analysis reveals the biosynthetic potential to produce novel types of compounds, but structures are not yet known (Challis, 2008). The basis of any genome-based pathway assignment or metabolite discovery is a thorough bioinformatic analysis. The computational analysis can be guided by precursor feeding studies that allow for a prediction of possible enzymatic features, or vice versa the bioinformatic analysis can suggest the composition of building blocks for a cryptic metabolite (Jenke-Kodama & Dittmann, 2009). Here, we compile a few methodological aspects that specifically apply to genomic mining strategies in cyanobacteria.

## 4.1. Mutagenesis of biosynthetic genes in cyanobacteria

Mutagenesis provides a definite possibility to assign a metabolite to a biosynthetic pathway and vice versa. The competence of cyanobacteria and the availability of genetic techniques differ from strain to strain even within the same genus. There are, however, a few characteristics that may help to select an appropriate strain for mutagenesis. Unicellular cyanobacterial strains including those of the genera *Synechococcus*, *Synechocystis*, and *Microcystis* are often naturally competent, that is, are capable of taking up free extracellular DNA (Koksharova & Wolk, 2002). The success of a genetic manipulation approach largely depends on the presence and number of extracellular and intracellular nucleases that break down foreign DNA. Methods of choice for unicellular cyanobacteria are the direct addition of DNA on suicide plasmids or the introduction of these constructs through electroporation (Koksharova & Wolk, 2002). The majority of the cyanobacteria can be manipulated via double-crossover recombination. For *Microcystis*, *Synechocystis*, and *Synechococcus*, the following protocol using chloramphenicol resistance as selective marker can be used.

### 4.1.1 Genetic manipulation of cyanobacteria

1. Select an *E. coli* host strain for knockout plasmid propagation that carries *dam* and *dcm* methylation genes (e.g., XL1 Blue). Plasmid constructs should contain your gene of interest and the selective resistance cassette flanked by 1 kb continuous sequence on both sides.
2. Prepare plasmid DNA using standard plasmid purification kits.
3. *In vitro* methylate DNA using 1 U *Sss*I methylase and 160 $\mu M$ S-adenyosyl-methionine (New England Biolabs) per $\mu g$ of plasmid DNA for 2 h at 37 °C (optional, especially recommended for *Microcystis*).
4. Harvest a 10-mL sample of a cyanobacterial log-phase culture ($6 \times 10^7$ cells mL$^{-1}$) and redissolve the pellet in 200 $\mu L$ BG-11 growth medium (Rippka, 1988).
5. Add 10 $\mu g$ of purified and *in vitro* methylated plasmid DNA and incubate for 1 h (40 $\mu E$ m$^{-2}$ s$^{-1}$ continuous white light).
6. Spread cells on nonselective BG-11 plates (1% Bacto-Agar, Difco Laboratories).
7. Grow cells for 48 h at 40 $\mu E$ m$^{-2}$ s$^{-1}$ continuous white light at 25 °C.
8. Add a chloramphenicol gradient to the plate from one side by slightly lifting the agar on one side (500 $\mu L$ of chloramphenicol at 10 $\mu g$ mL$^{-1}$). The gradient will form by diffusion.
9. Grow cyanobacteria at 40 $\mu E$ m$^{-2}$ s$^{-1}$ continuous white light for up to 5 weeks. Transformants will be obtained after 2–4 weeks depending on the selected cyanobacterial genus.

Electroporation is also the method of choice for filamentous motile cyanobacteria of the section *Oscillatoriales*, although only one genus, *Planktothrix*, was so far successfully manipulated (Christiansen, Fastner, Erhard, Borner, & Dittmann, 2003). Whereas mutants of unicellular strains can usually be selected on plates after a few weeks the motile mutant strains are more easily selected in liquid medium. The following protocol can be used for *Planktothrix* mutagenesis.

1. Prepare plasmid DNA from an *E. coli* host strain containing *dam* and *dcm* methylation genes (e.g., XL1 Blue) using a standard plasmid purification kit.
2. Linearize the vector in the multicloning site.
3. Prepare single-stranded vector DNA by incubating the plasmid at 95 °C for 10 min.
4. Harvest 50 mL of a mid-log-phase *Planktothrix* culture and wash the pellet three times with 1 m$M$ HEPES buffer.
5. Resuspend the cells in 200 µL of 1 m$M$ HEPES buffer and add 10 µg of single-stranded plasmid DNA.
6. Electroporate the cells (1.0 kV, 25 µF, 200 Ω).
7. Inoculate 100 mL of BG-11 medium and cultivate the cells at 20 µE m$^{-2}$ s$^{-1}$ continuous white light at 25 °C. After 3 days, add 50 µg of chloramphenicol to the culture medium.
8. Add fresh BG-11 medium (Rippka, 1988) containing 0.5 µg mL$^{-1}$ after 4–6 weeks.
9. Purify transformants by stepwise increasing the chloramphenicol concentration up to 5 µg mL$^{-1}$.

Nonmotile filamentous cyanobacterial strains of the section *Nostocales* can differentiate distinct cell types for nitrogen fixation, the so-called heterocysts. A few strains of this section, including *Nostoc* PCC7120, *Anabaena variabilis* ATCC29413, and *Nostoc punctiforme* PCC73102, were successfully manipulated. The commonly used technique is conjugation. The efficiency of the manipulation can be enhanced by using conjugative vectors carrying genes for endogenous restriction methylases. The conjugation vectors also carry genes that allow for a discrimination of single- and double-crossover events, such as the levansucrase gene *sacB* that facilitates growth of strains on sucrose (Koksharova & Wolk, 2002). Examples of biosynthetic pathways that could be successfully assigned by mutagenesis include the anabaenopeptilide pathway in *Anabaena* str. 90 (Rouhiainen et al., 2000), the microcystin pathway in *M. aeruginosa* PCC7806 and K81 (Dittmann, Neilan, Erhard, von Döhren, & Borner, 1997; Nishizawa, Asayama, Fujii, Harada, & Shirai, 1999), the aeruginosin pathway in *M. aeruginosa* PCC7806 (Ishida et al., 2009), the aeruginoside pathway in *P. agardhii* Niva-CYA126 (Ishida et al., 2007), and the microviridin pathway in the same strain (Philmus, Christiansen, Yoshida, & Hemscheidt, 2008).

If an assignment of biosynthetic genes is not possible via bioinformatic analysis a random mutagenesis approach might be suitable. There are, however, only a few cyanobacterial strains for which random mutagenesis has been established, namely, *Synechococcus* PCC7002, *Synechococcus* PCC7942, *Synechocystis* PCC6803, and *Nostoc* PCC7120 (Koksharova & Wolk, 2002). The most common technique used is transposon mutagenesis. There is currently no example for an assignment of a secondary metabolite pathway via a random mutagenesis approach.

## 4.2. Transcriptional activation of biosynthetic genes in cyanobacteria

Problems with the detection of secondary metabolites from cyanobacteria might be due to the fact that a given biosynthetic gene cluster is silent or transcribed at a very low level. Mining of the corresponding metabolites can be facilitated by finding conditions that lead to induction of transcription of biosynthetic genes or by introducing strong promoters. In any case, quantitative analysis of transcript amounts under different environmental conditions and analysis of the transcriptional organization are a good starting point. For a few model strains, such as *Synechocystis* PCC6803, *Nostoc* PCC7120, *N. punctiforme* PCC73102, *Cyanothece* ATCC 51142, *Prochlorococcus* spp., and *Microcystis* PCC7806 microarray data and RNA deep-sequencing data are available for various environmental conditions (e.g., Christman, Campbell, & Meeks, 2011; Mitschke, Vioque, Haas, Hess, & Muro-Pastor, 2011; Murata & Suzuki, 2006; Shi, Ilikchyan, Rabouille, & Zehr, 2010; Steglich, Futschik, Rector, Steen, & Chisholm, 2006; Straub, Quillardet, Vergalli, Tandeau de Marsac, & Humbert, 2011). Analysis of known secondary metabolite biosynthetic gene clusters from these datasets revealed that the clusters are typically not silent, although some of them are transcribed at a very low level. Conditions that were shown to influence the transcriptional level of biosynthetic gene clusters in cyanobacteria include iron limitation, light quantity and quality, and nitrogen availability (Neilan, Pearson, Muenchhoff, Moffitt, & Dittmann, 2012). For the well-characterized cyanobacterial hepatotoxin microcystin, an autoinduction could be shown (Neilan et al., 2012). Nevertheless, the factor by which the transcription is induced typically does not exceed a factor of five (Neilan et al., 2012).

Strong cyanobacterial promoters are extensively tested for the metabolic engineering of cyanobacteria toward enhanced biofuel production (Ruffing, 2011). To date, promoter exchange has not yet been applied for an enhanced production of cyanobacterial secondary metabolites. Promoters that have been very successfully applied for overexpression approaches in cyanobacteria include the light-dependent *psbA2* promoter that regulates transcription of the gene encoding the D1 protein of photosystem II in cyanobacteria, the IPTG-inducible lacZ promoter (Ruffing, 2011) and the $Cu^{2+}$-dependent *petJ* promoter regulating synthesis of a respiratory cytochrome in cyanobacteria (Tous, Vega-Palas, & Vioque, 2001).

## 4.3. Heterologous production of natural products from cyanobacteria

An alternative possibility for the mining of cyanobacterial secondary metabolites and the assignment of biosynthetic pathways is production of the respective compounds by heterologous expression in other bacterial hosts. *E. coli* has been proven as an appropriate host for the expression of cyanobacterial biosynthetic gene clusters (Li et al., 2010; Long, Dunlap, Battershill, & Jaspars, 2005; Schmidt et al., 2005; Ziemert, Ishida, Liaimer et al., 2008; Ziemert, Ishida, Quillardet, et al., 2008). Both groups of bacteria are Gram-negative and share a very similar codon usage, allowing the transfer of genes without adapting the codon sequences. Moreover, cyanobacterial promoters are often recognized in *E. coli*. Hence, expression of genes and production of metabolites are possible without applying laborious cloning procedures and utilizing *E. coli* expression systems. Patellamides and microviridins, for example, were effectively produced from *E. coli* strains carrying cyanobacterial fosmid clones (Long et al., 2005; Ziemert, Ishida, Liaimer et al., 2008; Ziemert, Ishida, Quillardet, et al., 2008). Heterologous production was also shown for a number of further peptides of the cyanobactin family and prochlorosins, a family of lantipeptides from *Prochlorococcus* (Donia et al., 2008; Li et al., 2010). The advantage of heterologous production in *E. coli* is not only the correlation of biosynthetic genes and compound production but also the much faster growth of these bacteria compared to cyanobacteria and the accessibility to genetic manipulation.

*Streptomyces coelicolor* and fast-growing cyanobacteria such as *Synechocystis* PCC6803 have been tested as alternative hosts for cyanobacterial biosynthetic pathways (Jones et al., 2012; Sielaff et al., 2003). Whereas heterologous expression of single biosynthetic genes was successful, no expression of a complete pathway was achieved yet. *Synechocystis* may nevertheless be an alternative for pathways that do not function in *E. coli* (e.g., because certain precursors are missing).

## 4.4. *In vitro* reconstitution of cyanobacterial biosynthetic pathways

An alternative possibility to assign a biosynthetic pathway to a product is the partial or complete cell-free reconstitution of the biosynthesis. Successful examples for cyanobacterial pathways established *in vitro* include the mycosporic acid pathways in *A. variabilis* ATCC 29413 and *N. punctiforme* ATCC 29133. Partial reconstitution could be shown for curacin A, jamaicamide, anatoxin A, scytonemin, cyanobactins, microviridins, and cylindrospermopsin (for an overview, see Kehr et al., 2011). Heterologous production of the enzymes is typically achieved in *E. coli*. The assembly of the product not only requires knowledge about precursors and cofactors but also allows for a thorough biochemical characterization of the enzymes.

## 5. SUMMARY

Cyanobacteria from various habitats produce a wide range of complex metabolites. Most prominent compounds originate from nonribosomal peptide and polyketide biosynthetic multienzyme systems, while cyanobactins are modified peptides of ribosomal origin. These oxygenic phototrophic bacteria may originate from aquatic or terrestrial habitats and in associations with invertebrates, plants, or fungi. They generally show low growth rates and biomass yields and require specific isolation and culturing procedures. Current advances in analytics, molecular engineering, and cultivation techniques have been discussed. A useful tool is the direct detection of metabolites in whole cells by MALDI-TOF mass spectrometry with a sensitivity to single filaments of about 50 bacterial cells. Rapid detection methods lead to the identification of large numbers of structural analogs. Genomic studies permit the description of various biosynthetic gene clusters, some understanding of the plasticity of synthetase genes, and successful molecular manipulations to generate knock out and overproduction mutants. Cell-free *in vitro* reconstitution of enzyme systems is another tool to establish the chemistry of metabolite biosynthesis. The heterologous expression of metabolite clusters has yet to be achieved.

## REFERENCES

Adams, D. G., & Duggan, P. S. (1999). Heterocyst and akinete differentiation in cyanobacteria. *The New Phytologist, 144*, 3–33.

Antal, O., Karisztl-Gacsi, M., Farkas, A., Kovacs, A., Acs, A., Toro, N., et al. (2011). Screening the toxic potential of Cylindrospermopsis raciborskii strains isolated from Lake

Balaton, Hungary. *Toxicon: Official Journal of the International Society on Toxinology, 57*, 831–840.

Ballot, A., Fastner, J., & Wiedner, C. (2010). Paralytic shellfish poisoning toxin-producing cyanobacterium *Aphanizomenon gracile* in Northeast Germany. *Applied and Environmental Microbiology, 76*, 1173–1180.

Beutler, J. A. (2012). Natural products as a foundation for drug discovery. *Current Protocols in Pharmacology, 46s*, 1–21.

Bigham, D. L., Hoyer, M. V., & Canfield, D. E. (2009). Survey of toxic algal (microcystin) distribution in Florida lakes. *Lake and Reservoir Management, 25*, 264–275.

Bindseil, K. U., Jakupovic, J., Wolf, D., Lavayre, J., Leboul, J., & van der Pyl, D. (2001). Pure compound libraries; a new perspective for natural product based drug discovery. *Drug Development Today, 6*, 840–847.

Bolch, C. J. S., & Blackburn, S. I. (1996). Isolation and purification of Australian isolates of the toxic cyanobacterium *Microcystis aeruginosa* Kutz. *Journal of Applied Phycology, 8*, 5–13.

Brennan, L., & Owende, P. (2010). Biofuels from microalgae—A review of technologies for production, processing, and extractions of biofuels and co-products. *Renewable and Sustainable Energy Reviews, 14*, 557–777.

Challis, G. L. (2008). Mining microbial genomes for new natural products and biosynthetic pathways. *Microbiology (Reading, England), 154*, 1555–1569.

Chlipala, G. E., Mo, S., & Orjala, J. (2011). Chemodiversity in freshwater and terrestrial cyanobacteria—A source for drug discovery. *Current Drug Targets, 12*, 1654–1673.

Christiansen, G., Fastner, J., Erhard, M., Borner, T., & Dittmann, E. (2003). Microcystin biosynthesis in planktothrix: Genes, evolution, and manipulation. *Journal of Bacteriology, 185*, 564–572.

Christman, H. D., Campbell, E. L., & Meeks, J. C. (2011). Global transcription profiles of the nitrogen stress response resulting in heterocyst or hormogonium development in Nostoc punctiforme. *Journal of Bacteriology, 193*, 6874–6886.

Czarnecki, O., Lippert, I., Henning, M., & Welker, M. (2006). Identification of peptide metabolites of *Microcystis* (Cyanobacteria) that inhibit trypsin-like activity in planktonic herbivorous *Daphnia* (Cladocera). *Environmental Microbiology, 8*, 77–87.

Dittmann, E., Neilan, B. A., Erhard, M., von Döhren, H., & Borner, T. (1997). Insertional mutagenesis of a peptide synthetase gene that is responsible for hepatotoxin production in the cyanobacterium Microcystis aeruginosa PCC 7806. *Molecular Microbiology, 26*, 779–787.

Dittmann, E., & Wiegand, C. (2006). Cyanobacterial toxins—Occurrence, biosynthesis and impact on human affairs. *Molecular Nutrition & Food Research, 50*, 7–17.

Donia, M. S., Ravel, J., & Schmidt, E. W. (2008). A global assembly line for cyanobactins. *Nature Chemical Biology, 4*, 341–343.

Fastner, J., Erhard, M., Carmichael, W. W., Sun, F., Rinehart, K. L., Rönicke, H., et al. (1999). Characterization and diversity of microcystins in natural blooms and strains of the genera *Microcystis* and *Planktothrix* from German freshwaters. *Archives of Hydrobiology, 145*, 147–163.

Forchert, A., Neumann, U., & Papendorf, O. (2001). New cyanobacterial substances with bioactive properties. In I. Chorus (Ed.), *Cyanotoxins: Occurrence causes consequences* (pp. 295–315). Berlin: Springer.

Gademann, K., Portmann, C., Blom, J. F., Zeder, M., & Jüttner, F. (2010). Multiple toxin production in the cyanobacterium Microcystis: Isolation of the toxic protease inhibitor cyanopeptolin 1020. *Journal of Natural Products, 73*, 980–984.

Haddix, P. L., Hughley, C. J., & Lechevallier, M. W. (2007). Occurrence of microcystins in 33 US water supplies. *Journal of American Water Works Association, 99*, 118–125.

Huisman, J. (1999). Population dynamics of light-limited phytoplankton: Microcosm experiments. *Ecology, 80*, 202–210.

Ishida, K., Christiansen, G., Yoshida, W. Y., Kurmayer, R., Welker, M., Valls, N., et al. (2007). Biosynthesis and structure of aeruginoside 126A and 126B, cyanobacterial peptide glycosides bearing a 2-carboxy-6-hydroxyoctahydroindole moiety. *Chemistry and Biology, 14*, 565–576.

Ishida, K., Matsuda, H., & Murakami, M. (1998). Micropeptins 88-A to 88-F, chymotrypsin inhibitors from the cyanobacterium *Microcystis aeruginosa* (NIES-88). *Tetrahedron, 54*, 5545–5556.

Ishida, K., Matsuda, H., Murakami, M., & Yamaguchi, K. (1997). Micropeptins 478-A and -B, plasmin inhibitors from the cyanobacterium *Microcystis aeruginosa*. *Journal of Natural Products, 60*, 184–187.

Ishida, K., Welker, M., Christiansen, G., Cadel-Six, S., Bouchier, C., Dittmann, E., et al. (2009). Plasticity and evolution of aeruginosin biosynthesis in cyanobacteria. *Applied and Environmental Microbiology, 75*, 2017–2026.

Jenke-Kodama, H., & Dittmann, E. (2009). Bioinformatic perspectives on NRPS/PKS megasynthases: Advances and challenges. *Natural Product Reports, 26*, 874–883.

Jones, A. C., Gu, L., Sorrels, C. M., Sherman, D. H., & Gerwick, W. H. (2009). New tricks from ancient algae: Natural products biosynthesis in marine cyanobacteria. *Current Opinion in Chemical Biology, 13*, 216–223.

Jones, A. C., Ottilie, S., Eustaquio, A. S., Edwards, D. J., Gerwick, L., Moore, B. S., et al. (2012). Evaluation of Streptomyces coelicolor A3(2) as a heterologous expression host for the cyanobacterial protein kinase C activator lyngbyatoxin A. *The FEBS Journal, 279*, 1243–1251.

Kehr, J. C., Gatte Picchi, D., & Dittmann, E. (2011). Natural product biosyntheses in cyanobacteria: A treasure trove of unique enzymes. *Beilstein Journal of Organic Chemistry, 7*, 1622–1635.

Koksharova, O. A., & Wolk, C. P. (2002). Genetic tools for cyanobacteria. *Applied Microbiology and Biotechnology, 58*, 123–137.

Li, L., Garden, R. W., Romanova, E. V., & Sweedler, J. V. (1999). *In situ* sequencing of peptides from biological tissues and single cells using MALDI-PSD/CID analysis. *Analytical Chemistry, 71*, 5451–5458.

Li, B., Sher, D., Kelly, L., Shi, Y., Huang, K., Knerr, P. J., et al. (2010). Catalytic promiscuity in the biosynthesis of cyclic peptide secondary metabolites in planktonic marine cyanobacteria. *Proceedings of the National Academy of Sciences of the United States of America, 107*, 10430–10435.

Liu, H., & Scott, P. M. (2011). Determination of the cyanobacterial toxin cylindrospermopsin in algal food supplements. *Food Additives and Contaminants: Part A, Chemistry, Analysis, Control, Exposure and Risk Assessment, 28*, 786–790.

Long, P. F., Dunlap, W. C., Battershill, C. N., & Jaspars, M. (2005). Shotgun cloning and heterologous expression of the patellamide gene cluster as a strategy to achieving sustained metabolite production. *Chembiochem, 6*, 1760–1765.

Luesch, H., Williams, P. G., Yoshida, W. Y., Moore, R. E., & Paul, V. J. (2002). Ulongamides A–F, new beta-amino acid-containing cyclodepsipeptides from palauan collections of the marine cyanobacterium Lyngbya sp. *Journal of Natural Products, 65*, 996–1000.

Mitschke, J., Vioque, A., Haas, F., Hess, W. R., & Muro-Pastor, A. M. (2011). Dynamics of transcriptional start site selection during nitrogen stress-induced cell differentiation in Anabaena sp. PCC7120. *Proceedings of the National Academy of Sciences of the United States of America, 108*, 20130–20135.

Molinari, G. (2009). Natural products in drug discovery: Present status and perspectives. *Advances in Experimental Medicine and Biology, 655*, 13–27.

Murakami, M., Ishida, K., Okino, T., Okita, Y., Matsuda, H., & Yamaguchi, K. (1995). Aeruginosins 98-A and B, trypsin inhibitors from the blue-green alga *Microcystis aeruginosa* (NIES-98). *Tetrahedron Letters, 36*, 2785–2788.

Murakami, M., Itou, Y., Ishida, K., & Shin, H. J. (1999). Prenylagaramides A and B, new cyclic peptides from two strains of *Oscillatoria agardhii*. *Journal of Natural Products*, *62*, 752–755.

Murata, N., & Suzuki, I. (2006). Exploitation of genomic sequences in a systematic analysis to access how cyanobacteria sense environmental stress. *Journal of Experimental Botany*, *57*, 235–247.

Neilan, B. A., Pearson, L. A., Muenchhoff, J., Moffitt, M. C., & Dittmann, E. (2012). Environmental conditions that influence toxin biosynthesis in cyanobacteria. *Environmental Microbiology*, http://dx.doi.org/10.1111/j.1462-2920.2012.02729.x. [Epub ahead of print] Mar 20.

Nishizawa, T., Asayama, M., Fujii, K., Harada, K., & Shirai, M. (1999). Genetic analysis of the peptide synthetase genes for a cyclic heptapeptide microcystin in Microcystis spp. *Journal of Biochemistry*, *126*, 520–529.

Okumura, H. S., Philmus, B., Portmann, C., & Hemscheidt, T. K. (2009). Homotyrosine-containing cyanopeptolins 880 and 960 and anabaenopeptins 908 and 915 from Planktothrix agardhii CYA 126/8. *Journal of Natural Products*, *72*, 172–176.

Parmar, A., Singh, N. K., Pandey, A., Gnansounou, E., & Madamwar, D. (2011). Cyanobacteria and microalgae: A positive prospect for biofuels. *Bioresource Technology*, *102*, 10163–10172.

Patterson, G. M. L., Larsen, L. K., & Moore, R. E. (1994). Bioactive natural-products from blue-green-algae. *Journal of Applied Phycology*, *6*, 151–157.

Philmus, B., Christiansen, G., Yoshida, W. Y., & Hemscheidt, T. K. (2008). Post-translational modification in microviridin biosynthesis. *Chembiochem*, *9*, 3066–3073.

Reshef, V., & Carmeli, S. (2001). Protease inhibitors from a water bloom of the cyanobacterium *Microcystis aeruginosa*. *Tetrahedron*, *57*, 2885–2894.

Rippka, R. (1988). Isolation and purification of cyanobacteria. *Methods in Enzymology*, *167*, 3–27.

Rohrlack, T., Edvardsen, B., Skulberg, R., Halstvedt, C. B., Utkilen, H. C., Ptacnik, R., et al. (2008). Oligopeptide chemotypes of the toxic freshwater cyanobacterium Planktothrix can form subpopulations with dissimilar ecological traits. *Limnology and Oceanography*, *53*, 1279–1293.

Rouhiainen, L., Paulin, L., Suomalainen, S., Hyytiainen, H., Buikema, W., Haselkorn, R., et al. (2000). Genes encoding synthetases of cyclic depsipeptides, anabaenopeptilides, in Anabaena strain 90. *Molecular Microbiology*, *37*, 156–167.

Ruffing, A. M. (2011). Engineered cyanobacteria: Teaching an old bug new tricks. *Bioengineered Bugs*, *2*, 136–149.

Schmidt, E. W., Nelson, J. T., Rasko, D. A., Sudek, S., Eisen, J. A., Haygood, M. G., et al. (2005). Patellamide A and C biosynthesis by a microcin-like pathway in Prochloron didemni, the cyanobacterial symbiont of Lissoclinum patella. *Proceedings of the National Academy of Sciences of the United States of America*, *102*, 7315–7320.

Shi, T., Ilikchyan, I., Rabouille, S., & Zehr, J. P. (2010). Genome-wide analysis of diel gene expression in the unicellular N(2)-fixing cyanobacterium Crocosphaera watsonii WH 8501. *The ISME Journal*, *4*, 621–632.

Shin, H. J., Murakami, M., Matsuda, H., Ishida, K., & Yamaguchi, K. (1995). Oscillapeptin, an elastase and chymotrypsin inhibitor from the cyanobacterium *Oscillatoria agardhii* (NIES-204). *Tetrahedron Letters*, *36*, 5235–5238.

Sielaff, H., Christiansen, G., & Schwecke, T. (2006). Natural products from cyanobacteria: Exploiting a new source for drug discovery. *IDrugs*, *9*, 119–127.

Sielaff, H., Dittmann, E., Tandeau De Marsac, N., Bouchier, C., von Döhren, H., Börner, T., et al. (2003). The mcyF gene of the microcystin biosynthetic gene cluster from Microcystis aeruginosa encodes an aspartate racemase. *The Biochemical Journal*, *373*, 909–916.

Spengler, B. (1997). Post-source decay analysis in matrix-assisted laser desorption/ionization mass spectrometry of biomolecules. *Journal of Mass Spectrometry, 32*, 1019–1036.

Steglich, C., Futschik, M., Rector, T., Steen, R., & Chisholm, S. W. (2006). Genome-wide analysis of light sensing in Prochlorococcus. *Journal of Bacteriology, 188*, 7796–7806.

Straub, C., Quillardet, P., Vergalli, J., Tandeau de Marsac, N., & Humbert, J. F. (2011). A day in the life of microcystis aeruginosa strain PCC 7806 as revealed by a transcriptomic analysis. *PLoS One, 6*, e16208.

Taniguchi, M., Nunnery, J. K., Engene, N., Esquenazi, E., Byrum, T., Dorrestein, P. C., et al. (2010). Palmyramide A, a cyclic depsipeptide from a Palmyra Atoll collection of the marine cyanobacterium Lyngbya majuscula. *Journal of Natural Products, 73*, 393–398.

Tillett, D., Dittmann, E., Erhard, M., von Döhren, H., Börner, T., & Neilan, B. A. (2000). Structural organization of microcystin biosynthesis in *Microcystis aeruginosa* PCC7806: An integrated peptide-poliketide synthetase system. *Chemistry and Biology, 7*, 753–764.

Tous, C., Vega-Palas, M. A., & Vioque, A. (2001). Conditional expression of RNase P in the cyanobacterium Synechocystis sp. PCC6803 allows detection of precursor RNAs. Insight in the in vivo maturation pathway of transfer and other stable RNAs. *The Journal of Biological Chemistry, 276*, 29059–29066.

van Apeldoorn, M. E., van Egmond, H. P., Speijers, G. J., & Bakker, G. J. (2007). Toxins of cyanobacteria. *Molecular Nutrition & Food Research, 51*, 7–60.

van Hannen, E. J., Zwart, G., van Agterveld, M. P., Gons, H. J., Ebert, J., & Laanbroek, H. J. (1999). Changes in bacterial and eukaryotic community structure after mass lysis of filamentous cyanobacteria associated with viruses. *Applied and Environmental Microbiology, 65*, 795–801.

Wan, F., & Erickson, K. L. (2001). Georgamide, a new cyclic depsipeptide with alkynoic acid residue from an Australian cyanobacterium. *Journal of Natural Products, 64*, 143–146.

Welker, M. (2008). Cyanobacterial hepatotoxins: Microcystins, nodularins, and cylindrospermopsins. In L. M. Botana (Ed.), *Seafood and freshwater toxins: Pharmacology, physiology and detection* (pp. 825–844). (2nd ed.). London: Taylor & Francis Group.

Welker, M., Christiansen, G., & von Döhren, H. (2004). Diversity of coexisting *Planktothrix* (Cyanobacteria) chemotypes deduced by mass spectral analysis of microcystins and other oligopeptides. *Archives of Microbiology, 182*, 288–298.

Welker, M., Marsalek, B., Sejnohova, L., & von Döhren, H. (2006). Detection and identification of oligopeptides in *Microcystis* (cyanobacteria) colonies: Toward an understanding of metabolic diversity. *Peptides, 27*, 2090–2103.

Welker, M., & von Döhren, H. (2006). Cyanobacterial peptides—Nature's own combinatorial biosynthesis. *FEMS Microbiology Reviews, 30*, 530–563.

Westrick, J. A., Szlag, D. C., Southwell, B. J., & Sinclair, J. (2010). A review of cyanobacteria and cyanotoxins removal/inactivation in drinking water treatment. *Analytical and Bioanalytical Chemistry, 397*, 1705–1714.

WHO, (2006). *Guidelines for drinking-water quality* (3rd ed.). *Recommendations* (Vol. 1). Geneva: World Health Organisation.

Williamson, R. T., Singh, I. P., & Gerwick, W. H. (2004). Taveunamides: New chlorinated toxins from a mixed assemblage of marine cyanobacteria. *Tetrahedron, 60*, 7025–7033.

Wolk, C. P. (1988). Purification and storage of nitrogen-fixing filamentous cyanobacteria. In L. Packer & A. N. Glazer (Eds.), *Cyanobacteria* (pp. 93–95). San Diego: Academic Press.

Zegura, B., Straser, A., & Filipic, M. (2011). Genotoxicity and potential carcinogenicity of cyanobacterial toxins—A review. *Mutation Research, Reviews in Mutation Research, 727*, 16–41.

Ziemert, N., Ishida, K., Liaimer, A., Hertweck, C., & Dittmann, E. (2008). Ribosomal synthesis of tricyclic depsipeptides in bloom-forming cyanobacteria. *Angewandte Chemie (International Ed. in English), 47*, 7756–7759.

Ziemert, N., Ishida, K., Quillardet, P., Bouchier, C., Hertweck, C., de Marsac, N. T., et al. (2008). Microcyclamide biosynthesis in two strains of *Microcystis aeruginosa*: From structure to genes and vice versa. *Applied and Environmental Microbiology*, *74*, 1791–1797.

Zurawell, R. W., Chen, H. R., Burke, J. M., & Prepas, E. E. (2005). Hepatotoxic cyanobacteria: A review of the biological importance of microcystins in freshwater environments. *Journal of Toxicology and Environmental Health*, *8*, 1–37.

CHAPTER THREE

# Isolating Antifungals from Fungus-Growing Ant Symbionts Using a Genome-Guided Chemistry Approach

## Ryan F. Seipke*, Sabine Grüschow[†], Rebecca J.M. Goss[†], Matthew I. Hutchings*,[1]

*School of Biological Sciences, University of East Anglia, Norwich Research Park, Norwich, United Kingdom
[†]School of Chemistry, University of East Anglia, Norwich, United Kingdom
[1]Corresponding author: e-mail address: m.hutchings@uea.ac.uk

## Contents

## Abstract

We describe methods used to isolate and identify antifungal compounds from actinomycete strains associated with the leaf-cutter ant *Acromyrmex octospinosus*. These ants use antibiotics produced by symbiotic actinomycete bacteria to protect themselves and their fungal cultivar against bacterial and fungal infections. The fungal cultivar serves as the sole food source for the ant colony, which can number up to tens of

*Methods in Enzymology*, Volume 517
ISSN 0076-6879
http://dx.doi.org/10.1016/B978-0-12-404634-4.00003-6

47

thousands of individuals. We describe how we isolate bacteria from leaf-cutter ants collected in Trinidad and analyze the antifungal compounds made by two of these strains (*Pseudonocardia* and *Streptomyces* spp.), using a combination of genome analysis, mutagenesis, and chemical isolation. These methods should be generalizable to a wide variety of insect-symbiont situations. Although more time consuming than traditional activity-guided fractionation methods, this approach provides a powerful technique for unlocking the complete biosynthetic potential of individual strains and for avoiding the problems of rediscovery of known compounds. We describe the discovery of a novel nystatin compound, named nystatin P1, and identification of the biosynthetic pathway for antimycins, compounds that were first described more than 60 years ago. We also report that disruption of two known antifungal pathways in a single *Streptomyces* strain has revealed a third, and likely novel, antifungal plus four more pathways with unknown products. This validates our approach, which clearly has the potential to identify numerous new compounds, even from well-characterized actinomycete strains.

# 1. INTRODUCTION

The spread of drug resistance amongst bacteria and fungi makes the isolation and development of new and improved antimicrobials an essential goal. In addition, the adverse side–effects of conventional antifungal drugs (antimycotics) and the increasing numbers of immunocompromised patients makes the need for new antimycotics even more urgent. The vast majority of useful antibacterial compounds, as well as numerous anthelmintic, anticancer, and antimycotic compounds, are derived from the high–GC branch of Gram-positive bacteria, known as actinomycetes, most notably from the genus *Streptomyces* (Challis & Hopwood, 2003). To avoid autotoxicity, these actinomycetes also carry resistance genes to the antibiotics that they produce and these spread widely to other bacteria through horizontal gene transfer (Palumbi, 2001). The prevalence of these natural resistance genes is one reason why major pharmaceutical companies are withdrawing from natural product research, which is perhaps a premature decision, given that many sequenced microbial genomes encode "silent" or "cryptic" biosynthetic pathways for potentially useful secondary metabolites yet to be isolated (Challis, 2008). Importantly, actinomycetes do not encode genes conferring resistance to antimycotics, as these compounds typically target features specific to fungi, with no equivalent in bacteria. The aim of our research is to exploit a remarkable symbiotic relationship between actinomycete bacteria and fungus-growing ants to discover new antinfective therapeutics.

Many insects engage in fungiculture, including ants, termites, beetles, and gall midges, but probably the best studied fungus-growing insects are the attine ants (tribe *Attini*). Presently, 13 genera and more than 210 species of attine ants have been described and they differ enormously in their ecology and the way they cultivate their fungus. We have focused our efforts on a branch of the higher attine ants known as leaf-cutters, which includes the genera *Atta* and *Acromyrmex* (Schultz & Brady, 2008), and specifically on the species *Acromyrmex octospinosus*. These ants harvest leaves and feed them to the coevolved fungus *Leucoagaricus gongylophorus*, which produces specialized hyphae (gongylidia) rich in fats and sugars, which serve as the sole food source for the colony (a live video feed of a captive colony of *A. octospinosus* ants is available at http://www.uea.ac.uk/~b247/antcam.html). The fungus garden can be parasitized by a coevolved fungus in the genus *Escovopsis* (Currie, 2001; Currie, Mueller, & Malloch, 1999) and the ants have evolved a symbiotic relationship with actinomycete bacteria to protect their fungus garden as well as themselves against infection (Currie, Bot, & Boomsma, 2003; Mattoso, Moreira, & Samuels, 2012). This is known as a protective mutualism (a mutually beneficial symbiosis) in which the ants provide the bacteria with food from specially evolved crypts (Currie, Poulsen, Mendenhall, Boomsma, & Billen, 2006), and in return the actinomycetes provide antibiotics that inhibit unwanted bacterial and fungal growth. There is good evidence that attine ants have coevolved with *Pseudonocardia* bacteria (Currie et al., 2006). More recently, evidence has accumulated that attine ants can also acquire *Pseudonocardia* spp. horizontally from other attine ant colonies and perhaps even from the soil environment (Barke et al., 2010; Barke, Yu, Seipke, & Hutchings, 2011; Sen et al., 2009). Furthermore, many attines, including *A. octospinosus*, are associated with *Streptomyces* species (Barke et al., 2010; Haeder, Wirth, Herz, & Spiteller, 2009; Kost et al., 2007; Schoenian et al., 2011; Sen et al., 2009). A strain of *Streptomyces albus* has been isolated by different research groups, including our own, from ants collected in Panama and Trinidad, leading to speculation that this strain is a common mutualist of attine ants (Barke et al., 2010; Haeder et al., 2009; Seipke, Barke, & Brearley, et al., 2011; Seipke, Kaltenpoth & Hutchings, 2011). Genome analysis of a strain we have named *S. albus* S4 revealed that, although it is 100% identical to type *S. albus* at the 16S rDNA level, it encodes additional secondary metabolic gene clusters (Ryan F Seipke & Matthew I Hutchings, unpublished).

In order to document methods for isolating antibiotics from fungus-growing ants, we here describe specific examples of the isolation and identification of antifungal compounds from two actinomycete strains taken from a single colony of *A. octospinosus* collected in Trinidad, using a combination of genome analysis, mutagenesis, and chemical isolation. We expect that these methods, with suitable adjustment, are adaptable to a wide range of such studies. Although more time consuming than traditional activity-guided fractionation methods, this provides a powerful technique for unlocking the complete biosynthetic potential of individual strains and for avoiding the problems of rediscovery of known compounds.

## 2. STRAIN ISOLATION AND GENOME SEQUENCING

Attine ants are widespread agricultural pests in South and Central America and the Southern USA and they are relatively easy to collect. Typically, scientists sample in Trinidad, French Guiana, and Panama, the latter being home to the Smithsonian Tropical Research Institute. We have collected leaf-cutter ants from Trinidad and here we describe how we isolated and analyzed actinomycete bacteria from these *A. octospinosus* ants.

### 2.1. Selectively isolating actinomycetes

Ants are typically collected and stored in 20% glycerol at $-20\ ^\circ$C until use. Either the ants are streaked or serial dilutions of the 20% glycerol stock are plated onto mannitol-soya flour (MS) agar (20 g soya flour, 20 g mannitol, 1 L tap water) (Kieser, Bibb, Buttner, Chater, & Hopwood, 2000) or hydrolyzed chitin (HC) agar is prepared as follows. Briefly, 40 g chitin (Sigma-Aldrich) is hydrolyzed by stirring in 300 mL concentrated hydrochloric acid for 30 min and then washed repeatedly in distilled water; the pH is adjusted to 7.0 using NaOH and the HC filtered and dried (Hsu & Lockwood, 1975). HC agar is prepared by adding 0.4 g of HC and 2 g agar to 100 mL aliquots of tap water in 250-mL flasks and autoclaved. Autoclaved HC agar is stored until needed and then melted in a microwave oven. For both MS and HC agar, 100 mL agar is used to pour four 25-mL plates (such thick plates dry out more slowly during long incubation periods than standard plates). Plates are incubated at 25 $^\circ$C for up to 6 weeks to allow growth of *Pseudonocardia* species and other slow-growing actinomycetes. *Streptomyces* species typically appear within a week. Filamentous actinomycetes are identified initially by eye, by their fluffy appearance, and then further analyzed by light microscopy and 16S rDNA sequencing with universal 16S rDNA primers 533F (5′-GTGCCAGCMGCCGCGGTAA) and 1492R (5′-GGTTACCTTGTTACGACTT) (Hugenholtz, Goebel, & Pace, 1998;

Lane, 1991). Spore or hyphal fragment suspensions of each strain are stored at − 20 °C in 20% glycerol until use.

## 2.2. Bioassaying for antifungal activity

Inoculate the center of an MS agar plate with 10 μL of unquantified spore or hyphal fragment stocks of each actinomycete strain and incubate at 30 °C for 10 days to allow a good-sized colony to grow. Actinomycete isolates are challenged with either the human pathogen *Candida albicans* or the leaf-cutter ant nest parasite *Escovopsis weberi* (Barke et al., 2010). For *C. albicans* challenges, soft nutrient agar (SNA, 8 g Difco nutrient broth powder, 5 g agar, 1 L deionized water) is prepared, autoclaved, and stored until needed. Upon use, SNA is melted using a microwave oven and allowed to cool to 50 °C in a water bath. Five millimeter of SNA is inoculated with 200 μL of *C. albicans* culture grown overnight at 37 °C in Lennox broth (10 g Bacto tryptone, 5 g yeast extract, 5 g NaCl, 1 L deionized water) (Kieser et al., 2000). *C. albicans*-containing SNA (5 mL) is pipetted onto the MS agar plate containing the actinomycete culture and gently swirled until the SNA is evenly distributed. (The hydrophobic nature of actinomycete aerial hyphae prevents submersion of the colony in SNA.) After inoculation with *C. albicans*, plates are incubated at room temperature and inspected daily for 2–3 days before photographs are taken. For challenges with *E. weberi*, a loop of spores is taken from a plate of *E. weberi* (maintained on MS agar) and used to inoculate a pea-sized spot on a far edge of the MS agar plate containing the actinomycete culture. Plates are wrapped in Parafilm, incubated at room temperature for 2–3 weeks and photographed when zones of inhibited growth become apparent or when the negative control plates are over-grown. *Streptomyces lividans* ZX1 strain 66 (stock number 1326 from the John Innes Center collection) serves as a negative control for bioactivity with both *C. albicans* and *E. weberi*. Two actinomycete isolates, *Pseudonocardia* P1 and *S. albus* S4, showed clear bioactivity in these assays and were further analyzed with respect to their secondary metabolite potential as described in the following paragraphs.

## 2.3. Genome sequencing and analysis of *Pseudonocardia* P1 and *S. albus* S4

Bacterial genomes were sequenced by The Genome Analysis Centre (TGAC) on the Norwich Research Park using the GS FLX Sequencer (Roche) and the GS FLX Titanium series chemistry kit. Reads were assembled into contigs by the Newbler Assembly v2 or v2.3 software (Roche) using the default settings. Contigs were annotated using the Web-based Rapid Annotation Seed Technology Server (RAST) (Aziz et al., 2008). The stand-alone command line version of the Basic Local Alignment Search Tool (BLAST) 2.2.23+ (Altschul, Gish, Miller, Myers, & Lipman, 1990) was used to create BLAST-compatible databases of the genomes (command: makeblastdb -dbtype nucl -in genome.fasta -out database_name).

## 3. IDENTIFYING ANTIFUNGALS MADE BY *Pseudonocardia* P1

If, as suggested, the *Pseudonocardia* symbiont strains have coevolved with the attine ants over tens of millions of years, we might expect them to produce novel antibiotics. In fact, a *Pseudonocardia* symbiont of the lower attine ant *Apterostigma dentigerum* makes a novel peptide antibiotic named dentigerumycin, adding some support for this hypothesis (Oh, Poulsen, Currie, & Clardy, 2009). It was therefore of primary interest to analyze the antifungal compounds made by *Pseudonocardia* strains isolated from *A. octospinosus*, and here we describe our partial success in identifying a new nystatin from the first strain we analyzed, named *Pseudonocardia* P1. Analysis of the secondary metabolome of *S. albus* S4 is discussed in Section 4.

## 3.1. Searching for antifungal pathways in the P1 genome

The P1 genome was sequenced using shotgun 454 sequencing. Assembly of ~95.7 Mb of sequence resulted in 876 large contigs (>500 bp) representing a genome of ~6.38 Mb (GenBank accession number ADUJ00000000.1) (Barke et al., 2010). Analysis of the RAST-annotated contigs revealed several polyketide synthase (PKS) gene fragments with >90% amino acid identity to proteins encoding the biosynthesis of an antifungal compound named nystatin-like *Pseudonocardia* polyene (NPP) produced by *Pseudonocardia autotrophica* (Kim et al., 2009). NPP is related to nystatin, a polyene antifungal made by *Streptomyces noursei*, which is widely used to treat fungal skin infections in humans (Brautaset et al., 2000; Brown & Hazen, 1957).

In order to determine whether or not *Pseudonocardia* P1 contained the entire biosynthetic gene cluster for a nystatin-like compound, contigs were aligned against the characterized *npp* biosynthetic gene cluster from *P. autotrophica* using NUCmer (Kurtz et al., 2004) run using the default settings (command: nucmer -p npp_alignment npp_cluster.fasta contigs.fasta).

Next, the show-tiling utility was used to extract information about the aligned contigs (command: show-tiling –i 80 npp_alignment.delta > npp_alignment.tiling). (The –i 80 option sets the nucleotide cutoff value to 80%, a value that we found to be suitable for our purposes.) Microsoft Excel was used to convert the output of the show-tiling utility to Gene Finder Format (see http://www.sanger.ac.uk/resources/software/gff/ spec.html) and visualized and compared to the *npp* cluster using Artemis (release 11.22) (Rutherford et al., 2000). Table 3.1 displays URLs to software programs used for genomic analyses.

The aligned contigs spanned the entire *npp* cluster (Fig. 3.1) including the six PKS genes that assemble the aglycone, the non-sugar-containing backbone of nystatin. Full-length coding sequences were captured for 11 genes (*nypF*, *nypH*, *nypDIII*, *nypL*, *nypN*, *nypDII*, *nypDI*, *nypE*, *nypO*, *nypRIV*, *nypM*), which are proposed to be primarily involved in the post-PKS modification of the nystatin aglycone, and two new genes, *nypY* and

**Table 3.1** URLs to software programs described in this chapter

| Software program | Web URL |
| --- | --- |
| RAST | http://rast.nmpdr.org |
| BLAST 2.2.23+ | ftp://ftp.ncbi.nlm.nih.gov/blast/executables/blast+/ LATEST/ |
| NUCmer (part of MUMmer3.23) | http://sourceforge.net/projects/mummer/ |
| Artemis | http://www.sanger.ac.uk/resources/software/artemis/ |

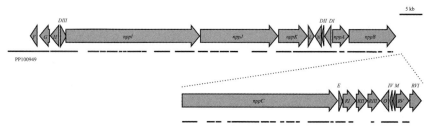

**Figure 3.1** Identification of the nystatin P1 biosynthetic gene cluster. Alignment of *Pseudonocardia* P1 contigs (represented by horizontal black lines, GenBank accession number ADUJ00000000.1) to the NPP biosynthetic gene cluster from *P. autotrophica* (GenBank accession number EU108007). Contig PP100949 contains the glycosyltransferase gene, *nypY*, presumed to be involved in attachment of the additional hexose moiety during the biosynthesis of nystatin P1.

*nypZ*, with unknown functions (Barke et al., 2010; Brautaset et al., 2000). Interestingly, a second glycosyltransferase gene (named *nypY*), absent from the nystatin producer *S. noursei* and in *P. autotrophica*, is present in the *nyp* gene cluster. NypY belongs to the same glycosyltransferase family as NypDI, the glycosyltransferase that adds a mycosamine sugar group to the nystatin aglycone. However, it displays only 42% amino acid identity to NypDI and is therefore unlikely to be a functionally redundant copy of NypDI. These bioinformatic analyses provided important clues that led to the chemical identification of a new nystatin analog that we have named nystatin P1.

## 3.2. Chemical identity of nystatin P1

Our early efforts to obtain antifungal activity during submerged growth of *Pseudonocardia* P1 using a variety of conditions failed. However, *Pseudonocardia* P1 reproducibly inhibited the growth of both *C. albicans* and *E. weberi* in agar-based bioassays, though the size of the zone of inhibited growth was small in comparison to the bioactivity of other isolates (Fig. 3.2; bioassays against *E. weberi* are not pictured). However, we found that adding sodium butyrate (150 m*M*) to the bioassay plates increased antifungal production,

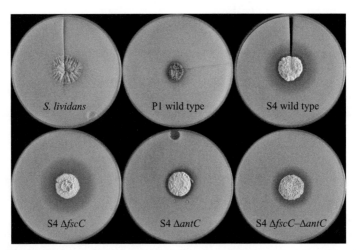

**Figure 3.2** Antifungal bioactivity of actinomycete strains against *Candida albicans*. Antifungal activity is evident by the zone of clearing present around the actinomycete colony in the center of the plate. A mutant that could not produce candicidin (S4 Δ*fscC*) was just as effective at killing *C. albicans* as the wild-type strain; a mutant unable to produce antimycins (S4 Δ*antC*) showed dramatically reduced bioactivity against *C. albicans*; a candicidin- and antimycin-deficient strain (Δ*fscC*–Δ*antC*) still inhibited the growth of *C. albicans*, suggesting that S4 produces an additional antifungal compound(s). *Modified from Seipke, Barke, Brearley, et al. (2011).*

presumably by increasing expression of the *nyp* biosynthetic genes or inducing the expression of another antifungal gene cluster. Sodium butyrate is commonly used as a histone deacetylase (HDAC) inhibitor in eukaryotes and such HDAC inhibitors can alter transcriptional activity and switch on genes that are usually silent, including antibiotic biosynthetic gene clusters in fungi (Davies, 2003). It appears that HDAC inhibitors have similar effects on actinomycete strains (see chapter 18 of this volume).

In order to extract the antifungal(s), *Pseudonocardia* P1 was allowed to sporulate fully on MS agar. The agar was subsequently sliced, placed in a flask, and extracted twice with methanol (200 mL). The solvent was evaporated under vacuum, and the residue resuspended in 50% aqueous methanol (150 µL) and subjected to LC–MS/MS analysis alongside a nystatin standard as previously described (Barke et al., 2010). An authentic nystatin standard (Sigma-Aldrich) was prepared at 0.1 mg/mL in 50% aqueous methanol. Immediately before analysis, the crude extract and the standard were diluted twofold with 20% aqueous methanol and spun in a microcentrifuge at maximum speed for 4 min to remove any insoluble matter. Only the supernatant was used for injection (5 µL). The samples were run on a Surveyor HPLC system attached to a LCQ DecaXPplus ion trap mass spectrometer (both Thermo Fisher). Separation of metabolites was performed on a Luna C18(2) 3 µm 100 × 2 mm column (Phenomenex) with 0.1% formic acid in water as solvent A and methanol as solvent B using the following gradient: 0–20 min 20–95% solvent B, 20–22 min 95% solvent B, 22–23 min 95–20% solvent B, and 23–30 min 20% solvent B. The flow rate was set to 260 µl/min and the column temperature was maintained at 30 °C. Detection was by PDA (full spectra from 200 to 600 nm) and by positive electrospray MS, using spray chamber conditions of 350 °C capillary temperature, 50 units sheath gas, 5 units auxiliary gas, and 5.2 kV spray voltage. Molecular ions for nystatin (*m/z* 926.5) or for NPP (*m/z* 1129.6) produced by *S. noursei* and *P. autotrophica*, respectively, were not detected. However, a molecular ion of *m/z* 1088.6 with retention time and UV absorbance characteristics similar to nystatin was identified (Fig. 3.3). Fragmentation of the molecular ion *m/z* 1088.6 in conjunction with the presence of an additional glycosyltransferase gene in the *nyp* gene cluster strongly suggested that the mass difference of 162 between nystatin and nystatin P1 corresponded to an additional hexose moiety (e.g., glucose) attached to the amino sugar mycosamine (Barke et al., 2010).

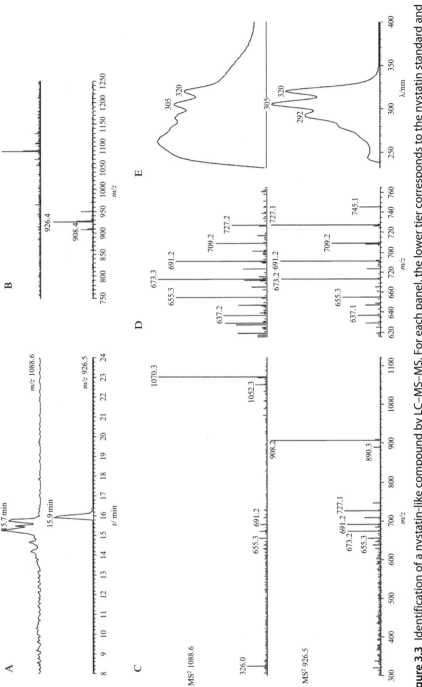

**Figure 3.3** Identification of a nystatin-like compound by LC–MS-MS. For each panel, the lower tier corresponds to the nystatin standard and the upper tier to the *Pseudonocardia* P1 extract. (A) Extracted ion chromatograms for *m/z* 926.5 (nystatin standard) and *m/z* 1088.6 (nystatin P1). (B) Mass spectra averaged across the retention times indicated in (A). (C) MS² analysis of the molecular ions identified above. The main mode of fragmentation is the loss of water molecules (*m/z* 18). (D) Enlarged region of the MS² spectra. These product ions arise from loss of the carbohydrate portion plus one to seven water molecules, and most are found in the nystatin standard as well as in nystatin P1. (E) UV spectra extracted at the retention times indicated in (A). *Reproduced from Barke et al. (2010).*

## 3.3. Attempts at heterologous expression

The slow growth of *Pseudonocardia* P1 and its relatively low-level production of nystatin P1 make heterologous expression of the nystatin P1 gene cluster attractive. Higher production in a faster growing strain would allow larger amounts of nystatin P1 to be purified for further structural and bioactivity studies and assessment in terms of potential utility in the clinic. Unfortunately, our attempts to produce nystatin P1 in an *S. noursei* strain engineered to overexpress the *nypY* glycosyl transferase from *Pseudonocardia* P1 were unsuccessful. Only nystatin was detected in extracts from this strain using LC–MS, with no sign of nystatin P1. Future efforts will require heterologous expression of the complete nystatin P1 gene cluster in a suitable *Streptomyces* host strain and/or use of promoter sequences native to the nystatin gene cluster of *S. noursei* rather than constitutively expressing *nypY* using the *ermE** promoter from a phage integration site.

## 4. IDENTIFYING ANTIFUNGALS MADE BY *S. albus* S4

While we were carrying out our research, Dieter Spiteller's group published the first report of an antifungal compound produced by actinomycete bacteria associated with attine ants. The compound, candicidin, is a well-known antifungal and the candicidin gene cluster is widely distributed amongst streptomycetes (Jørgensen et al., 2009). Furthermore, it is worth noting that the strains isolated by Spiteller and colleagues are identical to the *S. albus* S4 strain that we isolated from *A. octospinosus* ants at the 16S rDNA level (Haeder, 2009).

## 4.1. Searching for antifungal pathways in the S4 genome

The S4 genome was sequenced using a combination of shotgun, 3- and 8-kb paired-end 454 sequencing. After the assembly of ∼335 Mbp of sequence, 12 scaffolds containing 211 large contigs (>500 bp) were generated. The genome is composed of 7.47 Mb and consists of one large linear chromosome, one linear plasmid, and one circular plasmid (Seipke, Crossman, et al., 2011). Annotation and sequence analysis of the S4 genome was performed as described for the P1 genome. As expected, we identified a candicidin biosynthetic gene cluster in the S4 genome (Barke et al., 2010; Seipke, Barke, Brearley, et al., 2011). However, we could not identify any other pathways for the biosynthesis of known antifungals in the S4 genome.

## 4.2. Disrupting candicidin biosynthesis

To determine whether S4 can make antifungal(s) in addition to candicidin, we deleted *fscC*, a key gene in candicidin biosynthesis that encodes the PKS elongation modules 6–10 (Chen et al., 2003; Seipke, Barke, Brearley, et al., 2011). To mutagenize *fscC*, two 3-kb knockout arms were PCR-amplified using primers engineered at their 5′-end to contain appropriate restriction sites for cloning. The resulting PCR products were sequentially cloned into pKC1132, which contained the RK2 conjugal origin of transfer and an apramycin resistance gene for counterselection (Bierman et al., 1992; Kieser et al., 2000). After the upstream and downstream arms were cloned into pKC1132, a hygromycin B resistance cassette was PCR-amplified from pIJ10700 (Gust et al., 2004) and cloned between the upstream and downstream arms. The knockout plasmid was then introduced by electroporation into the methylation-deficient *E. coli* strain ET12567 containing the driver plasmid pUZ8002 (MacNeil et al., 1992) and transferred to *S. albus* S4 by conjugation as described (Kieser et al., 2000). Transconjugants were selected for apramycin resistance. An apramycin-resistant transconjugant was obtained and subsequently replica-plated to obtain hygromycin-resistant and apramycin-sensitive colonies, a phenotype indicating that the *fscC* gene had been entirely replaced by the hygromycin resistance cassette and that the plasmid backbone was no longer present. The integrity of the Δ*fscC* strain was confirmed by PCR.

In order to confirm that deletion of *fscC* abolished candicidin production, the LC–MS profiles of wild-type and mutant strains were compared. The S4 wild-type and Δ*fscC* mutant were cultivated in 50 mL of liquid MS medium (20 g soya flour, 20 g mannitol, 1 L tap water) in a 250-mL flask with shaking at 270 rpm for 10 days at 30 °C. Mycelium was removed by centrifugation and the supernatant of three biological replicates was combined. The combined supernatant (50 mL) was extracted three times with an equal volume of butanol. Butanol extracts were combined and evaporated to dryness under vacuum and the residue was resuspended in 0.5 mL of 50% aqueous methanol. Samples were spun in a microcentrifuge at maximum speed for 5 min immediately prior to analysis to remove any insoluble matter. Only the supernatant was injected into a Shimadzu quadrupole LC–MS-2010A mass spectrometer equipped with Prominence HPLC system. Metabolites were separated on a Waters XBridge™ C18 3.5 µm 2.1 × 100 mm column using 0.1% formic acid in water as solvent A and 0.1% formic acid in acetonitrile as solvent B. The flowrate was 0.35 mL/min and the gradient was as follows: 0.01–0.5 min 15% solvent B, 0.5–14 min 15–95% solvent B, 14–16 min 95% solvent B, 16–16.5 min 95–15% solvent B, and 16.5–19 min 15% solvent B. Mass spectra were acquired in positive-ion mode with the capillary voltage set to 1.3 kV. As expected, a metabolite with UV absorbance characteristics and a molecular ion consistent with candicidin was detected in the extract from the wild-type strain but was absent from the Δ*fscC* extract (Fig. 3.4).

To our surprise, bioassays against *C. albicans* (Fig. 3.2) and *E. weberi* using the wild-type and Δ*fscC* strain of S4 revealed no difference in antifungal activity, suggesting that S4 must make antifungals in addition to candicidin. While this work was in progress, it was reported that antimycins are produced by *Streptomyces* species associated with attine ants (Schoenian et al., 2011). Therefore, we investigated whether *S. albus* S4 makes antimycins and if this could account for the antifungal bioactivity retained by the Δ*fscC* mutant.

In order to determine if *S. albus* S4 can produce antimycins, the wild-type strain was fermented in liquid MS medium for 4 days at 30 °C. The clarified supernatant was extracted with XAD16 resin (Sigma-Aldrich) on a rotary platform overnight at 4 °C. The resin was washed twice with deionized water and metabolites were eluted from the resin with 1 mL of 100% methanol. Methanolic extracts were diluted to a methanol content of 50% with water and spun in a microcentrifuge for 5 min at maximum speed immediately prior to analysis in order to remove insoluble matter. Only the supernatant (10 μL) was analyzed by LC–MS (as described above for the detection of candicidin). The MS analysis identified eight compounds with molecular ions that match those reported for antimycins A1–A4 (Fig. 3.5). To determine if any of the eight compounds could be antimycins, we coinjected commercially available antimycin standards A1–A4 (Sigma-Aldrich) with our wild-type extract (5 μL extract mixed with 5 μL standards). Half of the S4 compounds were identical to the commercially available antimycin standards A1–A4 in both UV absorbance profile and LC retention time, while the remaining four compounds likely represent other members of the large antimycin family. Intriguingly, although antimycins were discovered >60 years ago (Dunshee, Leben, Keitt, & Strong, 1949) the biosynthetic pathway had never been identified, explaining why we could not predict antimycin biosynthesis through genome analysis of *S. albus* S4.

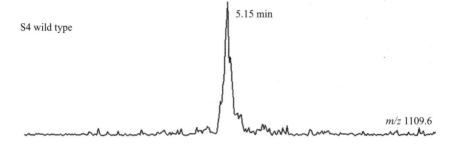

S4 wild type

5.15 min

*m/z* 1109.6

S4 Δ*fscC*

*m/z* 1109.6

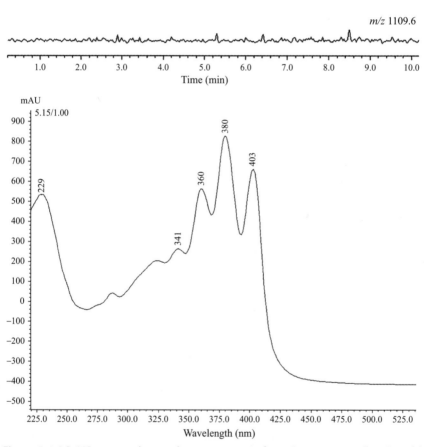

**Figure 3.4** LC–MS was used to analyze supernatant from *Streptomyces albus* S4 wild type and S4 Δ*fscC*. The extracted ion chromatogram for candicidin (*m/z* 1109.6) is shown and confirmed that only S4 wild type and not the Δ*fscC* mutant produced candicidin. The UV–Visible spectrum for the peak at RT 5.15 min displays absorption characteristics consistent with polyene compounds, as shown (bottom). *Reproduced from Seipke, Barke, Brearley, et al. (2011), Seipke, Barke, Ruiz-Gonzalez, et al. (2011), Seipke, Crossman, et al. (2011), and Seipke, Kaltenpoth, et al. (2011).*

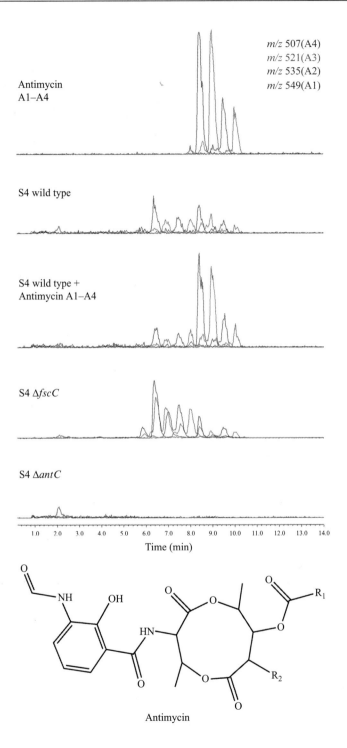

Antimycin

## 4.3. Identification of the antimycin gene cluster

The structure of antimycins (Fig. 3.5) suggested that they may be synthesized, at least in part, by a nonribosomal peptide synthetase (NRPS) that utilizes threonine as a substrate. In order to identify the gene cluster encoding the production of antimycins, BLAST was used to query the S4 genome with the threonine adenylation domain from the daptomycin biosynthetic protein DptA (Miao et al., 2005). This led to the discovery of a hybrid NRPS/PKS gene cluster that displayed significant homology to the DptA threonine adenylation domain. The Web-based software program NRPSpredictor (Rausch, Weber, Kohlbacher, Wohlleben, & Huson, 2005) further corroborated the amino acid selectivity. The hybrid NRPS/PKS gene cluster showed significant homology and synteny with another hybrid NRPS/PKS gene cluster from *Streptomyces ambofaciens* and *S. albus* J1074 (GenBank accession AM238663 and ABYC00000000, respectively) (Seipke, Barke, Brearley, et al., 2011).

To confirm that the observed antimycins were the product of the hybrid NRPS/PKS gene cluster, we disrupted *antC*, which encodes the NRPS that is predicted to utilize threonine. In order to disrupt the *antC* gene, a 1.5-kb internal fragment of *antC* was PCR-amplified and cloned into the commercially available vector pGEMT-EZ (Promega), followed by cloning of the apramycin resistance cassette containing a conjugal origin of transfer (*aac (3)IV*+oriT) from pIJ773 (Gust et al., 2003). This plasmid was introduced by electroporation into ET12567/pUZ8002 (MacNeil et al., 1992) and mobilized to S4 wild type as described previously (Kieser et al., 2000). Transconjugants were selected for apramycin resistance, indicating that the suicide plasmid had integrated into the *antC* gene. The integrity of the

---

**Figure 3.5** LC–MS analysis of *Streptomyces albus* S4 wild-type and mutant strains compared to antimycin standards. The extracted ion chromatograms for antimycins A1–A4 are shown. Eight compounds consistent with the mass of antimycin A1–A4 were produced by S4 wild type and S4 Δ*fscC*, but not by the Δ*antC* mutant. Coinjection of antimycin A1–A4 with the S4 wild-type extract demonstrated that antimycin A1–A4 have the same retention time as four of the eight compounds produced by S4 wild type. Antimycin A1: $R_1 = CH(CH_3)CH_2CH_3$, $R_2 = (CH_2)_5CH_3$. Antimycin A2: $R_1 = CH(CH_3)_2$, $R_2 = (CH_2)_5CH_3$. Antimycin A3: $R_1 = CH(CH_3)CH_2CH_3$, $R_2 = (CH_2)_3CH_3$. Antimycin A4: $R1 = CH(CH_3)_2$ $R_2 = (CH_2)_3CH_3$. Reproduced from Seipke, Barke, Brearley, et al. (2011); the UV–Visible spectra and positive mode electrospray ionization mass spectra for antimycin A1–A4 and the eight antimycin compounds produced by S4 wild type are presented there. (See Color Insert.)

$\Delta antC$ mutant was confirmed by PCR amplification of the upstream and downstream junctions of integration. The $\Delta antC$ mutant displayed dramatically reduced antifungal activity against *C. albicans* compared to that of the wild-type strain (Fig. 3.2). Confirmation that the hybrid NRPS/PKS encoded by *S. albus* S4 directs the biosynthesis of antimycins was obtained by comparing the extracted ion chromatograms of the wild-type and $\Delta antC$ mutant strains, which revealed that the $\Delta antC$ mutant does not produce the eight antimycins (Fig. 3.5).

## 4.4. Cloning the antimycin gene cluster

The antimycin biosynthetic pathway was unknown until we identified it in *S. albus* S4 and subsequently in the sequenced genomes of *S. albus* and *S. ambofaciens*. To facilitate future genetic and biochemical analysis of the antimycin gene cluster, we cloned the cluster into the cosmid Supercos1 (Stratagene). Standard molecular biology methods were used to prepare a large quantity of S4 genomic DNA (Kieser et al., 2000). One hundred micrograms of genomic DNA was partially digested with *Sau*3AI (New England Biolabs) for 3 min, a length of time that was empirically determined to yield DNA fragments > 20 kb. *Sau*3AI-digested genomic DNA was dephosphorylated with shrimp alkaline phosphatase (Roche) and ligated into Supercos1 DNA (prepared according to the manufacturer's instructions) using T4 DNA Ligase (Roche) and packaged into Gigapack III XL phage according to the manufacturer's instructions (Agilent Technologies). One thousand cosmid clones were patched onto Lennox agar (Lennox broth + 15 g/L agar) containing kanamycin (50 µg/mL) and carbenicillin (100 µg/mL). A PCR screen was used to identify cosmids containing the antimycin gene cluster. Briefly, 100 cosmid clones were inoculated into a single 250-mL flask containing 60 mL of Lennox broth containing kanamycin and carbenicillin and grown overnight while shaking at 37 °C. Bacterial cells were harvested by centrifugation and cosmids were isolated using standard alkaline lysis procedures (Kieser et al., 2000). Purified cosmid mixtures were screened by PCR using oligonucleotide primers that targeted a 1.5-kb DNA sequence of the antimycin gene cluster. The amount of total cosmid DNA used as template during PCR was such that each individual cosmid was assumed to be theoretically present in a concentration of ~30–40 µg/mL. This procedure was repeated first with pools of 20 from the original pool of 100 clones that yielded a positive result and subsequently with pools of 10 from the pool of 20 positive clones. At this stage of the

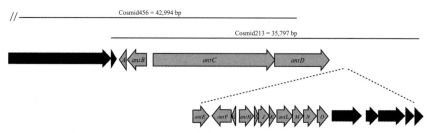

**Figure 3.6** Schematic representation of the antimycin gene cluster from *Streptomyces albus* S4. Genes in black have been experimentally determined not to be involved in the biosynthesis of antimycins (Seipke & Hutchings, unpublished). The two overlapping cosmids, cosmid213 and cosmid456, are shown; // denotes that the upstream border of the insert for cosmid456 falls outside the schematic presented.

deconvolution process, colony-PCR was used to identify the appropriate cosmid. This procedure identified two overlapping cosmids (cosmid213 and cosmid456), which were end-sequenced using primers RFS184 (5′-CCATTATTATCATGACATTAA) and RFS185 (5′-GTCCGTGG AATGAACAATGG) (Keyhani & Roseman, 1997). Sequences were mapped onto the S4 genome using BLAST. Cosmid213 contained what is presumed to be the entire antimycin gene cluster, based on comparative bioinformatics and mutagenesis studies of adjacent genes (Seipke & Hutchings, unpublished; Fig. 3.6).

As a first step toward analyzing antimycin biosynthesis, we attempted to heterologously express antimycins in both *S. lividans* and strains of *S. coelicolor* optimized for heterologous expression of secondary metabolite gene clusters (Gomez-Escribano & Bibb, 2011). The *bla gene* on cosmid213, conferring ampicillin resistance, was replaced with the *Ssp*I fragment from pMJCOS1 using lambda *red* recombination (Gust et al., 2003; Yanai, Murakami, & Bibb, 2006). The *Ssp*I fragment from pMJCOS1 contains an apramycin resistance gene, a conjugal origin of transfer, and the ϕC31 integrase and attachment site, which allow transfer and maintenance of the cosmid in the chromosome of the heterologous host. Unfortunately, we were unable to detect antifungal bioactivity from heterologous hosts containing the antimycin cosmid. We are currently attempting to express this cloned pathway in a strain more closely related to *S. albus* S4.

## 4.5. Other gene clusters in the S4 genome

Identification of the antimycin gene cluster leaves five putative antibiotic biosynthetic gene clusters with unknown products in the S4 genome (Table 3.2). As reported previously, an *S. albus* S4 *fscC-antC* double mutant,

**Table 3.2** Putative secondary metabolites encoded by *Streptomyces albus* S4

| Predicted biosynthetic system[a] | Genome coordinates[b] | Predicted metabolite or close relative | Biological properties |
|---|---|---|---|
| Hopene/squalene synthase | scaffold08: 588141–598581 | Hopanoids | Membrane stabilizers |
| NRPS-independent siderophore synthetase | scaffold05: 959198–972403 | Desferrioxamine | Siderophore |
| NRPS-independent siderophore synthetase | scaffold08: 1448607–1457963 | Unknown | Unknown |
| Ectoine synthase | scaffold05: 68880–72152 | Ectoine | Osmolyte |
| Phytoene/polyprenyl synthase | scaffold06: 410147–419826 | Carotenoids | Pigment |
| Terpene synthase | scaffold08: 1719586–1721871 | Geosmin | Unknown |
| Type III PKS | scaffold06: 295706–300701 | Tetrahydroxynaphthalene | Pigment |
| Type I PKS | scaffold06: 115150–253654 | Candicidin | Antifungal |
| Type I PKS/Type III PKS | scaffold05: 1001127–1064995 | Kendomycin | Anticancer |
| Type II PKS | scaffold08: 3878554–3911349 | Fredericamycin | Anticancer |
| Hybrid NRPS/PKS | scaffold06: 81953–106578 | Antimycin | Antifungal |
| Hybrid NRPS/PKS | scaffold06: 7264–45109 | Unknown | Unknown |
| Hybrid NRPS/PKS | scaffold08: 503983–520001 | Unknown | Unknown |
| NRPS | scaffold08: 4240081–4309220 | Gramicidin | Antibacterial |
| NRPS | scaffold08: 3002155–3042863 | Mannopeptimycin | Antibacterial |
| NRPS | scaffold06: 65083–81878 | Unknown | Unknown |

*Continued*

**Table 3.2** Putative secondary metabolites encoded by *Streptomyces albus* S4—cont'd

| Predicted biosynthetic system[a] | Genome coordinates[b] | Predicted metabolite or close relative | Biological properties |
|---|---|---|---|
| NRPS | scaffold08: 276268-301035 | Unknown | Unknown |
| NRPS | scaffold08: 3930113-3950474 | Unknown | Unknown |

[a]NRPS, nonribosomal peptide synthetase; PKS, polyketide synthase.
[b]*S. albus* S4 scaffolds are available under GenBank accession CADY00000000.1 and are also viewable at http://strepdb.streptomyces.org.uk.

which cannot make candicidin or antimycins, still has weak activity against *C. albicans* and *E. weberi*, suggesting that at least one of the five pathways might encode an additional antifungal agent (Fig. 3.2; Seipke, Barke, Brearley, et al., 2011). The lack of sequence homology to previously characterized secondary metabolite gene clusters implies that the pathway product is a structurally novel compound and/or of unknown biosynthetic origin. Work is underway to delete each of these five clusters in the wild-type and *fscC-antC* double mutant strains to try to identify the cluster responsible for the additional antifungal activity observed in bioassays and to identify the compound itself using comparative metabolomics.

# 5. SUMMARY AND PERSPECTIVES

In this chapter, we have described the genetic and chemical approaches we have taken to identify antifungal activities in two actinomycete strains that form stable interactions with leaf-cutter ants. We have used genome mining combined with mutagenesis to combat the problem of rediscovery, and by deleting pathways for known antifungal compounds in *S. albus* S4, we have uncovered a third antifungal which may be novel and which cannot be predicted from genome analysis alone. The approach we have outlined also allowed us to identify the biosynthetic pathway and coding gene cluster for a group of compounds named antimycins that were discovered more than 60 years ago. Recent advances in genome sequencing will make such discoveries commonplace and, with the identification of closely related pathways, such as those for nystatin, nystatin P1, and NPP, will facilitate genome engineering for further optimization of commonly used (and valuable) compounds such as nystatin. Although nystatin P1 was the only novel compound discovered in this study and is closely related to nystatin, the discovery of new chemical scaffolds seems close at hand with the new, as yet unexplored, gene clusters. Further analysis of these strains by gene disruption and comparative metabolomics, using tools such as the HDAC inhibitor, sodium butyrate, to unlock silent pathways, will no doubt facilitate these discoveries. In recent years, several new chemical scaffolds have been discovered in common streptomycete laboratory strains, suggesting that there is a lot of novel chemistry waiting to be discovered now that we have the genetic and chemical tools to unlock these pathways and compounds.

Thus far, efforts in natural product discovery from leaf-cutter ant symbionts as a source of new therapeutics have been met with mixed success. In the past 3 years, two known antibacterials (actinomycin and valinomycin), two known antifungals (candicidin and antimycin), one novel antifungal (dentigerumycin), and a new analog of nystatin have been identified (Barke et al., 2010; Haeder et al., 2009; Oh et al., 2009; Schoenian et al., 2011; Seipke, Barke, Brearley, et al., 2011). In order to increase the likelihood of identifying new antifungal therapeutics, we have begun to focus on antifungal-producing actinomycetes that have been isolated from fungus-growing ants outside the Attines, namely, the plant-ants belonging to the genera *Allomerus* and *Tetraponera* (Seipke, Barke, Ruiz-Gonzalez, et al., 2011; Seipke & Hutchings, unpublished). Nothing is known about the chemistry of these protective mutualisms, and there is potential to discover novel compounds. We have also changed the way that we screen for bioactivity. We now screen our isolates for bioactivity against an emerging, multidrug-resistant fungus, *Scedosporium prolificans* (Denning & Hope, 2010). Results from our screen against *S. prolificans* identified four actinomycetes in our strain library that show bioactivity against *S. prolificans* (Seipke & Hutchings, unpublished). Given the drug-resistant nature of *S. proflicans*, it is highly likely that the antifungal(s) produced by these strains has novel targets and will not only be useful in treating *S. prolificans* infections but will also be active against other fungal pathogens such as *C. albicans* and *Aspergillus fumigatus*.

## ACKNOWLEDGMENTS

We thank our colleagues who contributed to the research described here, most notably Jörg Barke and Doug Yu at the University of East Anglia and Darren Heavens and colleagues at the TGAC. Sequencing was carried out under a Capacity and Capability Challenge

Programme (project number CCC-1-12) with TGAC and our laboratory research was funded by a Milstein award from the Medical Research Council and by the University of East Anglia. We also thank Govind Chandra for advice on genome assembly and analysis and Sergey B. Zotchev for the gift of *S. noursei*.

## REFERENCES

Altschul, S. F., Gish, W., Miller, W., Myers, E. W., & Lipman, D. J. (1990). Basic local alignment search tool. *Journal of Molecular Biology, 215,* 403–410.

Aziz, R. K., Bartels, D., Best, A. A., DeJongh, M., Disz, T., Edwards, R. A., et al. (2008). The RAST Server: Rapid annotations using subsystems technology. *BMC Genomics, 9,* 75.

Barke, J., Seipke, R. F., Grüschow, S., Heavens, D., Drou, N., Bibb, M. J., et al. (2010). A mixed community of actinomycetes produce multiple antibiotics for the fungus farming ant *Acromyrmex octospinosus*. *BMC Biology, 8,* 109.

Barke, J., Yu, R. F., Seipke, D. W., & Hutchings, M. I. (2011). A mutualistic microbiome: How do fungus-growing ants select their antibiotic-producing bacteria. *Communicative & Integrative Biology, 4,* 41–43.

Bierman, M., Logan, R., O'Brien, K., Seno, E. T., Rao, R. N., & Schoner, B. E. (1992). Plasmid cloning vectors for the conjugal transfer of DNA from *Escherichia coli* to *Streptomyces* spp. *Gene, 116,* 43–49.

Brautaset, T., Sekurova, O. N., Sletta, H., Ellingsen, T. E., StLm, A. R., Valla, S., et al. (2000). Biosynthesis of the polyene antifungal antibiotic nystatin in *Streptomyces noursei* ATCC 11455: Analysis of the gene cluster and deduction of the biosynthetic pathway. *Chemistry & Biology, 7,* 395–403.

Brown, R., & Hazen, E. L. (1957). Present knowledge of nystatin, an antinfugal antibiotic. *Transactions of the New York Academy of Sciences, 19,* 447–456.

Challis, G. L. (2008). Mining microbial genomes for new natural products and biosynthetic pathways. *Microbiology, 154,* 1555–1569.

Challis, G. L., & Hopwood, D. A. (2003). Synergy and contingency as driving forces for the evolution of multiple secondary metabolite production by *Streptomyces* species. *Proceedings of the National Academy of Sciences of the United States of America, 100,* 14555–14561.

Chen, S., Huang, X., Zhou, X., Bai, L., He, J., Jeong, K. J., et al. (2003). Organizational and mutational analysis of a complete FR-008/candicidin gene cluster encoding a structurally related polyene complex. *Chemistry & Biology, 10,* 1065–1076.

Currie, C. R. (2001). Prevalence and impact of a virulent parasite on a tripartite mutualism. *Oecologia, 128,* 96–106.

Currie, C. R., Bot, A. N. M., & Boomsma, J. J. (2003). Experimental evidence of a tripartite mutualism: Bacteria protect ant fungus gardens from specialized parasites. *Oikos, 101,* 91–102.

Currie, C. R., Mueller, U. G., & Malloch, D. (1999). The agricultural pathology of ant fungus gardens. *Proceedings of the National Academy of Sciences of the United States of America, 96,* 7998–8002.

Currie, C. R., Poulsen, M., Mendenhall, J., Boomsma, J. J., & Billen, J. (2006). Coevolved crypts and exocrine glands support mutualistic bacteria in fungus-growing ants. *Science, 311,* 81–83.

Davies, J. R. (2003). Inhibition of histone deacetylase activity by butyrate. *The Journal of Nutrition, 133,* 2485S–2493S.

Denning, D. W., & Hope, W. W. (2010). Therapy for fungal diseases: Opportunities and priorities. *Trends in Microbiology, 18,* 195–204.

Dunshee, B. R., Leben, C., Keitt, G. W., & Strong, F. M. (1949). The isolation and properties of antimycin A. *Journal of the American Chemical Society, 71,* 2436–2437.

Gomez-Escribano, J. P., & Bibb, M. J. (2011). Engineering *Streptomyces coelicolor* for heterologous expression of secondary metabolite gene clusters. *Microbial Biotechnology*, *4*, 207–215.

Gust, B., Challis, G. L., Fowler, K., Kieser, T., & Chater, K. F. (2003). PCR-targeted Streptomyces gene replacement identifies a protein domain needed for biosynthesis of the sesquiterpene soil odor geosmin. *Proceedings of the National Academy of Sciences of the United States of America*, *100*, 1541–1546.

Gust, B., Chandra, G., Jakimowicz, D., Yuqing, T., Bruton, C. J., & Chater, K. F. (2004). Lambda Red-mediated genetic manipulation of antibiotic-producing *Streptomyces*. *Advances in Applied Microbiology*, *54*, 107–128.

Haeder, S., Wirth, R., Herz, H., & Spiteller, D. (2009). Candicidin-producing *Streptomyces* support leaf-cutting ants to protect their fungus garden against the pathogenic fungus *Escovopsis*. *Proceedings of the National Academy of Sciences of the United States of America*, *106*, 4742–4746.

Hsu, S. C., & Lockwood, J. L. (1975). Powdered chitin agar as a selective medium for enumeration of actinomycetes in water and soil. *Applied and Environmental Microbiology*, *29*, 422–426.

Hugenholtz, P., Goebel, B. M., & Pace, N. R. (1998). Impact of culture-independent studies on the emerging phylogenetic view of bacterial diversity. *Journal of Bacteriology*, *180*, 4765–4774.

Jørgensen, H., Fjaervik, E., Havag, S., Bruheim, P., Bredholt, H., Klinkenberg, G., et al. (2009). Candicidin biosynthesis gene cluster is widely distributed among *Streptomyces* spp. isolated from sediments and the neuston layer of the Trondheim Fjord, Norway. *Applied and Environmental Microbiology*, *75*, 3296–3303.

Keyhani, N. O., & Roseman, S. (1997). Wild-type Escherichia coli grows on the chitin disaccharide, N, N'-diacetylchitobiose, by expressing the cel operon. *Proceedings of the National Academy of Sciences of the United States of America*, *94*, 14367–14371.

Kieser, T., Bibb, M. J., Buttner, M. J., Chater, K. F., & Hopwood, D. A. (2000). *Practical Streptomyces genetics*. Norwich, UK: The John Innes Foundation.

Kim, B.-G., Lee, M.-J., Seo, J., Hwang, Y.-B., Lee, M.-Y., Han, K., et al. (2009). Identification of functionally clustered nystatin-like biosynthetic genes in a rare actinomycetes, *Pseudonocardia autotrophica*. *Journal of Industrial Microbiology & Biotechnology*, *36*, 1425–1434.

Kost, C., Lakatos, T., Bottcher, I., Arendholz, W.-R., Redenbach, M., & Wirth, R. (2007). Non specific association between filamentous bacteria and fungus-growing ants. *Die Naturwissenschaften*, *94*, 821–828.

Kurtz, B., Phillipy, A., Delcher, A. L., Smoot, M., Shumway, M., Antonescu, C., et al. (2004). Versatile and open software for comparing large genomes. *Genome Biology*, *5*, R12.

Lane, J. (1991). 16S/23S S rRNA sequencing. In E. Stackebrandt & M. Goodfellow (Eds.), *Nucleic acid techniques in bacterial systematics* (pp. 115–175). New York: John Wiley and Sons.

MacNeil, D. J., Gewain, K. M., Ruby, C. L., Dezeny, G., Gibbons, P. H., & MacNeil, T. (1992). Analysis of *Streptomyces avermitilis* genes required for avermectin biosynthesis utilizing a novel integrative vector. *Gene*, *111*, 61–68.

Mattoso, T. C., Moreira, D. D. O., & Samuels, R. I. (2012). Symbiotic bacteria on the cuticle of the leaf-cutting ant *Acromyrmex subterraneus subterraneus* protect workers from attack by entomopathogenic fungi. *Biology Letters*, *8*, 461–464. http://dx.doi.org/10.1098/rsbl.2011.0963.

Miao, V., Coeffet-LeGal, M.-F., Brian, P., Brost, R., Penn, J., Whiting, A., et al. (2005). Daptomycin biosynthesis in *Streptomyces roseosporus*: Cloning and analysis of the gene cluster and revision of peptide stereochemistry. *Microbiology*, *151*, 1507–1523.

Oh, D. C., Poulsen, M., Currie, C. R., & Clardy, J. (2009). Dentigerumycin: A bacterial mediator of an ant-fungus symbiosis. *Nature Chemical Biology*, *5*, 391–393.

Palumbi, S. R. (2001). Human's as the world's greatest evolutionary force. *Science*, *293*, 1786–1790.

Rausch, C., Weber, T., Kohlbacher, O., Wohlleben, W., & Huson, D. H. (2005). Specificity predcition of adenylation domains in nonribosomal peptide synthetases (NRPS) using transductive support vector machines. *Nucleic Acids Research*, *33*, 5799–5808.

Rutherford, K., Parkhill, J., Crook, J., Horsnell, T., Rice, P., Rajandream, M. A., et al. (2000). Artemis: Sequence visualization and annotation. *Bioinformatics*, *16*, 944–945.

Schoenian, I., Spiteller, M., Ghaste, M., Wirth, R., Herz, H., & Spiteller, D. (2011). Chemical basis of the synergism and antagonism in microbial communities in the nests of leaf-cutting ants. *Proceedings of the National Academy of Sciences of the United States of America*, *108*, 1955–1960.

Schultz, T. R., & Brady, S. G. (2008). Major evolutionary transitions in ant agriculture. *Proceedings of the National Academy of Sciences of the United States of America*, *105*, 5435–5440.

Seipke, R. F., Barke, J., Brearley, C., Hill, L., Yu, D. W., Goss, R. J. M., et al. (2011). A single *Streptomyces* symbiont makes multiple antifungals to support the fungus farming ant *Acromyrmex octospinosus*. *PLoS One*, *6*, e22028.

Seipke, R. F., Barke, J., Ruiz-Gonzalez, M. X., Orivel, J., Yu, D. W., & Hutchings, M. I. (2011). Fungus-growing *Allomerus* ants are associated with antibiotic-producing actinobacteria. *Antonie Van Leeuwenhoek*, *101*, 443–447.

Seipke, R. F., Crossman, L., Drou, D., Heavens, D., Bibb, M. J., Caccamo, M., et al. (2011). Draft genome sequence of *Streptomyces* S4, a symbiont of the leafcutter ant *Acromyrmex octospinosus*. *Journal of Bacteriology*, *193*, 4270–4271.

Seipke, R. F., Kaltenpoth, M., & Hutchings, M. I. (2011). *Streptomyces* as symbionts: An emerging and widespread theme? *FEMS Microbiology Reviews*, *36*, 1–15. http://dx.doi.org/10.1111/j.1574-6976.2011.00313.x.

Sen, R., Ishak, H. D., Estrada, E., Dowd, S. E., Hong, E., & Mueller, U. G. (2009). Generalized antifungal activity and 454-screening of *Pseudonocardia* and *Amycolatopsis* bacteria in nests of fungus-growing ants. *Proceedings of the National Academy of Sciences of the United States of America*, *106*, 17805–17810.

Yanai, K., Murakami, T., & Bibb, M. J. (2006). Amplification of the entire kanamycin biosynthetic gene cluster using emphirical strain improvement of *Streptomyces kanamyceticus*. *Proceedings of the National Academy of Sciences of the United States of America*, *103*, 9661–9666.

CHAPTER FOUR

# Gamma-Butyrolactone and Furan Signaling Systems in *Streptomyces*

**John D. Sidda, Christophe Corre[1]**
Department of Chemistry, University of Warwick, Coventry, United Kingdom
[1]Corresponding author: e-mail address: c.corre@warwick.ac.uk

## Contents

## Abstract

*Streptomyces* bacteria produce different classes of diffusible signaling molecules that trigger secondary metabolite production and/or morphological development within the cell population. The biosynthesis of gamma-butyrolactones (GBLs) and 2-alkyl-4-hydroxymethylfuran-3-carboxylic acids (AHFCAs) signaling molecules is related and involves an essential AfsA-like butenolide synthase. This chapter first describes the catalytic role of AfsA-like enzyme then provides details about methods for the discovery and characterization of potentially novel signaling molecules. In section 4, one approach for establishing the biological role of these signaling molecules is presented.

## 1. INTRODUCTION

*Streptomyces* bacteria are soil–dwelling, filamentous bacteria which, during their life cycle, undergo a morphological switch that involves activation of pathways involved in the biosynthesis of secondary metabolites, including a variety of antibiotics, antifungal, and anticancer agents (Bibb, 2005). In general, activation of these secondary metabolic biosynthetic

*Methods in Enzymology*, Volume 517
ISSN 0076-6879
http://dx.doi.org/10.1016/B978-0-12-404634-4.00004-8

71

pathways occurs with, or just before, the onset of development of aerial hyphae. Involved in the control of the morphological development of these bacteria are a range of small, diffusible molecules referred to as "microbial hormones," exemplified by autoregulatory factor (A-factor), which activates the secondary metabolite streptomycin and the morphological sporulation pathway in *Streptomyces griseus*.

Discovered in 1967, A-factor was the first example of a gamma-butyrolactone (GBL) to be isolated from *Streptomyces* species (Khokhlov et al., 1967). Because of the role of A-factor in the biosynthesis of the streptomycin antibiotic, A-factor has been the subject of much research in the field of bacterial signaling and has also indicated the wider functioning of A-factor in regulating morphological differentiation (Ohnishi, Kameyama, Onaka, & Horinouchi, 1999; Ohnishi, Yamazaki, Kato, Tomono, & Horinouchi, 2005). AfsA has been identified as the key biosynthetic enzyme required for A-factor production, and the intracellular target of A-factor—the transcriptional repressor protein ArpA—has also been identified (Kato, Funa, Watanabe, Ohnishi, & Horinouchi, 2007). ArpA controls expression of the gene encoding AdpA, a transcriptional activator implicated in the regulation of secondary metabolite production and morphological differentiation (Ohnishi et al., 1999).

Concurrent research over the past 30 years has led to the isolation and characterization of other GBLs and GBL signaling systems in a range of *Streptomyces* species: factor I from *S. viridochromogenes* (Gräfe, Schade, Eritt, Fleck, & Radics, 1982), three Gräfe factors isolated from *S. bikiniensis* (Gräfe et al., 1983), *Virginiae butanolides* VB-A to E isolated from *S. virginiae* (Yamada, Sugamura, Kondo, Yanagimoto, & Okada, 1987), factor IM-2 from *S. lavendulae* FRI-5 (Sato, Nihira, Sakuda, Yanagimoto, & Yamada, 1989), and SCB1-3 from *S. coelicolor* (Takano et al., 2000). The functions of these GBLs and the signaling pathways they regulate are diverse. The GBLs for which structures have been elucidated have been shown to regulate biosynthesis of different antibiotics in the *Streptomyces* species from which each of them was isolated. By contrast, not all GBLs are thought to be involved in morphological differentiation (Bibb, 2005; Takano, 2006).

GBLs have been the subject of several recent reviews (Bibb, 2005; Ohnishi et al., 2005; Takano, 2006; Willey & Gaskell, 2011). In addition, a previous chapter published in this series also focused on methods for isolation of several GBLs, and a discussion of reporter systems relating the presence of GBLs to pigment production or expression of kanamycin resistance using specially designed plasmids. The role of gel retardation assays to detect the DNA targets of GBL receptors were also discussed (Hsiao, Gottelt, & Takano, 2009).

Comparative genomic analysis has more recently led to the identification of several AfsA-like biosynthetic enzymes. ArpA-like putative GBL receptors and their proposed roles in the regulation of antibiotic biosynthesis and morphological differentiation have also been reviewed (Takano, 2006; Willey & Gaskell, 2011). However, many of the GBL or GBL-like signals proposed to be involved with these other putative pathways remain unidentified. Investigation of one such signaling pathway involved in regulation of methylenomycin biosynthesis has resulted in the identification and structural characterization of a new class of bacterial signaling molecules, the 2-alkyl-4-hydroxymethylfuran-3-carboxylic acids—AHFCAs—as exemplified by the methylenomycin furans (MMFs) (Corre, Song, O'Rourke, Chater, & Challis, 2008; O'Rourke et al., 2009). In this chapter, discussion will focus on the specific roles of biosynthetic enzymes required for biosynthesis of GBLs and AHFCAs. The biological function of the MMFs will also be discussed.

## 2. AfsA-LIKE ENZYMES AND THE BIOSYNTHESIS OF SIGNALING MOLECULES

Over the past decade, investigations into the biosynthesis of several GBLs and GBL-like molecules have revealed more details of the specific functions of AfsA and AfsA-like enzymes. For example, AfsA was shown to be the key enzyme required for A-factor biosynthesis by Kato et al. (2007), and MmfL was established to be the key enzyme for MMF biosynthesis by Corre et al. (2008).

## 2.1. A-factor biosynthesis by AfsA

AfsA has been known to be required for A-factor biosynthesis since the initial observation that A-factor production is restored by inserting the *afsA* gene into a *S. griseus* HH1 mutant lacking a DNA fragment that includes *afsA* (Horinouchi, Kumada, & Beppu, 1984). More recently Kato and coworkers confirmed *in vitro* that AfsA is the key enzyme for A-factor biosynthesis. These studies established that AfsA catalyzes the condensation of an intermediate from fatty acid metabolism—8-methyl-3-oxononanoyl-acyl carrier protein (ACP)—and dihydroxyacetone phosphate (DHAP) to form an 8-methyl-3-oxononanoyl-DHAP ester intermediate, shown in Fig. 4.1 (Kato et al., 2007).

An enzymatic assay utilizing His-tagged AfsA, $^{32}$P-labeled DHAP and the synthetic 8-methyl-3-oxononanoyl-ACP analog, 8-methyl-3-oxononanoyl-$N$-acetylcysteamine (SNAC) led to the detection of a radioactive ester intermediate on thin layer chromatography. This product was compared with an authentic synthetic standard of the ester intermediate. After dephosphorylation of the ester, analysis by liquid chromatography–mass spectrometry (LC–MS) confirmed the presence of the dephosphorylated butenolide that is expected to be formed by the nonenzymatically catalyzed intramolecular aldol reaction of the dephosphorylated ester intermediate (Fig. 4.1). This ester was also added to a lysate of an *S. griseus ΔafsA* mutant and resulted in production of A-factor, thus confirming that AfsA is required for A-factor biosynthesis, and that this ester intermediate could be converted into A-factor by other enzymes. Additionally, when testing other 3-carbon substrates in the AfsA assay, only on addition of DHAP did the initial concentration of 8-methyl-3-oxononanoyl-SNAC decrease with time, implying that the enzyme is specific for condensation of DHAP with the fatty acid derivatives to yield A-factor.

After the AfsA-catalyzed formation of the DHAP ester intermediate, three further steps are required to yield A-factor—dephosphorylation of the ester intermediate, reduction of the C2–C3 double bond in the butenolide intermediate and the nonenzymatically catalyzed intermolecular aldol reaction that precedes butenolide formation. Two initial pathways for these three steps were proposed by Sakuda and coworkers in which (i) dephosphorylation of the DHAP intermediate is followed by the aldol reaction and final reduction of the butenolide to A-factor and (ii) intramolecular aldol reaction is followed by the reduction step and dephosphorylation (Sakuda et al., 1993). These two pathways are shown in Fig. 4.1. Indeed, Kato and coworkers presented evidence for both of these pathways occurring during their *in vitro* investigations of AfsA and A-factor biosynthesis (Kato et al., 2007).

These results were also complemented by analysis of the metabolites produced by *E. coli* in which *afsA* had been introduced. This strain was found to produce GBLs that present the same fragmentation patterns as A-factor when analyzed by tandem mass spectrometry. These molecules that were different from A-factor reflected the differences in fatty acid metabolism between *Streptomyces* species and *E. coli*. Indeed, *Streptomyces* species mainly produce branched fatty acids such as 8-methyl-3-oxononanoyl-acyl chains, whereas fatty acid biosynthesis is primarily primed with acetyl CoA in *E. coli*. The GBL molecules produced by *E. coli* containing the *afsA* gene simply differed in their alkyl chains compared to A-factor. Because no other *Streptomyces* gene than *afsA* was expressed in this *E. coli* strain, downstream steps in GBL biosynthesis (reduction of the butenolide and dephosphorylation) were proposed to be catalyzed by enzymes that are nonspecific to the A-factor biosynthetic pathway. Such reductases and phosphatases may be present across a wider range of bacteria than just *Streptomyces*.

Interestingly, in *S. griseus* a gene coding for a putative oxidoreductase-like protein (BprA) is located next to *afsA*. This NADPH-dependent reductase, BprA, was shown to specifically catalyze the reduction of the butenolide phosphate to a butanolide phosphate but unable to catalyze the reduction of the dephosphorylated butenolide (Fig. 4.1). Both pathways initially proposed by Sakuda and coworkers for the downstream steps of A-factor biosynthesis after formation of the DHAP ester intermediate are still likely to occur.

Since the advent of next-generation sequencing techniques, genomic data for a number of *Streptomyces* strains is now available in the public domain and has led to the discovery of several *afsA* homologues in other streptomycetes, for example, *jadW1* from *S. venezuelae*, *sabA* from *S. acidiscabies*, and *tylP* from *S. fradiae*. However, it must be noted that these examples represent proposed biosynthetic enzymes whose products remain unknown (Willey & Gaskell, 2011). Within the genome of the streptomycetes model *S. coelicolor* A3(2), two *afsA*-like genes are present: *scbA* within the chromosomal DNA that has been shown to direct the biosynthesis of the *S. coelicolor* GBL (SCBs) and *mmfL* encoded within the natural plasmid SCP1. MmfL has been shown to direct the biosynthesis of AHFCAs as described in the next section (Corre et al., 2008).

## 2.2. The role of MmfL in MMF biosynthesis

In the wider context of signaling systems in streptomycetes, comparative genomics has revealed several genes that regulate secondary metabolites that, when characterized, led to the discovery of other classes of signaling molecules, most notably the MMFs isolated from *S. coelicolor* A3(2) (Corre et al., 2008).

After the discovery of the AHFCAs, Corre and coworkers proposed that the function of MmfL was to catalyze an analogous reaction to that of AfsA, by which the hydroxyl group of DHAP undergoes a condensation reaction with an intermediate from fatty acid metabolism, yielding an ester intermediate which then undergoes intramolecular aldol condensation to form a

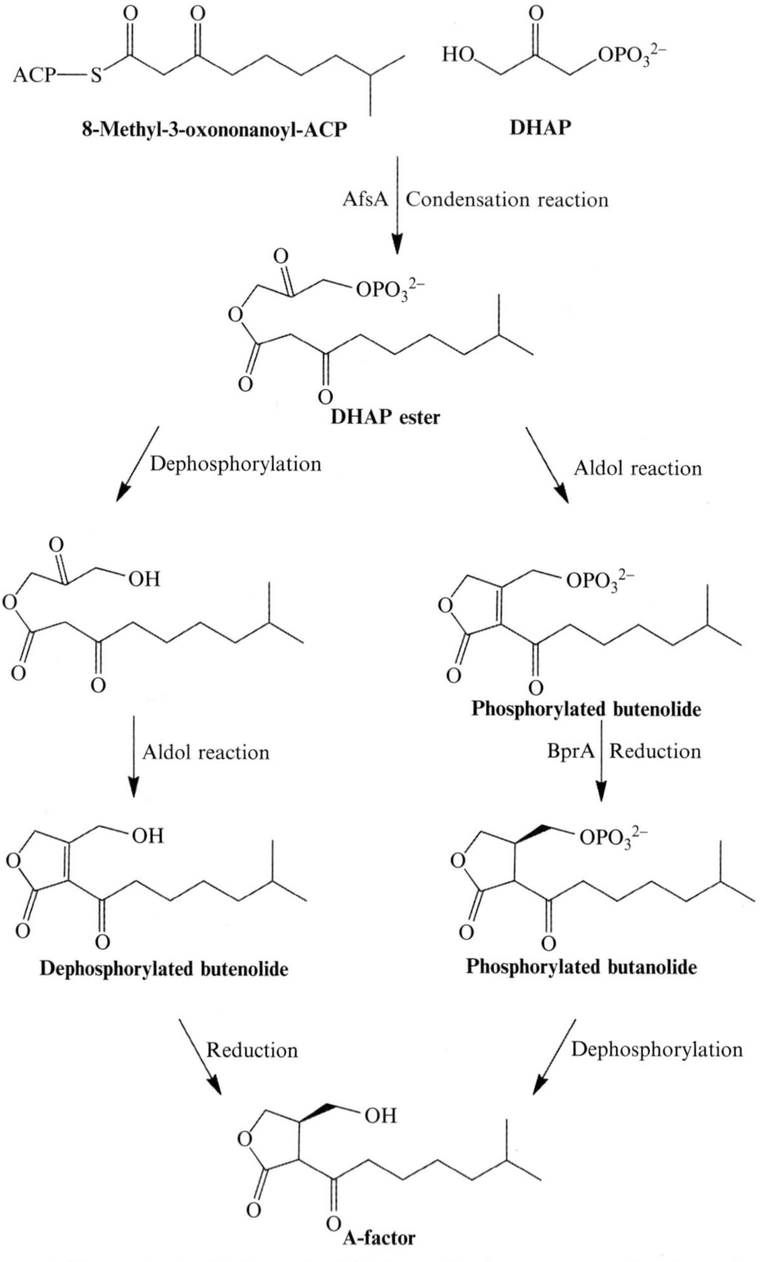

**Figure 4.1** Biosynthesis of A-factor by AfsA from dihydroxyacetone phosphate (DHAP) and 8-methyl-3-oxononanoyl-ACP. The two proposed pathways for the downstream steps of A-factor biosynthesis after the AfsA-catalyzed step are also shown.

butenolide intermediate. Subsequent dephosphorylation and reduction steps were proposed to be catalyzed by MmfP and MmfH, respectively, to yield the AHFCA molecules from the butenolide intermediate (Corre et al., 2008). Biosynthesis of one of these compounds, MMF2, is summarized in Fig. 4.2.

During initial structure elucidation studies of the MMF compounds, deuterium-labeled precursors of the starter units of fatty acid biosynthesis were found to be incorporated into MMF molecules. For example, all seven deuterium atoms from $D_7$-butyrate were incorporated into MMF2, via the 3-oxohexanoyl thioester shown in Fig. 4.2. The different alkyl chains found in the other MMF compounds originated from other precursor of fatty acid metabolism. This experiment provided the first evidence to support that the MMF resulted from the assembly of an intermediate derived from fatty acid metabolism with DHAP.

Cloning and expression of *mmfL* in a *Streptomyces* host (unable to produce AHFCA compounds) was shown to produce the same relative proportions of the MMF compounds as in the original *S. coelicolor* producer strain, establishing that MmfL is an essential enzyme for MMF biosynthesis. This study was complemented by cloning and expression of *mmfL* in *E. coli* BL21star via a high-copy number plasmid, which led to production of MMF2 and MMF5. Each of these two MMFs has nonbranched alkyl chains substitution in position 2, confirming that *mmfL* was the only *S. coelicolor* required enzyme for MMF biosynthesis and consistent with the observation that *E. coli* fatty acid metabolism produces only nonbranched chain molecules (Corre et al., 2008). In analogy with the observation that AfsA is the key enzyme required for A-factor biosynthesis, this result implies that there may be other enzymes present in the cells that allow the downstream steps of MMF biosynthesis to occur without the enzymes MmfH and MmfP. The precise biological functions of MmfH and MmfP have still to be established *in vitro*.

On completion of a successful biomimetic synthesis of the AHFCAs, Davis and coworkers proposed an alternative biosynthetic pathway for the AHFCAs that hypothesized a different function for MmfL that did not involve the formation of a butenolide intermediate (Davis, Bailey, & Sello, 2009). MmfL was proposed to catalyze a carbon–carbon bond formation instead of the esterification reaction that was evidenced by Kato and coworkers using recombinant AfsA. A key difference in the two proposed functions of MmfL had implications for the orientation of the DHAP molecule during the condensation reaction with the 3-oxohexanoyl-ACP.

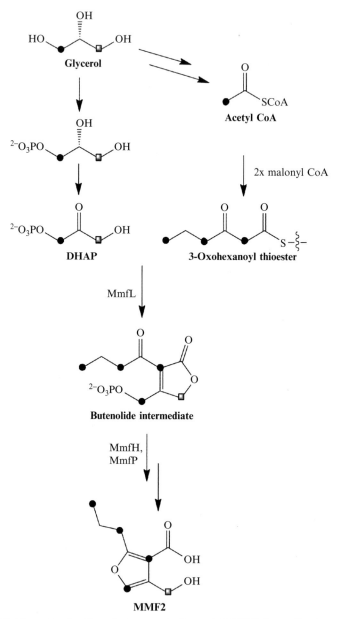

**Figure 4.2** Biosynthesis of methylenomycin furan 2 (MMF2) from dihydroxyacetone phosphate (DHAP) and a 3-oxohexanoyl thioester. The biosynthesis of the other MMF compounds differs only in the fatty acid precursor molecule. The black circles and gray squares represent the positions of $^{13}$C labels in the intermediates and products of this pathway when *Streptomyces* W74 is fed with *pro-R* and *pro-S* labeled glycerol, respectively.

Because *in vivo* DHAP is derived from glycerol, as is acetyl CoA (the precursor from which β-ketothioesters in fatty acid biosynthesis are derived), the incorporation of stable-isotope carbon atoms ($^{13}$C) into the MMF compounds using $^{13}$C-labeled glycerol can be hypothesized (Fig. 4.2).

Labeling of the *pro-S* carbon atom of glycerol (square in Fig. 4.2) was proposed to lead to incorporation of one $^{13}$C atom (from DHAP) into the hydroxymethyl group of MMFs. Labeling of the *pro-R* carbon atom of glycerol was proposed to lead to incorporation of up to four $^{13}$C atoms into the MMFs, as the *pro-R* carbon is retained in the biosynthetic steps that lead to formation of acetyl CoA from DHAP (Fig. 4.2). 3-Oxohexanoyl-ACP is derived from three units of acetyl CoA and therefore would contain three $^{13}$C labels, plus a $^{13}$C label from the DHAP molecule that undergoes the condensation reaction with the 3-oxohexanoyl thioester. By examining the mechanism of mass spectrometry fragmentation of MMFs, the fragment corresponding to the hydroxymethyl group could be identified, so from the fragmentation of the $^{13}$C-labeled MMFs the orientation of DHAP with respect to the ketothioester precursors could be deduced (Fig. 4.2).

The MMF compounds produced by *S. coelicolor* in presence of glycerol labeled at the *pro-S* carbon atom did contain one $^{13}$C atoms, as its molecular ion peak had the exact molecular weight and formula expected for MMFs that contain one $^{13}$C atoms (Corre, Haynes, Malet, Song, & Challis, 2010). On studying incorporation of glycerol labeled at the *pro-R* carbon atom, the molecular formulae of the MMFs generated were found to be consistent with the incorporation of up to four $^{13}$C atom.

MMF2 containing one $^{13}$C atom derived from the glycerol *pro-S* carbon lost its label during the fragmentation process, implying that the $^{13}$C atom was located on the hydroxymethyl group. However, the MMFs containing four $^{13}$C atoms (from DHAP and 3-oxohexanoyl thioester derived from *pro-R* labeled glycerol) did not lose any of the four $^{13}$C labels on fragmentation. These results provided evidence for the orientation of DHAP when incorporated into the MMFs, with the *pro-R* carbon from glycerol ultimately forming part of the furan ring in the MMFs, and the *pro-S* glycerol carbon atom providing the carbon for the hydroxymethyl group, as indicated in Fig. 4.2 (Corre et al., 2010).

These studies confirmed the initial proposed biosynthetic pathway (Corre et al., 2008) and are consistent with the enzymatic mechanism of AfsA itself that has a high homology to MmfL (27% amino acid identity, 41% amino acid sequence similarity over 274 amino acids).

## 3. HETEROLOGOUS EXPRESSION OF AN *AfsA*-LIKE-CONTAINING MINI-CLUSTER COMBINED WITH COMPARATIVE METABOLIC PROFILING

Mining the genome of sequenced *Streptomyces* species generally results in the identification of at least one *afsA* orthologue based on sequence homology analysis. The genetic context of this element can then be investigated. In *Streptomyces*, biosynthetic genes involved in the production of a natural product are most often clustered. Putative biosynthetic genes located adjacent to an *afsA*-like gene might be involved in modifying the butenolide intermediate that results from the catalytic activity of AfsA-like enzymes.

In the case of *S. coelicolor* A3(2), the *mmfLHP* minicluster (Fig. 4.3) was identified and was expressed in a *Streptomyces* host (Corre et al., 2008).

The protocol describing these steps is as follows:

1. Sub-cloning of the *afsA*-like containing biosynthetic gene cluster into an integrative vector.

Three strategies are possible:

**1a**. Digestion with carefully selected restriction enzymes (i.e., *Sex*AI/*Bgl*II 3554-bp fragment) and cloning into a pSET152-based integrative

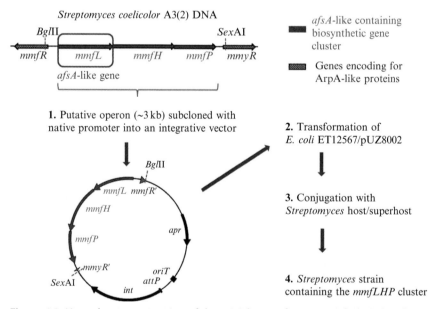

**Figure 4.3** Heterologous expression of the minicluster of genes *mmfL, H, P* that direct the biosynthesis of the methylenomycin furans (MMFs) in *Streptomyces*. (See Color Insert.)

vector. This plasmid contains a gene coding for an integrase and an oriT which will be utilized to introduce the engineered plasmid into a heterologous *Streptomyces* host.

*Note*: if the operon is smaller than 5000 bp, strategies 1b or 1c are more appropriate.

**1b.** PCR amplification of the operon. The native promoter can be included or replaced with a strong constitutive synthetic promoter. PCR primers include restriction sites at the 5′ end that are compatible with the restriction sites present on the integrative plasmid (pSET152).

*Note 1*: excluding the ARE region can be all-important to avoid possible transcriptional repression.

**1c.** Design of a synthetic operon. Similarly to strategy 1b, restriction sites are incorporated at the 5′ end of the synthetic operon to allow a straightforward cloning into the integrative vector. A strong constitutive synthetic promoter can be introduced in front of the synthetic operon. This approach is particularly convenient if the *afsA*-like gene-containing operon is too large to be amplified by PCR, or if the putative biosynthetic genes are not directly adjacent to each other.

**2.** Transformation of *E. coli* ET12567/pUZ8002. Using the apramycin resistance marker, the engineered integrative vector is used to transform *E. coli* ET12567/pUZ8002. *E. coli* ET12567 is a DNA methylase deficient strain. In addition, pUZ8002 encodes the tra functions required for mobilization of the integrative plasmid. This *E. coli* strain is used to circumvent the methyl-specific DNA restriction systems of *S. coelicolor*. This *E. coli* strain then allows transfer of the engineered vector into a *Streptomyces* host/superhost by intergenus conjugation.

**3.** Conjugation with *Streptomyces* host/superhost

The host initially used was *S. coelicolor* M512 (SCP1⁻, SCP2⁻, act⁻, red⁻) which cannot produce methylenomycin, actinorhodin, or prodiginine antibiotics; however, a series of improved *S. coelicolor* superhosts (M1152, M1154) is now available from Prof. Bibb's laboratory (Gomez-Escribano & Bibb, 2010) as well as a series of *S. avermitilis* superhosts from Prof. Ikeda's laboratory (Komatsu, Uchiyama, Ōmura, Cane, & Ikeda, 2010) (see also chapters by Gomez-Escribano and Bibb and by Yamada *et al.* in this volume and in the companion volume in this set).

pIJ6584 was integrated into the chromosome of *S. coelicolor* M512 *via* site-specific integration to generate *S. coelicolor* W74. These strains were then analyzed by PCR to confirm the presence of the *mmfLHP* operon.

**4.** Comparison of the metabolites produced by engineered *Streptomyces* strain containing the *mmfLHP* cluster and the *Streptomyces* host/superhost.

Both strains were grown in a supplemented minimum medium and diffusible small molecules were analyzed by collecting the supernatant after 5 days of incubation at 30 °C. Direct analysis of the supernatant by LC–MS is not very reliable as the sensitivity of the detection of small organic metabolites in the culture medium is relatively low.

The metabolites from both strains (W74 and M512) are therefore extracted from the supernatant using organic solvent such as EtOAc at different pH.

To identify the cryptic signaling molecule(s), the metabolite profiles are then compared by LC–MS. First, the base peak chromatograms are compared but unless the new compounds represent a significant proportion of the metabolites analyzed, no difference is visible. To pinpoint differences in metabolite profiles, extracted ion chromatograms can be compared using various mass ranges (i.e., $m/z = 150$–$250$). Five compounds present in extracts of the W74 strain but absent from extracts of the M512 strain were identified when extracting the ion chromatograms corresponding to a mass-over-charge ratio range from 150 to 250; these compounds were named MMFs 1–5.

*Note*: these analyses are facilitated if the medium in which metabolites are produced is minimal. Indeed, metabolites of interest might be difficult to identify if they are part of a complex mixture of components, they would also be more difficult to purify.

Fractions containing cryptic metabolites were then analyzed by accurate mass spectrometry. Thus, the molecular formula of each compound can be determined. These natural products can then be purified using HPLC and their chemical structure elucidated using a combination of mass spectrometry and NMR spectroscopy analyses. The new metabolites can then be screened for signaling activity.

## 4. GENETIC DISRUPTION OF *mmfLHP* AND CHEMICAL COMPLEMENTATION EXPERIMENTS

This section reports the methodology used to silence a specific biosynthetic gene cluster and reactivate this pathway using isolated or synthetic signaling molecules (or analogs).

The machinery that direct signaling molecule biosynthesis was first disrupted by replacing the *mmfLHP* genes by the apramycin resistance gene *apr* using PCR targeting technology (Fig. 4.4) (Gust, Challis, Fowler, Kieser, & Chater, 2003). The integrative cosmid C73-787, which contains the

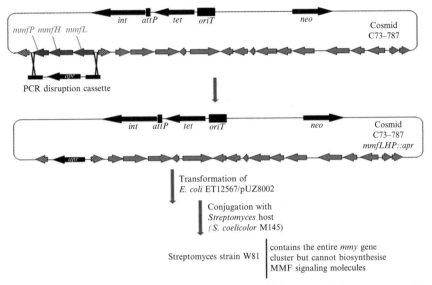

**Figure 4.4** Construction of a *Streptomyces* strain in which the methylenomycin antibiotic pathway has been silenced by disrupting the machinery that direct the biosynthesis of methylenomycin furan signaling molecules (MMFs). (See Color Insert.)

entire methylenomycin biosynthetic gene cluster, was used to generate the *S. coelicolor* MMF and methylenomycin non-producing strain W81 (Corre et al., 2008).

1. PCR amplification of an apramycin resistance cassette (*apr* gene under the control of a constitutive promoter from pCC60) using long PCR primers that contain homology to the sequence upstream of *mmfL* and downstream of *mmfP*.

2. Screening for apramycin and kanamycin resistant clones. Check by PCR and restriction digest.

3. As per the protocol described in the previous section: transformation of *E. coli* ET12567/pUZ8002 followed by conjugation with a *Streptomyces* host/superhost resulted in a new *Streptomyces* strain named W81. Indeed, the cosmid C73-787 also contained an integrative element which permits site-specific integration into the *Streptomyces* chromosome. Southern blot and PCR analyses are then used on the DNA isolated from the new *Streptomyces* strain to confirm integration of the mutated cosmid into the host.

4. Analysis of the production of furan signaling molecules (MMF) and methylenomycin antibiotics. Strain W81 was grown on SMM and, as

**Figure 4.5** Restoration of methylenomycin antibiotic production in the silenced strain by addition of 1 µg of synthetic MMF1. (For color version of this figure, the reader is referred to the online version of this chapter.)

expected, LC–MS analysis of the organic extract established that the new strain was unable to produce MMF compounds. In addition, no methylenomycin antibiotic could be detected; thus the biosynthetic pathway for the production of these antibiotics has been silenced. Feeding of synthetic signaling molecules (MMFs) to the culture medium restored production of the methylenomycin antibiotics.

5. Strain W81 was then used to assess the bioactivity of signaling molecules (MMFs and synthetic analogs) with regards to induction of methylenomycin production (Fig. 4.5). An SMMS plate (pH 5.0, this pH allows a better diffusion of methylenomycin antibiotic) is inoculated with W81 and incubated at 30 °C for 48 h. Plugs from this plate are then transferred onto an SMMS plate (pH 5.0) spread with a spore suspension of a methylenomycin sensitive strain (in this case *S. coelicolor* M145; however, any bacteria sensitive to methylenomycin could be used such as *Bacillus subtilis* or *Micrococcus luteus*). One microgram of pure MMF1 (in water) is then added on the top of an agar plug next to plug where W81 is growing. Alternatively, the compounds to be tested can be added directly onto the plug where W81 is growing. Following incubation at 30 °C for an additional 24 h, a zone of inhibition became apparent surrounding plugs where the methylenomycin antibiotic was produced.

## 5. CONCLUSION

The development of *Streptomyces* genome mining combined with the use of increasingly more sensitive analytical instrumentation (mass spectrometry and NMR spectroscopy) is of great interest with regards to the characterization of microbial signaling molecules that are naturally produced at concentrations in the nanomolar range (Fig. 4.6).

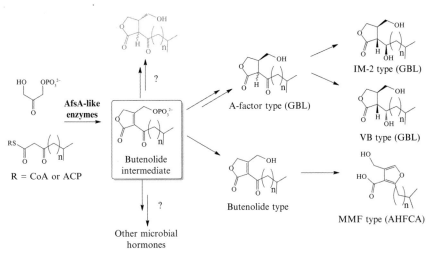

**Figure 4.6** Overview of diverse types of *Streptomyces* antibiotic production inducers that result from AfsA-like catalytic activities.

Thus, novel *Streptomyces* signaling molecules such as the MMFs have recently been identified and characterized using the strategies presented in this chapter. Strain W81 was then engineered to assess the biological role of the MMF signaling molecules. This strain is also currently used for assessing the structure–activity relationships involved in the interactions between transcriptional repressors and signaling molecules. The transcriptional repressors that interact with GBL and AHFCA signaling molecules are discussed in the chapter 17 by Aigle and Corre in this volume.

## ACKNOWLEDGMENT

We are grateful to The Royal Society (URF) for funding.

## REFERENCES

Bibb, M. J. (2005). Regulation of secondary metabolism in streptomycetes. *Current Opinion in Microbiology*, *8*, 208–215.

Corre, C., Haynes, S. W., Malet, N., Song, L., & Challis, G. L. (2010). A butenolide intermediate in methylenomycin furan biosynthesis is implied by incorporation of stereospecifically [13]C-labelled glycerols. *Chemical Communications*, *46*, 4079–4081.

Corre, C., Song, L., O'Rourke, S., Chater, K. F., & Challis, G. L. (2008). 2-Alkyl-4-hydroxymethylfuran-3-carboxylic acids, antibiotic production inducers discovered by *Streptomyces coelicolor* genome mining. *Proceedings of the National Academy of Sciences of the United States of America*, *105*, 17510–17515.

Davis, J. B., Bailey, J. D., & Sello, J. K. (2009). Biomimetic synthesis of a new class of bacterial signalling molecules. *Organic Letters*, *11*, 2984–2987.

Gomez-Escribano, J. P., & Bibb, M. J. (2010). Engineering *Streptomyces coelicolor* for heterologous expression of secondary metabolite gene clusters. *Microbial Biotechnology, 4,* 207–215.

Gräfe, U., Reinhardt, G., Schade, W., Eritt, I., Fleck, W. F., & Radics, L. (1983). Interspecific inducers of cytodifferentiation and anthracycline biosynthesis from *Streptomyces bikinensis* and *S. cyaneofuscatus*. *Biotechnology Letters, 5,* 591–596.

Gräfe, U., Schade, W., Eritt, I., Fleck, W. F., & Radics, L. (1982). A new inducer of anthracycline biosynthesis from *Streptomyces viridochromogenes*. *The Journal of Antibiotics, 35,* 1722–1723.

Gust, B., Challis, G. L., Fowler, K., Kieser, T., & Chater, K. F. (2003). PCR-targeted *Streptomyces* gene replacement identifies a protein domain needed for biosynthesis of the sesquiterpene soil odor geosmin. *Proceedings of the National Academy of Sciences of the United States of America, 100,* 1541–1546.

Horinouchi, S., Kumada, Y., & Beppu, T. (1984). Unstable genetic determinant of A-factor biosynthesis in streptomycin-producing organisms: Cloning and characterisation. *Journal of Bacteriology, 158,* 481–487.

Hsiao, N.-H., Gottelt, M., & Takano, E. (2009). Regulation of antibiotic production by bacterial hormones. *Methods in Enzymology, 458,* 143–157.

Kato, J., Funa, N., Watanabe, H., Ohnishi, Y., & Horinouchi, S. (2007). Biosynthesis of γ-butyrolactone autoregulators that switch on secondary metabolism and morphological development in *Streptomyces*. *Proceedings of the National Academy of Sciences of the United States of America, 104,* 2378–2383.

Khokhlov, A. S., Tovarova, I. I., Borisova, L. N., Pliner, S. A., Shevchenko, L. N., Kornitskaia, E., Ivkina, N. S., & Rapoport, I. A. (1967). The A-factor, responsible for streptomycin biosynthesis by mutant strains of *Actinomyces streptomycini*. *Doklady Akademii Nauk SSSR, 177,* 232–235.

Komatsu, M., Uchiyama, T., Ōmura, S., Cane, D. E., & Ikeda, H. (2010). Genome-minimized *Streptomyces* host for the heterologous expression of secondary metabolism. *Proceedings of the National Academy of Sciences of the United States of America, 107,* 2646–2651.

O'Rourke, S., Wietzorrek, A., Fowler, K., Corre, C., Challis, G. L., & Chater, K. F. (2009). Extracellular signalling, translational control, two repressors and an activator all contribute to the regulation of methylenomycin production in *Streptomyces coelicolor*. *Molecular Microbiology, 71,* 763–778.

Ohnishi, Y., Kameyama, S., Onaka, H., & Horinouchi, S. (1999). The A-factor regulatory cascade leading to streptomycin biosynthesis in *Streptomyces griseus*: Identification of a target gene of the A-factor receptor. *Molecular Microbiology, 34,* 102–111.

Ohnishi, Y., Yamazaki, H., Kato, J., Tomono, A., & Horinouchi, S. (2005). AdpA, a central transcriptional regulator in the A-factor regulatory cascade that leads to morphological development and secondary metabolism in *Streptomyces griseus*. *Bioscience, Biotechnology, and Biochemistry, 69,* 431–439.

Sakuda, S., Tanaka, S., Mizuno, K., Sukcharoen, O., Nihira, T., & Yamada, S. (1993). *Journal of the Chemical Society Perkin Transactions 1,* 2309–2315.

Sato, K., Nihira, T., Sakuda, S., Yanagimoto, M., & Yamada, Y. (1989). Isolation and structure of a new butyrolactone autoregulator from *Streptomyces* sp. FRI-5. *Journal of Fermentation and Bioengineering, 68,* 170–173.

Takano, E. (2006). γ-Butyrolactones: *Streptomyces* signalling molecules regulating antibiotic production and differentiation. *Current Opinion in Microbiology, 9,* 287–294.

Takano, E., Nihira, T., Hara, Y., Jones, J. J., Gershater, C. J., Yamada, Y., & Bibb, M. (2000). Purification and structural determination of SCB1, a γ-butyrolactone that elicits antibiotic production in *Streptomyces coelicolor* A3(2). *The Journal of Biological Chemistry, 275,* 11010–11016.

Willey, J. M., & Gaskell, A. A. (2011). Morphogenetic signalling molecules of the strepto-mycetes. *Chemical Reviews*, *111*, 174–187.

Yamada, Y., Sugamura, K., Kondo, K., Yanagimoto, M., & Okada, H. (1987). The structure of inducing factors for virginiamycin production in *Streptomyces virginiae*. *The Journal of Antibiotics*, *40*, 496–504.

CHAPTER FIVE

# Old Meets New: Using Interspecies Interactions to Detect Secondary Metabolite Production in Actinomycetes

## Mohammad R. Seyedsayamdost*[,1], Matthew F. Traxler[†,1], Jon Clardy*[,2], Roberto Kolter[†,2]

*Department of Biological Chemistry and Molecular Pharmacology, Harvard Medical School, Boston, Massachusetts, USA
†Department of Microbiology and Immunobiology, Harvard Medical School, Boston, Massachusetts, USA
[1]These authors contributed equally.
[2]Corresponding authors: e-mail address: jon_clardy@hms.harvard.edu; roberto_kolter@hms.harvard.edu

## Contents

## Abstract

Actinomycetes, a group of filamentous, Gram-positive bacteria, have long been a remarkable source of useful therapeutics. Recent genome sequencing and transcriptomic studies have shown that these bacteria, responsible for half of the clinically used antibiotics, also harbor a large reservoir of gene clusters, which have the potential to produce novel secreted small molecules. Yet, many of these clusters are not expressed under common culture conditions. One reason why these clusters have

*Methods in Enzymology*, Volume 517
ISSN 0076-6879
http://dx.doi.org/10.1016/B978-0-12-404634-4.00005-X
89

not been linked to a secreted small molecule lies in the way that actinomycetes have typically been studied: as pure cultures in nutrient-rich media that do not mimic the complex environments in which these bacteria evolved. New methods based on multispecies culture conditions provide an alternative approach to investigating the products of these gene clusters. We have recently implemented binary interspecies interaction assays to mine for new secondary metabolites and to study the underlying biology of interactinomycete interactions. Here, we describe the detailed biological and chemical methods comprising these studies.

# 1. INTRODUCTION

The majority of antibacterial compounds used clinically since the 1950s have been derived from actinomycetes, a group of filamentous, high $G + C$ Gram–positive bacteria long known as prolific producers of secondary metabolites (Berdy, 2005; Fischbach & Walsh, 2009; Newman & Cragg, 2012; Newman, Cragg, Holbeck, & Sausville, 2002). Recent genome sequencing efforts have brought this production capacity into focus by revealing dozens of easily identifiable secondary metabolite gene clusters per sequenced strain, indicating that actinomycetes contain vast reservoirs of potentially useful therapeutic agents (Bentley et al., 2002; Copeland et al., 2009; Nett, Ikeda, & Moore, 2009; Oliynyk et al., 2007; Zhao et al., 2010). However, the bulk of these recognizable clusters appear to be inactive under the conditions in which actinomycetes have been examined. As a consequence, these clusters are not associated with cognate small molecule products (Nett et al., 2009). Whether they represent gene clusters that are actively transcribed but not yet linked to a small molecule, or silent gene clusters that are inactive under the conditions examined, remains largely unknown. In some cases, transcriptomic or gene knockout studies have already allowed a distinction between these cluster types. The preponderance of biosynthetic gene clusters without known small molecule products found in *Saccharopolyspora erythraea* and *Streptomyces griseus*, the erythromycin and streptomycin producers, respectively, is especially astonishing (Nett et al., 2009; Ohnishi et al., 2008; Oliynyk et al., 2007). These strains have been studied under a large number of conditions for several decades and > 75% of their gene clusters still have no known small molecule product (Nett et al., 2009). Clearly, alternative approaches are needed to explore metabolite production in actinomycetes.

The number of secondary metabolite gene clusters revealed by genome sequencing and the growing appreciation that the exchange of small molecules mediates inter- and intraspecies communication suggest that bacterial interaction assays should be reassessed as a potential conduit for the discovery of new small molecules (Fischbach & Krogan, 2010; Schmidt, 2008; Shank & Kolter, 2009; Straight & Kolter, 2009). Several features of actinomycete biology lend themselves well to interaction screens, as previously demonstrated by Ueda, Beppu, and coworkers (Ueda et al., 2000; Yamanaka et al., 2005). Actinomycete colonies often have striking morphologies that are visible to the naked eye, including pigmentation (by one or more compounds), complex colony structure (i.e., wrinkles or other features), aerial hyphae, and pigmented spores (Chater, 1998). Alterations in any of these visible phenotypes can be due to compounds secreted by a nearby colony. Thus, these phenotypic changes can serve as sensors for small molecules which might not be found in screens that only search for inhibition of bacterial growth. At the same time, several new lines of evidence suggest that coculture may stimulate activation of silent gene clusters.

With these possibilities in mind, we began a series of interaction assays within a group of 20 sequenced actinomycetes to look for molecules whose effects could be recognized through altered morphology as well as molecules whose production might be stimulated through transcriptional activation of otherwise silent gene clusters (Seyedsayamdost, Traxler, Zheng, Kolter, & Clardy, 2011). Here, we describe detailed procedures for the methods we have recently used to gain insight into novel bioactive molecules that affect developmental changes within cocultured actinomycetes. These screens are described on a scale that allows a single researcher to evaluate hundreds to thousands of interspecies interactions for new secondary metabolites.

## 2. SETTING UP BINARY ACTINOMYCETE INTERACTIONS

Actinomycetes have traditionally been isolated from soil, which typically contains hundreds or thousands of different strains of these bacteria per gram (Curtis, Sloan, & Scannell, 2002). With such staggering biological diversity, the soil environment has been vastly undersampled with regards to its total actinomycete capacity (Baltz, 2008; Lazzarini, Cavaletti, Toppo, & Marinelli, 2000). Nonetheless, the rediscovery rate of secondary metabolites produced by streptomycetes is relatively high (Baltz, 2008; Watve, Tickoo, Jog, & Bhole, 2001). This high rate in part reflects the inherent bias in most commonly used isolation techniques, which rely on two important features, (1) sporulation and (2) utilization of complex mixtures of starch and proteins for growth. Most isolation protocols employ heat and desiccation steps to select only for microbes capable of forming resistant spores, followed by growth on media where starch and/or proteins are the major substrates, thereby selecting for strains which produce extracellular amylases and proteases, a hallmark of actinomycetes. Efforts to maximize the diversity of cultured actinomycetes have included development of protocols for isolating strains outside the genus *Streptomyces*. These are the so-called rare actinomycetes (Lazzarini et al., 2000), including those from the genera *Actinoplanes*, *Saccharopolyspora*, *Amycolatopsis*, and *Streptosporangium*. More recently, isolation of actinomycetes from more varied sources has also led to discovery of new compounds. Important new environmental sources include the ocean (Goodfellow & Fiedler, 2010), both from sediments (Jensen, Gontang, Mafnas, Mincer, & Fenical, 2005; Jensen, Mincer, Williams, & Fenical, 2005) and bacterial symbionts of marine organisms (Jiang et al., 2007; Kurahashi, Fukunaga, Sakiyama, Harayama, & Yokota, 2010), and insects, especially ants (Oh, Poulsen, Currie, & Clardy, 2009), beetles (Oh, Scott, Currie, & Clardy, 2009; Scott et al., 2008), and wasps (Oh, Poulsen, Currie, & Clardy, 2011; Poulsen, Oh, Clardy, & Currie, 2011). A number of culture conditions for isolating a variety of actinomycetes are listed in Table 5.1. As a starting point, we include here a basic protocol for the isolation of actinomycetes using starch–casein agar.

## 2.1. Actinomycete isolation medium

1. To 1 L of distilled water, add the following:
     10.0 g soluble starch (Fisher #S-516)
     0.30 g Casein hydrolysate (Sigma #C-9386)
     2.0 g $KNO_3$
     2.0 g NaCl
     2.0 g $K_2HPO_4$
     0.05 g $MgSO_4 \cdot 7H_2O$
     0.02 g $CaCO_3$
     0.01 g $FeSO_4 \cdot 7H_2O$
     15.0 g Granulated agar
2. Boil the agar solution, allow to cool slightly, and adjust to pH 7 with HCl or NaOH.
3. Autoclave at 121 °C for 15 min. After cooling to approximately 45 °C, add 5 mg cycloheximide in 1.0 mL methanol, mix thoroughly, and pour plates.

**Table 5.1** Culture media for isolation of various clades of actinomycetes

| References | Target actinobacterial clade | Agar type | Selection criteria |
|---|---|---|---|
| Wellington and Toth (1994) | General actinomycete | Starch–casein | Starch and protein as primary growth substrates |
| Difco | General actinomycete | Proprietary | Proprietary, includes glycerol |
| Hamaki et al. (2005) | Streptosporangium, Actinomadura | Soil extract | Growth on soil extract, small colonies picked under microscopic evaluation |
| Hop et al. (2011) | Actinoplanes, Kineosporia, Cryptosporangia | Humic acid-vitamin agar, nalidixic acid, kabicidin | Rehydration-centrifugation to select for species with motile spores |
| Hop et al. (2011) | Micromonospora, Nonomuraea, Streptomyces | Humic acid-vitamin agar, nalidixic acid, kabicidin | Sodium dodecyl sulfate-yeast extract dilution method |
| Tan, Ward, and Goodfellow (2006) | Amycolatopsis | SM1, SM2, SM3 | Combinations of C-sources and antibiotics: sorbitol, melezitose, neomycin, naladixic, acid, novobiocin |
| Kim, Kshetrimayum, and Goodfellow (2011) | Dactylosporangium | Streptomyces isolation agar | Gentamicin or oxytetracycline resistance |

## 2.2. Desiccation of soil

1. Place 1–5 g of soil in an empty, sterile Petri dish.
2. Place dish at 50 °C for 48 h or until soil is completely dry.
3. Begin a dilution series by adding 0.1 g of soil to 0.9 mL sterile water. Mix by vortexing and continue dilutions to $10^{-6}$.
4. Plate 100 μL of the $10^{-4}$–$10^{-6}$ dilutions on the starch–casein plates and allow to dry.
5. Incubate plates at 30 °C for 5 days or longer.
6. Restreak colonies that exhibit actinomycete morphology (i.e., tough or brittle colony consistency, visible white aerial hyphae, etc.) for isolation.

7. Once pure cultures are obtained, proceed with making spore stocks for use in interaction screens.

## 2.3. Generation of spore stocks for long-term storage of actinomycetes

For most actinomycetes, it is possible to make frozen stocks of vegetative mycelium or spores. Vegetative mycelial stocks are more prone to loss of viability with repeated freeze-thaw cycles than spore stocks. In addition to remaining viable almost indefinitely if stored at $-20\,^{\circ}\mathrm{C}$ or $-80\,^{\circ}\mathrm{C}$, spore stocks give more reliable colony forming unit counts than mycelial stocks, which tend to be composed of clumps of biomass.

1. Once a pure culture of an actinomycete has been obtained, a densely covered plate is desired to provide ample spore numbers for production of a highly concentrated spore stock. This can be achieved by taking an initial inoculum of spores from the pure culture plate and suspending it in 100–500 μL sterile water.

2. Vortex vigorously and spread 100 μL of this solution onto the sporulation medium of choice. If robust development was observed on starch–casein (Section 2.2, step 7), then this medium may be used in step 1. Alternatively, ISP2 (Becton Dickinson #277010) and Oatmeal agar (Becton Dickinson #255210) are also common media that can promote robust actinomycete sporulation.

3. Incubate the plates at 30 °C until sporulation occurs, often marked by the production of powdery gray material after the formation of fuzzy white aerial hyphae. Some actinomycetes do not produce pigmented spores, and in these cases, the plates will remain white even though mature spores have formed. Microscopic examination of the eventual spore stock can be useful in verifying the presence of spores as opposed to only aerial hyphae.

4. Place 500 μL of sterile 20% glycerol into a cryovial.

5. Pipette 1 mL of sterile 20% glycerol onto the spore-covered plate surface. In most cases, the glycerol solution will quickly form beads as a result of the hydrophobic surface layer covering spores and aerial hyphae.

6. Using a sterile cotton swab, gently rub the surface of the plate until the swab is completely covered with spores.

7. Transfer spores by twisting the swab against the interior wall of the cryovial until a dense spore solution is obtained. Repeat as necessary.

8. The concentration of the spore stock may be determined by serial dilution techniques followed by counting the numbers of the resulting colonies.

## 2.4. Calculation and setup of interaction arrays

Once spore stocks of an actinomycete have been obtained and quantitated, the layout of the interaction array must be considered. The equation for calculating unique combinations from a larger set of items is

$$\text{Number of combinations} = \frac{n!}{(n-r)! \times (r!)} \qquad [5.1]$$

where $n$ = total number of items and $r$ = number of items in each combination. For example, if a total of eight actinomycetes will be assayed in pairs, then there will be

$$\frac{8!}{(8-2)! \times (2!)} = 28 \text{ unique combinations} \qquad [5.2]$$

It is recommended to conduct screens in duplicate so that any observed interaction phenotypes can be verified. The simplest way to set up an interaction array on this scale is to spot 1 µL drops of dense spore solutions 0.5 cm apart on an agar surface. Square 12-in. Petri dishes can accommodate many interactions on this scale when each row of interactions is spaced 1 cm apart, laterally. The procedure for setting up screens in an 8 × 8 format, where all interactions within a set of eight actinomycete strains are monitored, follows:

1. Make 1 mL of a 1:100 spore stock dilution in 20% sterile glycerol.
2. Array 100 µL aliquots of each spore stock into the first eight columns in a 96-well plate as shown in the top panel of Fig. 5.1.
3. Using an eight-lane multichannel pipettor, place 1 µL drops from the first column onto the agar surface.
4. Place 1-µL drops from the first row onto the agar surface in parallel (0.5 cm away) to the drops from the previous step.
5. Repeat in the pattern shown in the bottom panel of Fig. 5.1. This yields an array that includes each of the 28 interactions twice and each strain versus itself once.
6. A row containing each of the strains alone is offset to the right to serve as a control and to check for interactions that may be occurring vertically.
7. Incubate plates at 30 °C for 4–5 days or until developmental progress is visible in control colonies.

Interactions between actinomycetes can alter many phenotypes that are visible on a macroscale, including development (i.e., aerial hyphae formation and sporulation), pigmentation, and growth. A typical set of interactions is

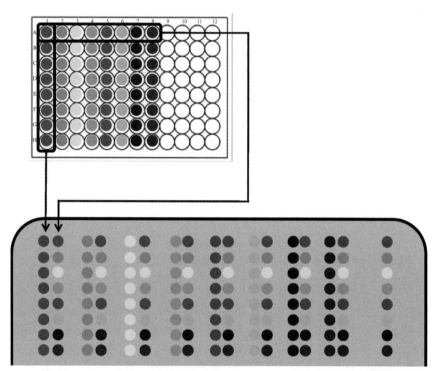

**Figure 5.1** Cartoon outline for generating screens within a group of eight bacterial strains. The strains of interest are arrayed in a 96-well plate as shown in the top panel. These are then transferred to an agar plate as shown in the bottom panel yielding an array that includes each of the 28 possible interactions twice and each strain versus itself once. A column to the right includes each strain by itself for comparison. (See Color Insert.)

shown using R2YE agar (Fig. 5.2). An example of an altered developmental phenotype can be seen in Fig. 5.2A (red outline), where *Amycolatopsis* sp. AA4 (hereafter AA4) inhibited aerial hyphae formation on the adjacent colony of *Streptomyces coelicolor* M145 (John Innes Centre). This interaction was chosen for chemical analysis and prompted the discovery of a new *Amycolatopsis* siderophore, amychelin (Seyedsayamdost et al., 2011). The chemical and biological experiments which led to the isolation and characterization of amychelin are detailed in the following sections.

Antibiotic production may also be stimulated or repressed by an interaction. This can be observed by overlaying the entire interaction plate with a layer of agar containing a test organism such as *Staphylococcus aureus*, as seen in Fig. 5.2B.

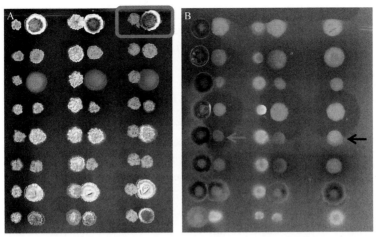

**Figure 5.2** Binary actinomycete interaction assays. (A) Shown are binary interactions of *Streptomyces tendae* (left column), *Streptomyces albus* (middle column), and *Amycolatopsis* sp. AA4 (right column) with eight actinomycete strains. The interaction between *S. coelicolor* and AA4, chosen for further study, is outlined at the top. (B) Binary actinomycete assays are overlaid with *S. aureus* to detect anti-Staphylococcal compounds that are produced as a function of coculture. The black arrow points to a colony that does not generate an anti-Staphylococcal compound in isolation. In coculture with another actinomycete, however, a halo in the overlaid *S. aureus* lawn is observed (red arrow), indicating production of an antibiotic. (For interpretation of the references to color in this figure legend, the reader is referred to the online version of this chapter.)

An overlay can be made by following these steps:

1. Grow an overnight culture of the test organism in LB (Becton Dickinson #244620) or other suitable broth (OD > 1.5).
2. Autoclave a 1% agar solution and cool to ~45 °C.
3. Add 1% inoculum (v/v) from the overnight culture and mix thoroughly.
4. Pour over interaction array and allow to cool.
5. Incubate overnight at 37 °C or appropriate temperature.
6. Check for altered zones of inhibition.

Alternatively, from step 3:

4. Pour into a 12-in. Petri plate at a thickness of ~2 mm and allow to cool.
5. Transfer this agar layer onto a sheet of plastic wrap by overturning the plate and separating the agar layer with a metal spatula.
6. Place the interaction array plate face down over the top of the agar layer on the plastic wrap.
7. Invert the entire assembly, firmly pressing the agar layer containing the test organism onto the surface of the interaction array.

This method has the advantage of not displacing spores that may be present on some actinomycete colonies. Spores picked up in poured agar may germinate overnight, obscuring or further altering zones of inhibition.

# 3. ACTIVITY-GUIDED FRACTIONATION FOR AMYCHELIN ISOLATION

As described above, our interpretation of the phenotypes observed in the binary assays is that small molecules produced by one actinomycete affect developmental or other cellular processes in the neighboring actinomycete. To identify the molecule responsible for these changes, activity-guided fractionation may be carried out. In this method, various purification steps are carried out concomitant with a specific activity assay to identify the fraction containing the active ingredient. This procedure is carried out repeatedly until a pure compound is obtained. The assay may vary depending on the interaction but is critical in guiding the purification of small quantities of a compound present in a pool of many other, more abundant, compounds. To identify the small molecule(s) that inhibited development in *S. coelicolor* in the interaction shown in Fig. 5.2A (red outline), we carried out activity-guided fractionation using the phenotypic effect on *S. coelicolor* as a readout.

## 3.1. Biological assays to guide purification of small molecules

1. To ensure that liquid AA4 cultures also produce the compound that inhibits development in *S. coelicolor*, prepare several 3-mL cultures of AA4 in R2YE medium (Seyedsayamdost et al., 2011) and incubate in a rotating drum incubator at 30 °C.
2. At various time points (3–8 days), isolate the supernatant by centrifugation (2 min, $15,000 \times g$, room temperature), followed by filtration through a 0.2-μm Nalgene membrane.
3. Prepare a lawn of *S. coelicolor* by plating $\sim 10^5$ spores, prepared as described in Section 2.3, on solid R2YE agar medium followed by incubating the plate for 16 h at 30 °C.
4. Punch 1-cm wells into the agar using the round end of a sterile, plastic 1-mL pipette tip. Then mix 50–100 μL of the supernatant with 200–250 μL of sterile, melted 0.7% agar solution and add the combined 300 μL mixture into the agar well.
5. Monitor the *S. coelicolor* morphology over the next 24–48 h.

## 3.2. Large-scale fermentation of AA4 and activity-guided fractionation

Here, purification of the bioactive small molecule amychelin, which inhibits development in *S. coelicolor* (Fig. 5.2A), is described. The procedure may be used as a general template for activity-guided purification, though compound-specific modifications will likely become necessary at each step.

1. For large-scale cultivation of AA4, prepare 2 L of a minimal medium consisting of 25 mM N-[tris(hydroxymethyl)methyl]-2-aminoethanesulfonic acid, pH 7.2, 2.34 mM $MgSO_4$, 0.5 mM $NaH_2PO_4$, 0.5 mM $K_2HPO_4$, 55.4 mM glucose, 85.5 mM NaCl, and 66.5 mM glycine. Add 1 mL of R2YE trace element solution to each liter of the minimal medium. The trace element solution includes the following components:

| Trace elements solution components (1 L) | Amount (mg) |
| --- | --- |
| $ZnCl_2$ | 40 |
| $FeCl_3 \cdot 6H_2O$ | 200 |
| $CuCl_2 \cdot 2H_2O$ | 10 |
| $MnCl_2 \cdot 4H_2O$ | 10 |
| $Na_2B_4O_7 \cdot 10H_2O$ | 10 |
| $(NH_4)_6Mo_7O_{24} \cdot 4H_2O$ | 10 |

2. Inoculate four 2.8-L baffled Fernbach flasks, each containing 0.5 L of the minimal medium described above, with 20 µL of a concentrated spore solution and incubate for 8 days shaking at 240 rpm and 30 °C.

3. After 8 days, isolate the supernatant by centrifugation ($10,000 \times g$, 15 min, room temperature) and further by filtration through a 0.2-µm Nalgene filter unit.

4. Extract the filtrate with an equal volume of ethyl acetate in a separatory funnel. Carry out the *S. coelicolor* agar assay described above on the aqueous and organic fractions. Continue with the fraction that contains the active component.

5. Analyze the active fraction, in this case the aqueous phase, by HPLC–MS using a Phenomenex Luna C18 analytical column (5 µm, 4.6 × 100 mm) operating at 0.7 mL/min with a gradient of 10% MeCN in $H_2O$ to 100% MeCN over 25 min. Water and MeCN each contain 0.1% formic acid.

6. Concentrate the aqueous phase to 0.35 L *in vacuo* and fractionate using solid-phase extraction: wash a 0.5 g C18 Sep-Pak cartridge (Waters) with 10 column volumes (CV, 20 mL) of methanol, followed by 10 CV of

10% MeOH in water. Pass the aqueous phase through the column and collect the flow-through.

7. Elute bound material using 10 CV of 20%, 40%, 60%, 80% MeOH in water, followed by MeOH. Carry out the assay described above on each of the fractions and continue purifying the fraction that contains the active component.

8. Fractionate the flow-through further by another solid-phase extraction step with a 0.5-g C8 Sep-Pak cartridge (Waters) using the same procedure as for the C18 cartridge. Carry out the *S. coelicolor* developmental assay on each of the fractions.

9. Complete the purification of the active component by HPLC using a Phenomenex Luna C18 column (5 μm, 9.4 × 250 mm) operating at 3 mL/min eluting with a gradient of 10% MeCN in $H_2O$ to 26% MeCN in $H_2O$ over 20 min. Water and MeCN contain 0.1% formic acid.

## 3.3. Generation and purification of Ga- or Fe-amychelin complexes

The interaction in Fig. 5.2A (red outline) is mediated by the novel siderophore, amychelin. Certain functional groups in siderophores, such as oxazolines, thiazolines, or cyclic hydroxamates, may not be stable under the conditions typically employed in small molecule isolations. In such cases, a siderophore may be stabilized by formation of its metal complexes. The Ga-complex is often generated to facilitate structural elucidation by NMR (Mohn, Koehl, Budzikiewicz, & Lefevre, 1994; Sharman, Williams, Ewing, & Ratledge, 1995). The ferric Fe complex may also be generated and used in structural elucidation by X-ray crystallography (Miller, Parkin, Fetherston, Perry, & Demoll, 2006; Van der Helm & Poling, 1976; Zalkin, Forrester, & Templeton, 1964).

1. Transfer an 8-mL aliquot of the mixture from step 6 in Section 3.2, which contains ~1 mg/mL amychelin, into a 40-mL scintillation vial equipped with a magnetic stir bar, and begin to gently stir at room temperature.

2. Slowly add a 10-fold excess of solid $GaBr_3$ (or $FeCl_3$) over 5 min. [Alternatively, Ga may be added from a stock solution as described in detail, previously (Sharman et al., 1995)].

3. Continue to stir for an additional 10 min at room temperature.

4. Incubate the reaction at 4 °C overnight.

5. Addition of Fe leads to formation of a charge transfer band (435 nm) which indicates Fe complexation (Fig. 5.3). A charge transfer band is not formed with Ga because of the fully occupied d orbitals of the $d^{10}$

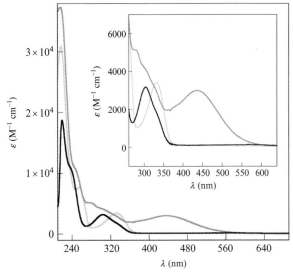

**Figure 5.3** UV–visible spectra of various forms of amychelin. Spectra for apo-amychelin (black trace), Ga-amychelin (light gray trace), and Fe-amychelin (dark gray trace) are shown. Ga- and Fe-amychelin were prepared as described in Section 3.3. The inset shows a magnified view of the charge transfer band for Fe-amychelin.

Ga$^{III}$ center. A red-shift is observed in the absorbance spectrum of Ga-amychelin, likely due to deprotonation of the hydroxyl group of the phenyloxazoline moiety.

6. Purify Ga- or Fe-amychelin by HPLC using a Phenomenex Luna C18 column (5 μm, 21.2 × 250 mm) operating at 13 mL/min eluting with a gradient of 10% MeCN in $H_2O$ to 26% MeCN in $H_2O$ over 14 min. Water and MeCN contain 0.1% formic acid. Monitor the elution of Fe-amychelin at 435 nm; Ga-amychelin may be monitored using its 335-nm absorption band.

## 4. DETERMINING THE AFFINITY OF AMYCHELIN FOR Fe

The scarcity of free Fe in the environment and its requirement in actinomycete development sets up a fierce competition for its acquisition (Challis & Hopwood, 2003; Hider & Kong, 2010; Miethke & Marahiel, 2007). Our studies on the AA4–*S. coelicolor* interactions have pointed to an arms race in which bacteria evolve siderophores with ever-increasing Fe affinities in order to ensure access to Fe and to prevent competitors from acquiring Fe. Studying the role of Fe in actinomycete development and competition requires determination of the affinity of siderophores for

Fe. Two different constants, $K_f$ and pFe, may be determined to quantitate the ability of a siderophore to bind Fe (Abergel, Zawadzka, & Raymond, 2008; Harris, Carrano, & Raymond, 1979a, 1979b; Miethke & Marahiel, 2007). $K_f$ is the formation constant of the Fe-siderophore complex, and it is generally reported as the fully deprotonated siderophore binding to Fe. Because these conditions are not physiologically relevant, determination of pFe may be preferred. pFe is defined as $-\log[Fe^{III}]$, when, according to convention, Fe and siderophore are present at 1 and 10 $\mu M$, respectively. It is usually determined at pH 7.5 and does not require knowledge of the $pK_a$s of the chelating groups.

1. The assay is carried out in Hepes buffer consisting of 10 m$M$ Hepes, 100 m$M$ KCl, pH 7.5.
2. Generate and purify Fe-amychelin according to the procedure described above. Exchange Fe-amychelin into Hepes buffer to a stock concentration of 1 m$M$.
3. Determine the extinction coefficient, $\varepsilon$, of Fe-amychelin in Hepes buffer as a function of wavelength, $\lambda$, using the Beer–Lambert equation.
4. Carry out steps 2 and 3 for Fe-EDTA.
5. Make a stock solution of EDTA at 10–100 m$M$.
6. The assay can be carried out in a 96-well plate format or in a series of Eppendorf tubes. In a 96-well format, each well will contain 100 µL buffer, Fe-amychelin at a concentration of 10 or 100 $\mu M$ and varied concentrations of EDTA ranging from 0.2-fold to 100-fold versus Fe-amychelin, including a sample that contains only Fe-amychelin in Hepes buffer and no EDTA (control). The volume of each well is kept at 150 µL using Hepes buffer.
7. Cover the plate with a lid or a nylon cover to prevent evaporation. Allow the reactions to equilibrate at room temperature for at least 24 h. Determine the UV–vis spectrum of each well. Each spectrum is a composite of UV–vis traces for Fe-amychelin, Fe-EDTA, free amychelin, and free EDTA (Fig. 5.4). Only Fe-amychelin has a charge transfer band at 435 nm.
8. For each sample, subtract the Fe-amychelin control spectrum from the composite spectrum. This gives the concentration of Fe-amychelin using the Beer–Lambert equation. Compute the concentration of free amychelin by subtracting the concentration of Fe-amychelin from total amychelin Eq. (5.4).
9. Use the subtraction spectrum to compute the concentration of Fe-EDTA and subsequently free EDTA using Eq. (5.5).

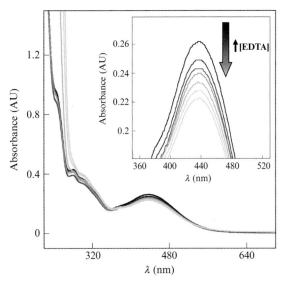

**Figure 5.4** EDTA competition assay to determine the pFe of amychelin. Fe-amychelin is incubated with various [EDTA]. Increasing EDTA concentrations result in disappearance of the Fe-amychelin charge transfer band, indicating conversion of Fe-amychelin into apo-amychelin and Fe-EDTA (inset). The data can be deconvoluted as described in Section 4 to obtain a pFe for amychelin.

10. To obtain the $\Delta pFe^{III}$ value of amychelin, plot log {[apo-EDTA]/[apo-amychelin]} versus log {[Fe-EDTA]/[Fe-amychelin]} and fit the data to Eq. (5.3), which has previously been derived, using nonlinear regression analysis. This equation gives $\Delta Fe^{III}$, the $pFe^{III}$ of amychelin minus that of EDTA. The $pFe^{III}$ of amychelin can be calculated using the known $pFe^{III}$ value of 23.42 for EDTA.

$$\log([Fe-EDTA]/[Fe-amychelin])$$
$$= \log([EDTA]/[amychelin]) + \Delta pFe^{III}) \qquad [5.3]$$
$$[amychelin]_T = [Fe-amychelin] + [apo-amychelin] \qquad [5.4]$$
$$[EDTA]_T = [Fe-EDTA] + [apo-EDTA] \qquad [5.5]$$

## 5. IDENTIFYING THE AMYCHELIN BIOSYNTHETIC GENE CLUSTER

The main advantage in using the Broad Institute actinomycete library (http://www.broadinstitute.org/annotation/genome/streptomyces_group) in the binary assays above is that the genome sequences of these strains have been determined, allowing us to connect the small molecules to their biosynthetic clusters. To link amychelin to its biosynthetic cluster, we first focused on the *N*-terminal hydroxybenzoyl group. The enzymes that incorporate this moiety into nonribosomal peptide synthetase (NRPS)-derived natural products have been well-studied. Performing a BLAST alignment of the protein sequence of one such enzyme from the actinomycete *Mycobacterium smegmatis* against the predicted proteins from the genome of AA4 led us to a gene cluster which appeared to contain the necessary genes for generating a hydroxybenzoyl-containing siderophore. Bioinformatic analysis of the genes in the cluster and gene interruption studies confirmed that this cluster was responsible for amychelin biosynthesis, allowing us to propose a model for the biosynthesis of amychelin. This general procedure allows rudimentary identification of possibly interesting biosynthetic steps that may be studied further by enzymology or in *in vivo* studies. Because these studies may reveal new and unusual enzymatic transformations, such biosynthetic gene clusters are not only reservoirs of new small molecules but also reservoirs of new enzymatic reactions. In Table 5.2, we summarize a number of online bioinformatic tools that may be used for initial analyses of secondary metabolite biosynthetic gene clusters.

## 6. TRANSCRIPTOMIC ANALYSIS USING NANOSTRING TECHNOLOGY

Gene expression at the level of transcription is among the most commonly measured experimental variables in modern biology. Methods for measuring transcriptomic output range from global, such as RNAseq and microarrays, down to a single gene using qRTPCR. Recently, a new methodology, called Nanostring, has become available (Geiss et al., 2008). Nanostring allows for the quantification of transcripts of up to 800 genes in a single reaction, placing it as an intermediate technology (between global and single gene techniques) in terms of the number of genes per assay. This technology works by annealing fluorescently bar-coded reporter oligonucleotides to gene transcripts in the test sample, and subsequently capturing and counting the complex of gene transcript and the labeled oligonucleotides. The result is a direct digital count of the number of transcripts in a sample, with no bias introduced from reverse transcriptase amplification. Perhaps even more appealing is the fact that traditional RNA isolation is not required, adding to the relative ease and speed of obtaining transcriptomic data. Here we provide a general method for preparing

**Table 5.2** Bioinformatic tools for functional prediction of genes in biosynthetic gene clusters

| Bioinformatic tool (website) | Brief tool description |
| --- | --- |
| ClustScan (http://bioserv.pbf.hr/cms/) | A database containing genetic and biochemical information on secondary metabolites derived from PKS, NRPS, and hybrid PKS/NRPS clusters. Allows identification and annotation of biosynthetic clusters in a genome or metagenome sequence. |
| BLAST (http://blast.ncbi.nlm.nih.gov/Blast.cgi) | Comparison of protein or DNA sequence to a database of available sequences. Allows identification of genes or proteins that most closely resemble the query sequence. |
| FASTA (http://www.ebi.ac.uk/Tools/sss/fasta/) | Homology search database, similar to BLAST. FASTA and BLAST are comparable for highly similar sequences, though BLAST is faster without significant loss of sensitivity. FASTA may be better for less similar sequences. |
| InterProScan (http://www.ebi.ac.uk/Tools/pfa/iprscan/) | InterPro contains a number of member databases, which provide amino acid sequence motifs for known protein families or domains. These identifiable features in known proteins are applied to a query sequence to provide a functional prediction. |
| PKS/NRPS Analysis (http://nrps.igs.umaryland.edu/nrps/) | Prediction of various NRPS and PKS domains (condensation/ketosynthase, adenylation/acyltransferase, etc.) within an NRPS or PKS protein sequence. Provides a predicted specificity for NRPS A-domains by comparison of the active site residues of known A-domains to those of a query sequence. |
| SBSPKS (www.nii.ac.in/sbspks.html) | Sequence and structure based analysis of PKSs. Query sequence is compared to a database of experimentally characterized PKSs with known reactions and protein structures. |
| Integrated Microbial Genomes (http://img.jgi.doe.gov/) | Platform for genome browsing and annotation including a number of useful and user-friendly tools for comparative analysis of genes, gene clusters, genomes, and protein function. |

actinomycete samples from colonies grown on solid medium for Nanostring analysis. More information regarding the method and data analysis can be found at the manufacturer's website: http://www.nanostring.com/.

1. Grow actinomycetes on solid medium until the desired time-point.
2. Scrape colony(s) from agar using a sterile razor blade, minimizing excess agar.
3. Transfer the colony onto the end of a clean ceramic pestle.
4. Pour a small amount of liquid nitrogen into a clean ceramic mortar.
5. Slowly submerge the tip of the pestle containing the colony into the liquid nitrogen, allowing the biomass to freeze.
6. Once frozen, grind the colony biomass thoroughly as the liquid nitrogen evaporates.
7. Add 150 µL buffer RLT (lysis buffer available from Qiagen) to the pestle.
8. Grind until an even-looking biomass crude lysate is obtained.
9. Use a pipette to transfer as much of the lysate as possible to a sterile 1.4-mL tube. 20–50 µL is more than enough for Nanostring analysis.
10. Freeze the samples immediately by dipping the tube into liquid nitrogen or by placing immediately at $-80\,^{\circ}\text{C}$. Store until use.
11. Add 1–2 µL of lysate to Nanostring reactions as indicated in the crude lysate manufacturer protocols.

## 7. SUMMARY

This chapter provides a guide for creating efficient screens to examine the biology and chemistry of binary interactions among morphologically complex actinomycetes. Application of a similar screen led to the observation that *Amycolatopsis* sp. AA4 inhibited development of *S. coelicolor* aerial hyphae. The chemical characterization of this interaction, which led to the discovery of a siderophore, amychelin, provides an example of how such interactions can drive novel compound discovery. While the format suggested here is focused on actinomycetes, one can imagine how these regimes could be adapted for other bacteria or fungi, for example, those of the human gut microbiota or from other varied and complex environments. Bacteria of many phylogenetic backgrounds are capable of producing pigments or have complex colony morphologies that could be altered through interspecies interactions. Genetically tractable test organisms can also be engineered such that fluorescent protein expression driven from a promoter of interest can be assayed in the context of chemically mediated interactions.

Creative variations of these screens may thus open a door into the previously unexplored chemistry of interactions between microorganisms; a chemical realm potentially occupied by many useful and unique metabolites.

## ACKNOWLEDGMENTS

This work was supported by the National Institutes of Health (Grant GM82137 to R. K., and Grants AI057159 and GM086258 to J. C.). M. R. S. is supported by a NIH K99 Pathway to Independence Award (Grant 1K99 GM098299-01). M. F. T. is a NIH Postdoctoral Fellow (Grant 5F32GM089044-02).

## REFERENCES

Abergel, R. J., Zawadzka, A. M., & Raymond, K. N. (2008). Petrobactin-mediated iron transport in pathogenic bacteria: Coordination chemistry of an unusual 3,4-catecholate/citrate siderophore. *Journal of the American Chemical Society*, *130*, 2124–2125.

Baltz, R. H. (2008). Renaissance in antibacterial discovery from actinomycetes. *Current Opinion in Pharmacology*, *8*, 557–563.

Bentley, S. D., Chater, K. F., Cerdeno-Tarraga, A. M., Challis, G. L., Thomson, N. R., James, K. D., Harris, D. E., Quail, M. A., Kieser, H., Harper, D., Bateman, A., Brown, S., et al. (2002). Complete genome sequence of the model actinomycete *Streptomyces coelicolor* A3(2). *Nature*, *417*, 141–147.

Berdy, J. (2005). Bioactive microbial metabolites. *The Journal of Antibiotics*, *58*, 1–26.

Challis, G. L., & Hopwood, D. A. (2003). Synergy and contingency as driving forces for the evolution of multiple secondary metabolite production by *Streptomyces* species. *Proceedings of the National Academy of Sciences of the United States of America*, *100*(Suppl. 2), 14555–14561.

Chater, K. F. (1998). Taking a genetic scalpel to the *Streptomyces* colony. *Microbiology*, *144*, 1465–1478.

Copeland, A., Lapidus, A., Glavina Del Rio, T., Nolan, M., Lucas, S., Chen, F., Tice, H., Cheng, J. F., Bruce, D., Goodwin, L., Pitluck, S., Mikhailova, N., et al. (2009). Complete genome sequence of *Catenulispora acidiphila* type strain (ID 139908). *Standards in Genomic Sciences*, *1*, 119–125.

Curtis, T. P., Sloan, W. T., & Scannell, J. W. (2002). Estimating prokaryotic diversity and its limits. *Proceedings of the National Academy of Sciences of the United States of America*, *99*, 10494–10499.

Fischbach, M. A., & Krogan, N. J. (2010). The next frontier of systems biology: Higher-order and interspecies interactions. *Genome Biology*, *11*, 208.

Fischbach, M. A., & Walsh, C. T. (2009). Antibiotics for emerging pathogens. *Science*, *325*, 1089–1093.

Geiss, G. K., Bumgarner, R. E., Birditt, B., Dahl, T., Dowidar, N., Dunaway, D. L., Fell, H. P., Ferree, S., George, R. D., Grogan, T., James, J. J., Maysuria, M., et al. (2008). Direct multiplexed measurement of gene expression with color-coded probe pairs. *Nature Biotechnology*, *26*, 317–325.

Goodfellow, M., & Fiedler, H. P. (2010). A guide to successful bioprospecting: Informed by actinobacterial systematics. *Antonie Van Leeuwenhoek*, *98*, 119–142.

Hamaki, T., Suzuki, M., Fudou, R., Jojima, Y., Kajiura, T., Tabuchi, A., Sen, K., & Shibai, H. (2005). Isolation of novel bacteria and actinomycetes using soil-extract agar medium. *Journal of Bioscience and Bioengineering*, *99*, 485–492.

Harris, W. R., Carrano, C. J., & Raymond, K. N. (1979a). Coordination chemistry of microbial iron transport compounds. 16. Isolation, characterization, and formation constants of ferric aerobactin. *Journal of the American Chemical Society, 101*, 2722–2727.

Harris, W. R., Carrano, C. J., & Raymond, K. N. (1979b). Spectrophotometric determination of the proton-dependent stability constant of ferric enterobactin. *Journal of the American Chemical Society, 101*, 2213–2214.

Hider, R. C., & Kong, X. (2010). Chemistry and biology of siderophores. *Natural Product Reports, 27*, 637–657.

Hop, D. V., Sakiyama, Y., Binh, C. T., Otoguro, M., Hang, D. T., Miyadoh, S., Luong, D. T., & Ando, K. (2011). Taxonomic and ecological studies of actinomycetes from Vietnam: Isolation and genus-level diversity. *The Journal of Antibiotics, 64*, 599–606.

Jensen, P. R., Gontang, E., Mafnas, C., Mincer, T. J., & Fenical, W. (2005). Culturable marine actinomycete diversity from tropical Pacific Ocean sediments. *Environmental Microbiology, 7*, 1039–1048.

Jensen, P. R., Mincer, T. J., Williams, P. G., & Fenical, W. (2005). Marine actinomycete diversity and natural product discovery. *Antonie Van Leeuwenhoek, 87*, 43–48.

Jiang, S., Sun, W., Chen, M., Dai, S., Zhang, L., Liu, Y., Lee, K. J., & Li, X. (2007). Diversity of culturable actinobacteria isolated from marine sponge Haliclona sp. *Antonie Van Leeuwenhoek, 92*, 405–416.

Kim, B. Y., Kshetrimayum, J. D., & Goodfellow, M. (2011). Detection, selective isolation and characterisation of Dactylosporangium strains from diverse environmental samples. *Systematic and Applied Microbiology, 34*, 606–616.

Kurahashi, M., Fukunaga, Y., Sakiyama, Y., Harayama, S., & Yokota, A. (2010). *Euzebya tangerina* gen. nov., sp. nov., a deeply branching marine actinobacterium isolated from the sea cucumber Holothuria edulis, and proposal of Euzebyaceae fam. nov., Euzebyales ord. nov. and Nitriliruptoridae subclassis nov. *International Journal of Systematic and Evolutionary Microbiology, 60*, 2314–2319.

Lazzarini, A., Cavaletti, L., Toppo, G., & Marinelli, F. (2000). Rare genera of actinomycetes as potential producers of new antibiotics. *Antonie Van Leeuwenhoek, 78*, 399–405.

Miethke, M., & Marahiel, M. A. (2007). Siderophore-based iron acquisition and pathogen control. *Microbiology and Molecular Biology Reviews, 71*, 413–451.

Miller, M. C., Parkin, S., Fetherston, J. D., Perry, R. D., & Demoll, E. (2006). Crystal structure of ferric-yersiniabactin, a virulence factor of Yersinia pestis. *Journal of Inorganic Biochemistry, 100*, 1495–1500.

Mohn, G., Koehl, P., Budzikiewicz, H., & Lefevre, J. F. (1994). Solution structure of pyoverdin GM-II. *Biochemistry, 33*, 2843–2851.

Nett, M., Ikeda, H., & Moore, B. S. (2009). Genomic basis for natural product biosynthetic diversity in the actinomycetes. *Natural Product Reports, 26*, 1362–1384.

Newman, D. J., & Cragg, G. M. (2012). Natural products as sources of new drugs over the 30 years from 1981 to 2010. *Journal of Natural Products, 75*, 311–335.

Newman, D. J., Cragg, G. M., Holbeck, S., & Sausville, E. A. (2002). Natural products and derivatives as leads to cell cycle pathway targets in cancer chemotherapy. *Current Cancer Drug Targets, 2*, 279–308.

Oh, D. C., Poulsen, M., Currie, C. R., & Clardy, J. (2009). Dentigerumycin: A bacterial mediator of an ant-fungus symbiosis. *Nature Chemical Biology, 5*, 391–393.

Oh, D. C., Poulsen, M., Currie, C. R., & Clardy, J. (2011). Sceliphrolactam, a polyene macrocyclic lactam from a wasp-associated *Streptomyces sp. Organic Letters, 13*, 752–755.

Oh, D. C., Scott, J. J., Currie, C. R., & Clardy, J. (2009). Mycangimycin, a polyene peroxide from a mutualist *Streptomyces sp. Organic Letters, 11*, 633–636.

Ohnishi, Y., Ishikawa, J., Hara, H., Suzuki, H., Ikenoya, M., Ikeda, H., Yamashita, A., Hattori, M., & Horinouchi, S. (2008). Genome sequence of the streptomycin-producing microorganism Streptomyces griseus IFO 13350. *Journal of Bacteriology, 190*, 4050–4060.

Oliynyk, M., Samborskyy, M., Lester, J. B., Mironenko, T., Scott, N., Dickens, S., Haydock, S. F., & Leadlay, P. F. (2007). Complete genome sequence of the erythromycin-producing bacterium *Saccharopolyspora erythraea* NRRL23338. *Nature Biotechnology, 25,* 447–453.

Poulsen, M., Oh, D. C., Clardy, J., & Currie, C. R. (2011). Chemical analyses of wasp-associated streptomyces bacteria reveal a prolific potential for natural products discovery. *PloS One, 6,* e16763.

Schmidt, E. W. (2008). Trading molecules and tracking targets in symbiotic interactions. *Nature Chemical Biology, 4,* 466–473.

Scott, J. J., Oh, D. C., Yuceer, M. C., Klepzig, K. D., Clardy, J., & Currie, C. R. (2008). Bacterial protection of beetle-fungus mutualism. *Science, 322,* 63.

Seyedsayamdost, M. R., Traxler, M. F., Zheng, S. L., Kolter, R., & Clardy, J. (2011). Structure and biosynthesis of amychelin, an unusual mixed-ligand siderophore from *Amycolatopsis sp.* AA4. *Journal of the American Chemical Society, 133,* 11434–11437.

Shank, E. A., & Kolter, R. (2009). New developments in microbial interspecies signaling. *Current Opinion in Microbiology, 12,* 205–214.

Sharman, G. J., Williams, D. H., Ewing, D. F., & Ratledge, C. (1995). Isolation, purification and structure of exochelin MS, the extracellular siderophore from *Mycobacterium smegmatis. The Biochemical Journal, 305,* 187–196.

Straight, P. D., & Kolter, R. (2009). Interspecies chemical communication in bacterial development. *Annual Review of Microbiology, 63,* 99–118.

Tan, G. Y., Ward, A. C., & Goodfellow, M. (2006). Exploration of *Amycolatopsis* diversity in soil using genus-specific primers and novel selective media. *Systematic and Applied Microbiology, 29,* 557–569.

Ueda, K., Kawai, S., Ogawa, H., Kiyama, A., Kubota, T., Kawanobe, H., & Beppu, T. (2000). Wide distribution of interspecific stimulatory events on antibiotic production and sporulation among *Streptomyces* species. *The Journal of Antibiotics, 53,* 979–982.

Van der Helm, D., & Poling, M. (1976). The crystal structure of ferrioxamine E. *Journal of the American Chemical Society, 98,* 82–86.

Watve, M. G., Tickoo, R., Jog, M. M., & Bhole, B. D. (2001). How many antibiotics are produced by the genus *Streptomyces*? *Archives of Microbiology, 176,* 386–390.

Wellington, E. M. H., & Toth, I. K. (1994). Actinomycetes. In R. W. Weaver, S. Angle, P. Bottomley, S. Bezdicek, S. Smith, A. Tabatabai & A. Wollum (Eds.), *Methods of soil analysis* (pp. 269–290). *SSSA Book SeriesNo. 5,* (pp. 269–290). Madison, WI: Soil Science Society of America.

Yamanaka, K., Oikawa, H., Ogawa, H. O., Hosono, K., Shinmachi, F., Takano, H., Sakuda, S., Beppu, T., & Ueda, K. (2005). Desferrioxamine E produced by *Streptomyces griseus* stimulates growth and development of *Streptomyces tanashiensis. Microbiology, 151,* 2899–2905.

Zalkin, A., Forrester, J. D., & Templeton, D. H. (1964). Crystal and molecular structure of Ferrichrome A. *Science, 146,* 261–263.

Zhao, W., Zhong, Y., Yuan, H., Wang, J., Zheng, H., Wang, Y., Cen, X., Xu, F., Bai, J., Han, X., Lu, G., Zhu, Y., et al. (2010). Complete genome sequence of the rifamycin SV-producing *Amycolatopsis mediterranei* U32 revealed its genetic characteristics in phylogeny and metabolism. *Cell Research, 20,* 1096–1108.

# Analyzing Gene Clusters

CHAPTER SIX

# Finding and Analyzing Plant Metabolic Gene Clusters

## Anne Osbourn*,[1], Kalliopi K. Papadopoulou[†], Xiaoquan Qi[‡], Ben Field[§], Eva Wegel[¶]

*Department of Metabolic Biology, John Innes Centre, Norwich Research Park, Norwich, United Kingdom
[†]Department of Biochemistry & Biotechnology, University of Thessaly, Larissa, Greece
[‡]Key Laboratory of Plant Molecular Physiology, Institute of Botany, Chinese Academy of Sciences, Beijing, PR China
[§]Laboratoire de Génétique et de Biophysique des Plantes, Unité Mixte de Recherche 7265 Centre National de la Recherche Scientifique-Commissariat à l'Energie Atomique et aux Energies Alternatives-Aix-Marseille Université, Marseille, France
[¶]Department of Biochemistry, University of Oxford, South Parks Road, Oxford, United Kingdom
[1]Corresponding author: e-mail address: anne.osbourn@jic.ac.uk

## Contents

*Methods in Enzymology*, Volume 517
ISSN 0076-6879
http://dx.doi.org/10.1016/B978-0-12-404634-4.00006-1

113

## Abstract

Plants produce an array of diverse secondary metabolites with important ecological functions, providing protection against pests, diseases, and abiotic stresses. Secondary metabolites are also a rich source of bioactive compounds for drug and agrochemical development. Despite the importance of these compounds, the metabolic diversity of plants remains largely unexploited, primarily due to the problems associated with mining large and complex genomes. It has recently emerged that genes for the synthesis of multiple major classes of plant-derived secondary metabolites (benzoxinones, diterpenes, triterpenes, and cyanogenic glycosides) are organized in clusters reminiscent of the metabolic gene clusters found in microbes. Many more secondary metabolic clusters are likely to emerge as the body of sequence information available for plants continues to grow, accelerated by high-throughput sequencing. Here, we describe approaches for the identification of secondary metabolic gene clusters in plants through forward and reverse genetics, map-based cloning, and genome mining and give examples of methods used for the analysis and functional confirmation of new clusters.

## 1. INTRODUCTION

Plants produce a huge array of diverse secondary metabolites that protect them against pests, diseases, and abiotic stresses. Plant secondary metabolites determine important crop traits and are also a rich source of bioactives for drug and agrochemical discovery. Around 20% of plant genes have predicted functions in secondary metabolism, but even in *Arabidopsis thaliana*, the functions of most of these are still unknown. The number of sequenced plant genomes now exceeds 30 (e.g., http://www.phytozome. net/), and a massive amount of transcriptome data is available for diverse plant species, including medicinal plants (Chen, Xiang, Guo, & Li, 2011; Yonekura-Sakakibara & Saito, 2009). This body of information is growing exponentially as sequencing methodologies develop. However, the metabolic diversity of plants remains largely unexploited due to the problems associated with mining large and complex genomes.

The finding that the genes for the synthesis of a number of major classes of plant-derived secondary metabolite are organized in clusters, reminiscent of the operons and metabolic gene clusters found in microbes, is now opening up new opportunities for pathway discovery (Chu, Wegel, & Osbourn, 2011; Osbourn, 2010; Osbourn & Field, 2009). Although these clusters have operon-like features (physical clustering and coregulation), they are clearly distinct from bacterial operons because the genes within each cluster are independently transcribed and separated by significant stretches of noncoding DNA. The first example of a plant secondary metabolic gene cluster was for the synthesis of the cyclic hydroxamic acids 2,4-dihydroxy-1,4-benzoxazin-3-one (DIBOA) and 2,4-dihydroxy-7-methoxy-1,4-benzoxazin-3-one (DIMBOA) in maize (*Zea mays*) (Frey et al., 1997). We have since discovered and characterized a further three clusters, one in oat and two in *A. thaliana* (for the synthesis of different triterpenes) (Field et al., 2011; Field & Osbourn, 2008; Qi et al., 2006). Gene clusters for other types of plant secondary metabolite have also been reported from diverse plant species. These include the phytocassane and momilactone diterpenes in rice (Shimura et al., 2007; Wilderman, Xu, Jin, Coates, & Peters, 2004) and cyanogenic glucosides in *Lotus japonicus*, cassava (*Manihot esculenta*) and sorghum (*Sorghum bicolor*) (Takos et al., 2011). These gene clusters have been identified by genetic analysis of mutants and map-based cloning, or through whole-genome sequencing, gene cluster annotation, and functional confirmation. Here, we give examples of different approaches for the discovery and analysis of gene clusters for new secondary metabolic pathways in plants.

## 2. DISCOVERY OF PLANT METABOLIC GENE CLUSTERS THROUGH GENETIC ANALYSIS

### 2.1. Discovery of secondary metabolic clusters for the synthesis of defense compounds in maize and oat

The cyclic hydroxamic acids DIBOA and its methylated derivative DIMBOA are present in maize and a variety of grasses and confer resistance to pests and pathogens. The maize mutation *bx1* (*benzoxazineless*) abolishes DIBOA synthesis. *Bx1* was cloned through a reverse genetics approach using the maize Mutator (Mu) transposon tagging system, and shown by heterologous expression in *Escherichia coli* to encode a tryptophan synthase α homolog that catalyzes the formation of indole (the precursor for DIBOA and DIMBOA) rather than tryptophan (Frey et al., 1997). The Mu system also enabled the cloning of another gene in the pathway, *Bx3*. *Bx3* is one of the four closely related cytochrome P450 (CYP) genes (*Bx2–5*) that map close to *Bx1* on maize chromosome 4 (Frey et al., 1997; Frey, Kliem, Saedler, & Gierl, 1995). Expression of Bx2–5 in yeast showed that these four CYPs catalyze successive steps in the conversion of indole to

DIBOA (Frey et al., 1997). Thus, *Bx1–5* were identified and functionally defined using a combination of genetics and biochemical analysis. These five genes, which form part of a metabolic gene cluster, are necessary and sufficient for the conversion of indole-3-glycerol phosphate to DIBOA. Other experiments have identified several other genes in this pathway, for the conversion of DIBOA to DIMBOA and for glucosylation of these compounds (Frey et al., 2003; Frey, Schullehner, Dick, Fiesselmann, & Gierl, 2009; Jonczyk et al., 2008).

The second example is the avenacin cluster for the synthesis of antimicrobial triterpene glycosides (saponins) that are produced in the roots of oat (Mugford et al., 2009; Qi et al., 2004, 2006). This gene cluster was initially defined by a forward screen to identify mutants of diploid oat (*Avena strigosa*) that were unable to produce avenacins (known as *saponin deficient*, or *sad*, mutants) (Papadopoulou, Melton, Leggett, Daniels, & Osbourn, 1999). Extensive genetic analysis revealed that six of the *Sad* loci defined by mutation were genetically linked and mapped to a 3.6-cM region of linkage group D of the diploid oat genome (Qi et al., 2004). One locus (*Sad4*, which is required for glucosylation) was unlinked (Mylona et al., 2008; Qi et al., 2004). The first of these genes to be cloned was *Sad1*, which encodes β-amyrin synthase, the enzyme that catalyzes the first committed step in avenacin synthesis (Haralampidis et al., 2001). Having cloned and characterized *Sad1* and the corresponding gene product by functional expression in yeast, we constructed a bacterial artificial chromosome (BAC) library for diploid oat (*A. strigosa*). This enabled us to retrieve BAC clones containing *Sad1* to construct a 400-kb BAC contig spanning the region of the genome around this gene. Sequence analysis of this BAC contig revealed further four genes in the immediate vicinity of *Sad1* that form part of a metabolic gene cluster and that have all subsequently been shown to have a role in avenacin synthesis (Chu et al., 2011; Mugford et al., 2009; Qi et al., 2006). Thus, the avenacin gene cluster was discovered and defined through a forward mutant screen, genetic analysis, map-based cloning, and analysis of biochemical function.

## 2.2. Example: identification of the avenacin cluster in oat

1. *Generation of oat mutants.* Seeds of diploid oat *A. strigosa* accession no. S75 (Papadopoulou et al., 1999) were presoaked with water for 4 h and mutagenized with 10 m*M* sodium azide in 0.1 *M* sodium phosphate buffer (pH 3.2) for 1 h in a fume hood. Following thorough washing,

the seeds were dried and then sown in the field. The $M_2$ seed from each individual $M_1$ plant was harvested and bagged separately. The efficiency of mutagenesis was assessed by monitoring the frequency of chlorophyll-deficient seedlings, which was $\sim 4.6\%$. *Note*: Sodium azide is extremely toxic. It is essential to consult your health and safety officer, and ensure that a full risk assessment and appropriate operating procedures are in place before using it.

2. *Screening of sad mutants.* The major oat triterpene avenacin A-1 contains *N*-methyl anthranilic acid and so has strong autofluorescence under ultraviolet (UV) illumination. Avenacin A-1 causes the roots of young oat seedlings to fluoresce bright blue, enabling root fluorescence to be used as a preliminary screen to identify *sad* mutants. Seeds of individual $M_2$ families were germinated in Petri dishes on moist filter paper and seedlings assessed for root fluorescence. In the first screen, 10 independent mutants with reduced fluorescence were identified after screening seedlings representing 1289 $M_2$ families (Papadopoulou et al., 1999). A further 82 mutants were identified in a subsequent large-scale screen (Mugford et al., 2009).

3. *Definition of Sad loci.* Test crosses were performed between *sad* mutants and the wild-type parent line to determine whether the *sad* phenotype of each mutant was likely to be due to a single mutation. For each cross, the ratio of the numbers of seedlings with wild-type root fluorescence to those with reduced fluorescence in the $F_2$ progeny was generally consistent with a 3:1 segregation, suggesting that the mutations were all recessive alleles of single genes. Allelism tests were conducted by crossing the mutants with each other and analyzing the $F_2$ generations. Where the $F_2$ progeny all had reduced fluorescence, the two mutant parents were affected at the same locus. Where a mixture of seedlings with wild-type and reduced root fluorescence was observed in the $F_2$ progeny, the two mutant parents were affected at different *Sad* loci. In this way, mutants could be grouped into allelic series as follows: *Sad1* (mutants 109 and 610); *Sad2* (mutants 283, 500, 638, 698, 791, 1027, 1325, and 1412); *Sad3* (mutants 105, 368, 891, and 1139); *Sad4* (mutants 9 and 933); *Sad6* (mutant 825); *Sad7* (mutants 19, 376, and 616); and *Sad8* (mutant 1243) (Haralampidis et al., 2001; Mugford et al., 2009; Mylona et al., 2008; Papadopoulou et al., 1999; Qi et al., 2004; Qin et al., 2010).

4. *Genetic linkage analysis.* In order to obtain a preliminary map position for *Sad1*, a single-nucleotide polymorphism in intron 17 of the *Sad1* gene that conferred presence or absence of a *Pac*I site was identified in

*A. strigosa* CI3815 × *A. wiestii* CI1994 recombinant inbred lines (Yu & Wise, 2000). For genotyping, PCR fragments for digestion were amplified with the primer pair: TGGGCAATGTTGGCTTTAATTT/ TGATGACATCGGTAGGAA. This enabled *Sad1* to be mapped to linkage group D of diploid oat (Qi et al., 2004).

5. *Mapping other Sad genes.* To map the other *Sad* loci defined by mutation, representative mutants for different loci (*sad2*: 1027; *sad3*: 1139; *sad4*: 9; *sad5*: 616; *sad6*: 825; *sad7*: 376; *sad8*: 1243) were crossed with the mapping line *A. strigosa* CI386. All *sad* mutations behaved as recessive alleles of single genes within these $F_2$ populations. The $F_2$ populations were scored for the *sad* phenotype and then genotyped for *Sad1* by single-nucleotide polymorphism analysis as described in the above section. Maps were constructed using JOINMAP 2.0 (Stam, 1993). Genetic analysis indicated that six of the seven other *Sad* loci were genetically linked to the *Sad1* gene. There was no recombination between *Sad1* and *Sad2* in a population of 2000 $F_2$ individuals. Three other *Sad* loci (*Sad5, 6, 7,* and *8*) also showed absolute cosegregation with *Sad1* (in populations of 150–170 $F_2$ individuals); *Sad3* was less closely linked, resolving the gene cluster to 3.6 cM around the *Sad1* locus (Qi et al., 2004).

6. *Construction of a BAC library and a BAC contig spanning Sad1.* A BAC library of *A. strigosa* accession number S75 was constructed using established methods (Allouis et al., 2003). Approximately 150,000 colonies with an average insert size of ∼110 kb (ca. 4.2 × genome coverage) were stored in 384-well microtitre plates and gridded onto high-density filters. Filters were screened with [32]P-labeled cDNA probes for *Sad1* and also with unique BAC end sequences. Hybridization and washing were carried out at stringencies of either 60° C or 65° C. BAC fingerprinting was conducted by digestion of BAC DNA with *Hind*III and *Bam*H1. The construction of the BAC contig was carried out by manual comparison of restriction fragments following agarose gel electrophoresis (Fig. 6.1). The resulting BAC contig covered a region of >300 kb spanning five of the genes in the pathway (Chu et al., 2011; Mugford et al., 2009; Qi et al., 2006).

7. *Expression analysis of genes in the cluster.* For Northern blot analysis, total RNA was extracted using TRI-REAGENT (Sigma). Hybridizations with biotin-labeled (Biotin-16-dUTP; Roche) antisense RNA probes for the cloned avenacin biosynthetic genes were carried out at high stringency (68° C) with signal detection using BrightStar® BioDetect™ (Ambion). For mRNA *in situ* analysis, biotin-labeled sense and antisense

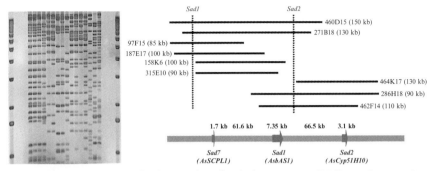

**Figure 6.1** Construction of a bacterial artificial chromosome (BAC) contig spanning *Sad1*. Left: BAC fingerprinting by digestion with *Hind*III. Right: BAC contig and sequenced genome region encompassing three genes in the avenacin pathway (Haralampidis et al., 2001; Mugford et al., 2009; Qi et al., 2004, 2006). (For color version of this figure, the reader is referred to the online version of this chapter.)

RNA probes were used. Tissue preparation and hybridization were carried out as described (Mao, Buschmann, Doonan, & Lloyd, 2006). The genes in the avenacin gene cluster are expressed preferentially in the roots, specifically in the epidermal cells of the root tip, the site of synthesis and accumulation of avenacin A-1 (shown for *Sad1*, *Sad2*, and *Sad7* in Fig. 6.2) (Haralampidis et al., 2001; Mugford et al., 2009; Qi et al., 2004, 2006).

## 3. MINING GENOMES FOR SECONDARY METABOLIC GENE CLUSTERS

### 3.1. Predicting candidate metabolic gene clusters

Secondary metabolic pathways consist of genes for "signature enzymes" that make the scaffold of the secondary metabolite, along with genes for "tailoring enzymes" that carry out subsequent modifications to this scaffold (Osbourn, 2010). Examples of plant signature enzymes are terpene synthases (for terpenes), chalcone synthases (for flavonoids), and CYP79 family enzymes for cyanogenic glucosides (D'Auria & Gershenzon, 2005; Osbourn, 2010). Examples of tailoring enzymes include oxidoreductases, methyltransferases, acyltransferases, and glycosyltransferases. A good candidate gene cluster will contain genes encoding a signature enzyme and tailoring enzymes. Given that these genes contribute to a common pathway, they may be expected to be tightly coexpressed, although this is not always the case (e.g., Takos et al., 2011). Coexpression can be tested by querying gene expression databases or by performing RT-PCR experiments on samples from

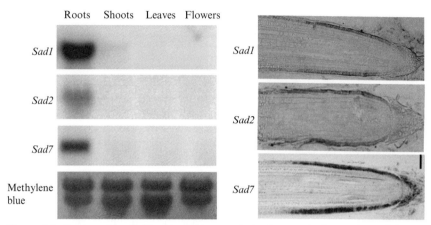

**Figure 6.2** Northern blot (left) and mRNA *in situ* hybridization (right) analysis of the expression of *Sad1*, *Sad2*, and *Sad7* (Haralampidis et al., 2001; Mugford et al., 2009; Qi et al., 2004, 2006). The methylene blue stained gel (bottom, left) indicates the RNA loading for the Northern blot analysis. (For color version of this figure, the reader is referred to the online version of this chapter.)

different plant tissues and from plants grown under different growth conditions. In our laboratory, the *A. thaliana* thalianol and marneral gene clusters were initially identified as candidate secondary metabolic gene clusters consisting of genes for a signature enzyme (an oxidosqualene cyclase) and tailoring enzymes (CYPs and acyltransferases) (Field et al., 2011; Field & Osbourn, 2008). In a similar manner, the momilactone and phytocassane gene clusters in rice were first implicated through a search of the rice genome for genes encoding the two signature enzymes ent-copalyl diphosphate synthase and ent-kaurene synthase (Sakamoto et al., 2004).

## 3.2. Defining the scaffold

Once a candidate gene cluster has been identified, the next step is to define the cognate metabolic pathway. A successful approach is to first identify the product of the putative signature enzymes using heterologous expression in bacteria or yeast, or by transient expression in *Nicotiana benthamiana*. The appropriate expression system is selected according to the predicted function of the signature enzyme. For example, the function of an oxidosqualene cyclase can be tested by expression in a strain of yeast such as GIL77 that accumulates the substrate 2,3-oxidosqualene (Kushiro, Shibuya, & Ebizuka, 1998). The product of the signature enzyme will usually accumulate to high levels provided that the enzyme is active. High product levels and the

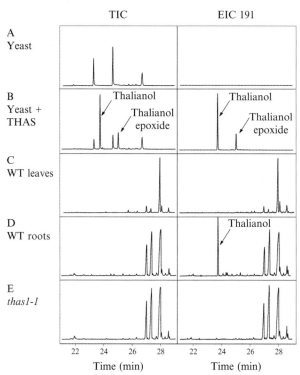

**Figure 6.3** Identification of thalianol in plant extracts using signature-ion fragments. Extracts from yeast and *A. thaliana* were analyzed for triterpene content by GC–MS: TIC, total-ion chromatograms; EIC 191, extracted-ion chromatograms at *m/z* 191. (A) Yeast empty vector control; (B) yeast expressing the thalianol synthase (THAS) cDNA; (C) leaf and (D) root extracts from wild-type *A. thaliana*; and (E) root extracts from a THAS knockout line (*thas1-1*). The chromatograms are scaled to the highest peak. Unlabeled peaks are sterols. *Adapted from Field and Osbourn (2008).*

relatively simple metabolite profile of microbes together facilitate product identification. A metabolite identified in this way serves as a standard for the subsequent identification of the metabolite (and derivatives) at lower levels in wild-type and mutant plant extracts (Fig. 6.3).

## 3.3. Example: analysis of an oxidosqualene cyclase by expression in yeast

1. *Cloning of an oxidosqualene cyclase cDNA.* A high-accuracy polymerase is used to amplify the coding sequence (CDS) of the gene of interest from a cDNA sample; this should be synthesized from RNA extracted from a tissue where the gene is known to be expressed. The Gateway cloning

system is useful for downstream flexibility. Our preferred approach is to directly introduce the CDS into the pCR8/GW/TOPO TA or pENTR/D-TOPO Gateway entry vectors from Invitrogen. The entire CDS is then sequenced to check for errors.

2. *Transfer to an expression vector.* The CDS is transferred to an expression vector using the Gateway LR recombination reaction and verified by restriction digestion. For expression in yeast, the galactose-inducible pYES/DEST-52 vector from Invitrogen can be used. Other Gateway-compatible vectors exist for expression in other systems, for example, the arabinose-inducible pBAD/DEST-49 vector in bacteria, or the pEAQ Dest overexpression vector (Sainsbury, Thuenemann, & Lomonossoff, 2009) in *N. benthamiana.* These vectors often include the possibility of adding an N- or C-terminal tag, which can be invaluable for confirming protein production and for potential purification applications. However, we find that tags sometimes interfere with enzymatic activity. Therefore, we recommend cloning an untagged version of the enzyme in parallel.

3. *Transformation, induction, and extraction of yeast GIL77.* Yeast strain GIL77 (*gal2 hem3−6 erg7 ura3−167*) is an ergosterol auxotroph that accumulates 2,3-oxidosqualene and must be maintained at 30 °C on a medium supplemented with 20 μg/mL ergosterol, 13 mg/mL hemin, and 5 mg/mL Tween80. GIL77 can be transformed with pYES2/DEST-52 using the standard lithium acetate method. For induction, a 2-mL starter culture is used to inoculate 100 mL of yeast peptone medium with 3% glucose and supplements (above) and incubated for 16–24 h at 30 °C with shaking (200 rpm). The culture is then centrifuged ($3000 \times g$ for 10 min), the pellet is washed with 1 volume of distilled water and centrifuged again before induction by resuspension in synthetic complete medium with 2% glucose and 1% raffinose plus supplements for 16–24 h. After induction, the culture is centrifuged and the pellet resuspended in 1 volume of 0.1 *M* potassium phosphate, pH 7.0, for 16–24 h. Finally, the culture is centrifuged and the pellet hydrolyzed at 70 °C for 2 h in 40 mL 10% KOH (w/v) in 80% EtOH (v/v) with 0.5 mg/mL butylated hydroxytoluene. After hydrolysis, the mixture is diluted with water (10 mL) and extracted three times with an equal volume of hexane. The sample is then evaporated under reduced pressure and transferred to an analysis vial using small quantities of hexane. The sample can be stored at − 20 °C prior to analysis.

4. *Metabolite analysis (suitable for nonpolar metabolites).* Samples can be analyzed directly or in a dervatized form by GC–MS. For analysis of triterpene alcohols, we derivatize the extract with Tri-Sil Z and analyze on an Agilent GC–MS fitted with a Phenomenex Zebron ZB-5 column. Raw GC–MS data is analyzed using the AMDIS spectral deconvolution software (Stein, 1999), and the target metabolites are saved into a user-defined AMDIS search library.

5. *Identification of the molecular scaffold in wild-type plant extracts.* A defined quantity of lyophilized plant tissue (100–500 mg) is hydrolyzed and analyzed by GC–MS as in steps 3 and 4. The chromatograms are then deconvoluted in AMDIS and automatically searched for components that match components previously added to the search library.

## 4. VALIDATING CANDIDATE METABOLIC GENE CLUSTERS

To demonstrate that a group of genes form a metabolic gene cluster, it is essential to show that the gene products act together in the same metabolic pathway. There are two principal ways in which this has been achieved: (i) by heterologous expression of the genes in plant or microbial systems; and/or (ii) using genetic approaches that include gene knockouts and knockdowns in the plant of origin. The use of these and similar approaches for the discovery of nonclustered secondary metabolism pathways is also discussed elsewhere (Dixon, Achnine, Deavours, & Naoumkina, 2006). Here, we will focus on the different approaches that have been used successfully in the functional analysis of plant gene clusters. Downstream metabolite analysis and identification approaches will also be discussed.

### 4.1. Heterologous expression approaches

A wide variety of different expression approaches has been used for the analysis of functional gene clusters in plants. Expression of cluster gene products in *E. coli*, followed by protein purification and assays with likely substrate molecules, has been successful for the characterization of a number of different enzymes, including glycosyl transferases, oxoglutarate-dependent dioxygenases, and methyltransferases (Frey et al., 2009; Jonczyk et al., 2008; Shimura et al., 2007; von Rad, Hüttl, Lottspeich, Gierl, & Frey, 2001; Wilderman et al., 2004). However, some enzymes, such as oxidosqualene cyclases (discussed above), are difficult to recover in an active form from prokaryotic expression systems. CYPs, encoded by all plant functional gene clusters reported to date, are particularly recalcitrant because they require an electron donor protein for catalytic activity, and both the CYP and donor are membrane-anchored (Bak et al., 2011). Catalytically active CYPs can be isolated in microsomes made from strains of yeast expressing an electron donor, for example, WAT11, which expresses Arabidopsis CYP reductase 1 (ATR1) from the chromosome (Frey et al., 1997). CYP activity can also be tested in insect cells using a baculovirus expression system (Swaminathan, Morrone, Wang, Fulton, & Peters, 2009). However, these approaches are not always successful, and we discuss some alternative approaches below.

The simultaneous expression of two or more enzymes from a metabolic pathway is a powerful approach for enzyme characterization and pathway discovery. The reconstitution of partial metabolic pathways removes the problem of substrate availability and the need to purify the enzymes or develop *in vitro* activity assays. The rapidity of the coexpression approach also permits the sampling of several different enzyme combinations. A recent example is the development of a bacterial system for the analysis of diterpenoid pathways in which the enzymes for the early steps of diterpene synthesis are coexpressed with an enzyme of interest (Cyr, Wilderman, Determan, & Peters, 2007). Combined with the coexpression of a CYP reductase, this system has been successfully used to analyze the function of several CYPs in the rice momilactone and phytocassane/oryzalide and oryzadione gene clusters (Swaminathan et al., 2009; Wang, Hillwig, & Peters, 2011; Wu, Hillwig, Wang, & Peters, 2011). The scalability of this system allows the purification of sufficient quantities of product to perform full structural characterization.

Plant cells can provide a cellular environment very similar to that experienced by a native plant enzyme. Therefore, expression of recombinant plant proteins in plants is likely to increase the likelihood of recovering accurate enzymatic activity. *Agrobacterium*-mediated transient expression in *N. benthamiana* has been used for the expression and purification of the acyltransferase SCPL1 from the avenacin gene cluster in oats (Mugford et al., 2009). The transient coexpression of multiple enzymes in *N. benthamiana* has also been used to reconstitute an entire cyanogenic glucoside pathway and gene cluster from *Lotus japonicus* (Takos et al., 2011). This new pathway consists of two CYPs and a glucosyl transferase that together synthesize the cyanogenic glucosides linamarin and lotaustralin. Similar results can also be obtained by the generation of stable transgenic plants overexpressing one or more enzymes, as was used for probing the function of OSCs and CYPs in the Arabidopsis marneral and thalianol gene clusters (Field et al., 2011; Field & Osbourn, 2008). Stable transgenics can also provide clues as to the function of pathway intermediates. We found, for example, that accumulation of hydroxylated triterpenes in Arabidopsis had severe consequence for plant growth and development (Field et al., 2011; Field & Osbourn, 2008). However, overexpression of recombinant enzymes in plants does have potential disadvantages. For example, endogenous plant enzymes may modify the metabolites of interest, and the identification and purification of target metabolites can be complicated by copurification of other compounds from the complex plant metabolome.

The major caveat that should accompany results generated from the above expression strategies is that the activity of the recombinant protein may not reflect the activity of the native protein (Dixon et al., 2006). Metabolic enzymes are frequently multispecific and/or promiscuous (i.e., they may accept many different substrates and produce multiple products). Therefore, an activity identified for a recombinant or ectopically expressed enzyme may not reflect the principal activity of the corresponding native enzyme. Investigation of plant mutants affected in different steps in the pathway can therefore provide important confirmation of predictions based on heterologous expression experiments.

## 4.2. Reverse genetic approaches

Genetic approaches have the potential to provide unambiguous evidence for the requirement of a specific gene product (or products) in a specific metabolic pathway. Forward genetic strategies, such as that employed in the identification of the avenacin pathway in oats, enable the *de novo* discovery of multiple genes required for a specific pathway. Reverse genetic approaches, as already mentioned for the maize DIBOA/DIMBOA pathway, require knowledge of candidate genes that can then be targeted for mutagenesis. A knockout will usually cause accumulation of the substrate of the missing enzyme, and loss of the end product of the metabolic pathway. Gene knockouts can be obtained for plant species where comprehensive mutant collections are available such as the T-DNA insertion collections available for Arabidopsis, or EMS mutagenized populations that can be screened using TILLING technology to identify mutations in a gene of interest (McCallum, Comai, Greene, & Henikoff, 2000). In certain species, such as in the moss, *Physcomitrella patens*, gene knockouts can be obtained using homologous recombination-mediated gene-targeting approaches.

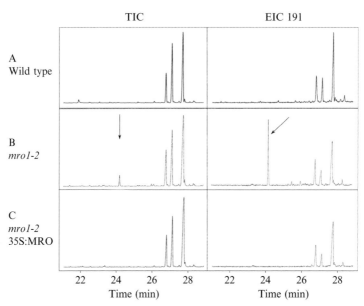

**Figure 6.4** Complementation of a marnerol oxidase (MRO) mutant. Extracts from *A. thaliana* were analyzed for triterpene content by GC–MS: TIC, total-ion chromatogram; EIC 191, extracted-ion chromatogram at *m/z* 191. Root extracts from (A) wild-type *A. thaliana*, (B) the marnerol oxidase mutant *mro1-2*, and (C) *mro1-2* complemented by overexpression of the MRO cDNA. Arrows show peaks representing trimethylsilyated marnerol. Unlabeled peaks are trimethylsilyated sterols. *Adapted from Field et al. (2011).*

A recent example of a reverse genetic approach to cluster discovery is the demonstration that the *THAS*, *THAH*, and *MRO* genes are required for the thalianol and marneral pathways in Arabidopsis using T-DNA insertion mutants (Field et al., 2011; Field & Osbourn, 2008). Gene knockouts generated by insertional mutagenesis may have insertions or mutations elsewhere in the genome. Therefore, to unequivocally link the knockout phenotype to the disrupted gene, it is necessary to complement the mutant by the expression of the wild-type gene (Fig. 6.4). If a knockout mutant for a gene of interest cannot be identified, it is possible to employ gene knockdown strategies using hairpin RNAs or artificial micro-RNA mediated gene silencing. This strategy has been employed to analyze the roles of *CYP99A2* and *CYP99A3* in the rice momilactone cluster (Shimura et al., 2007) and for *THAD* in the Arabidopsis thalianol cluster (Field & Osbourn, 2008). Care must be taken when using RNA silencing strategies because complementation is less trivial, and there is the potential for silencing of related nontarget genes. Nontarget gene silencing can be difficult to avoid when working with tandem duplicates. Indeed, in the above example of the rice momilactone cluster, double knockdowns could be obtained for the tandem duplicates *CYP99A2* and *CYP99A3*, but single-gene knockdown lines could not be isolated, despite the use of gene-specific target sequences.

Finally, chemical genetic strategies can be employed to narrow down the list of candidate genes for a gene cluster pathway. For example, the CYP inhibitor uniconazole-P was used to test the involvement of CYPs in the momilactone pathway (Shimura et al., 2007) and the 2-oxoglutarate-dependent dioxygenase inhibitor Prohexadion-Ca was used to help identify Bx6 in the DIMBOA pathway in maize (Frey et al., 2003).

## 4.3. Metabolite identification

The new metabolites generated through expression of recombinant enzymes, *in vitro* assays, or the creation of knockout/knockdown plants must be identified for effective delineation of a putative metabolic pathway. Metabolite identification can be difficult when attempted in complex mixtures such as plant extracts. Here, we briefly focus on methods for analyzing more complex extracts. The extraction procedure must be tailored towards the molecules of interest. For example, polar saponins and cyanogenic glucosides can be extracted from plant samples using aqueous methanol. In special cases, selective extraction procedures can be used. These have the advantage of simplifying complex extracts. For example, alkaline hydrolysis followed by hexane extraction is an effective approach for separating sterols and triterpenes from other lipids.

Analysis of metabolite extracts is most typically by chromatographic separation followed by mass spectrometry. New metabolites can be identified by comparing control samples (e.g., wild-type plant extract) with modified samples (e.g., mutant plant extract). In complex mixtures, new low-abundance metabolite peaks are often not immediately obvious in the raw chromatograms. However, prior information about the core structure of the target molecule and its ion fragmentation pattern can lead directly to the identification of new peaks representing related compounds. In the absence of prior information, the total-ion chromatograms can be manually or automatically deconvoluted and compared using software to identify peaks that are present in or absent from the control sample. In both cases, we stress the importance of establishing robust and repeatable growth and extraction procedures; otherwise, new peaks may simply reflect differences in growth conditions or sample preparation.

After the identification of a new peak, mass spectrometry techniques can be further used to gather valuable structural information. For example, tandem mass spectrometry techniques with accurate mass are ideal for identifying mass-shifts associated with specific modifications such as hydroxylation, acylation, and glycosylation (Prasad, Garg, Takwani, & Singh, 2011). Potential modifications can be predicted using prior knowledge about the molecular scaffold of the pathway under investigation (see Sections 3.1 and 3.2) and also the predicted general function of the enzyme being studied. In addition to improving separation properties, derivatization of extracts can be used for obtaining information about the functional groups displayed by a target molecule.

## 4.4. Example: analysis of knockouts and knockdowns for the thalianol gene cluster

1. *Plant growth.* Seeds of the mutant and wild-type genotypes are germinated in a growth chamber with a 16-h light/8-h dark photoperiod at 23 °C. After 8 days, seedlings are transplanted to pots of a Terra-Green (Oil-Dri UK Ltd., Wisbech, UK) and 16/30 sand (WBB Minerals, Sandbach, UK) mix (1:1). The plants are fed twice a week with Sangral 3:1:1 fertilizer solution (William Sinclair Horticulture Ltd., Lincoln, UK).
2. *Plant harvesting.* After 6 weeks growth, roots (where the thalianol pathway is expressed) are separated from the Terra-Green/sand mix in a tray of water and immediately frozen in liquid nitrogen. The root material is then ground to a fine powder in a mortar and pestle and lyophilized in a freeze drier.
3. *Extraction of triterpenes.* 100–500 mg of freeze-dried material is hydrolyzed at 70 °C for 2 h in 40 mL 10% KOH (w/v) in 80% EtOH (v/v) with 0.5 mg/mL butylated hydroxytoluene. At the beginning of hydrolysis, an internal standard such as betulin can be added. After hydrolysis, the mixture is diluted with water (10 mL) and extracted three times with an equal volume of hexane. The sample is then evaporated under reduced pressure and transferred to an analysis vial using small quantities of hexane. The sample can be stored at −20 °C prior to analysis.
4. *Metabolite analysis.* Samples are derivatized with Tri-Sil Z, and the trimethyl silyl ethers are analyzed on an Agilent GC–MS fitted with a Phenomenex Zebron ZB-5 column. Raw GC–MS data is analyzed using the AMDIS spectral deconvolution software (Stein, 1999).

## 5. IMAGING GENE CLUSTERS

Cytogenetic methods offer a route to determining whether cloned genes are clustered in unsequenced or only partially sequenced genomes. This can be achieved using DNA fluorescence *in situ* hybridization (DNA FISH) on metaphase spreads, pachytene spreads, or extended DNA fibers. Metaphase spreads are easiest to make but show the least spatial resolution (4–5 MB, de Jong, Fransz, & Zabel, 1999) and signal sensitivity (minimum of 15-kb probe length in our experience). They can, however, in conjunction with chromosome-specific probes identify the chromosome on which the cluster is localized and where it is located with reference to centromeres and telomeres. Pachytene spreads from pollen mother cells resolve down to 120 kb in euchromatic chromosome regions but are laborious to prepare and not successful in all species (de Jong et al., 1999). DNA fibers are difficult to produce but can show a resolution of 1 kb/$\mu$m (de Jong et al., 1999), which poses problems if genes are interspersed with more than 20 kb of repetitive sequences. DNA FISH in tissue can show how close genes are at the nuclear level (Wegel, Koumproglou, Shaw, & Osbourn, 2009). Here, the minimum target length is 5 kb, and a resolution of 50–100 kb is possible (de Jong et al., 1999). This technique can also be used to show differences in chromatin conformation in the cluster region depending on whether the cluster is transcriptionally active or not (Fig. 6.5; Wegel et al., 2009).

Once a cluster has been identified, fluorescence *in situ* hybridization can be used to show expression of the genes down to the single-cell level (Wegel et al., 2009). Using RNA FISH, it is possible to identify whether two or three genes are expressed in the same nucleus at the same time, how close they are spatially and also where the transcripts are localized in the cytoplasm (Fig. 6.6). Up to three probes can be used, for green, red, and far red; there are no strong dyes for blue/UV. More than three probes will cause imaging problems because very narrow bandpass filters are necessary to distinguish between the different fluorochromes, which reduces signal that may already be weak if copy numbers are low.

### 5.1. RNA probe preparation

DNA templates are synthesized by PCR using primers that start with a T7 or SP6 promoter sequence; linear PCR products are precipitated and resuspended in $H_2O$ to a concentration of 0.5–1 $\mu$g/$\mu$L. Transcripts are synthesized using an SP6/T7 Transcription Kit (Roche 10999644001) and can

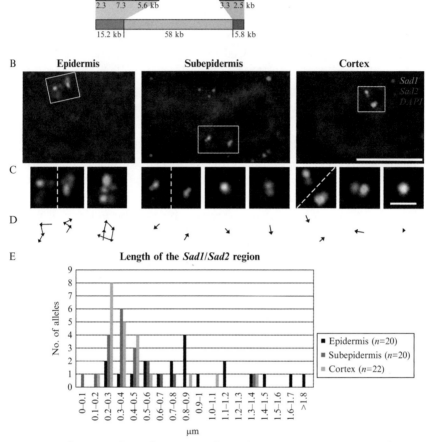

**Figure 6.5** Chromatin decondensation in the *Sad1/Sad2* region. (A) Diagram showing the *Sad1* (green) and *Sad2* (red) probes used for DNA *in situ* hybridization (coding regions in lighter shades of green and red). (B) The *Sad1/Sad2* region is decondensed in the nuclei of epidermal cells (left) compared with those of the subepidermis (center) and the cortex (right). In each panel, a single nucleus is shown. Bar = 5 μm. (C) Detailed views of individual gene regions. Dashed lines separate loci on two adjacent chromosomes. Bar = 1 μm. The images shown in (B) and (C) are overlays of several optical sections, section spacing 0.2 μm. Blue, chromatin stained with DAPI. (D) Line drawings showing the path length from the center of each fluorescence focus in (C) to the center of its nearest neighbor, starting from *Sad1*. (E) Length distributions of the *Sad1/Sad2* region in nuclei of epidermal, subepidermal, and cortical cells (reproduced with permission from Wegel et al., 2009). (See Color Insert.)

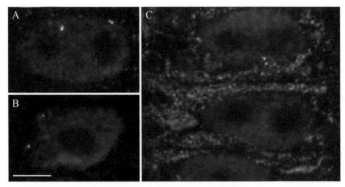

**Figure 6.6** Visualization of *Sad1* and *Sad2* transcripts in nuclei of oat root-tip cells. Nascent *Sad1* (red) and *Sad2* (green) transcripts; blue, chromatin stained with 4′,6-diamidino-2-phenylindole (DAPI). (A) Detection of nascent *Sad1* and *Sad2* transcripts in the nucleus of an epidermal root-tip cell. (B) In contrast to the epidermis, nuclei of the subepidermis, like the one shown, only express *Sad1*. (C) Distinctive localization patterns of *Sad1* and *Sad2* transcripts in the cytoplasm of root epidermal cells. All images are overlays of several optical sections with section spacing of 0.2 mm. Bar = 5 mm. Adapted from Wegel et al. (2009). (See Color Insert.)

be labeled with biotin-16-UTP (Roche 11388908910), digoxigenin-11-UTP (Roche 11209256910), or dinitrophenol-11-UTP (PerkinElmer NEL555001EA).

1. After the DNase I digest in the manufacturer's protocol, add 2 μL of 200 m$M$ EDTA (pH 8.0), 2 μL of LiCl (4 $M$), and 75 μL of ice-cold EtOH. Mix and precipitate at −20 °C overnight.
2. Centrifuge for 15 min at 4 °C at 15,000 rpm. Wash with ice-cold 70% EtOH and spin again.
3. Resuspend pellet in 50 μL H$_2$O (RNase-free), add 50 μL of fresh, filter-sterilized 200 m$M$ carbonate buffer (pH 10.2, made up of 80 m$M$ NaHCO$_3$ + 120 mM Na$_2$CO$_3$) and mix.
4. Incubate at 60 °C for required time, which depends on probe length:

$$t = \frac{L_i - L_f}{K \times L_i \times L_f},$$

where $t$, time in minutes; $K$, rate constant (=0.11 Kb/min), $L_i$, initial length (kb), and $L_f$, final length (optimal $L_f$=0.15 kb).

5. Stop the reaction. Add 5 μL of acetic acid (10%), 10 μL of sodium acetate (3 $M$, pH 5.5), 288 μL of EtOH (ice-cold) and precipitate for 2 h to overnight at −20 °C.

6. Spin, wash, and dry as above.
7. Resuspend pellet in 50 µL TE.
8. Check probe concentration by loading 3 µL on a 1% agarose gel.

## 5.2. DNA probe preparation

Label probes using the nick-translation mix (Roche 11745808910) according to the manufacturer's instructions with digoxigenin-11-dUTP (Roche 11093088910) or dinitrophenol-11-dUTP (PerkinElmer NEL551001EA) or biotin-16-dUTP (Roche 11093070910).

## 5.3. Preparation of sections for RNA and DNA fluorescence *in situ* hybridization (FISH)

Embed your tissue in wax after fixation with 4% formaldehyde in PBS, in our case, root tips from 3-day-old seedlings of diploid oat (*A. strigosa*). Prepare 12-µm thin longitudinal sections on a Leica RM2055 microtome (Leica, Nussloch, Germany) and transfer them to polylysine-coated slides (BDH).

1. Cover dewaxed and rehydrated sections with 1% driselase (Sigma-Aldrich, D9515), 0.5% cellulase "Onozuka" R10 (Yakult Honsha, Tokyo, Japan), 0.025% pectolyase Y23 (Kikkoman, Seishin Corp., Tokyo, Japan) in PBS and incubate in a humid chamber for 1 h at 37 °C. Aliquots of the mix can be kept frozen at $-20$ °C. Wash for 5 min in PBS.
2. Treat sections for RNA FISH with 10 µg/mL proteinase K (Sigma) for 30 min at 28 °C in a Coplin jar in proteinase K buffer (100 m$M$ Tris–HCl (pH 8.0), 50 m$M$ EDTA). Prewarm buffer and container, and add proteinase K just before use from a 10 mg/mL stock in distilled $H_2O$. Continue with step 10.
   2a. Incubate sections for DNA FISH in a Coplin jar in 0.1 mg/mL RNase A (Sigma) in 2× SSC for 1 h at 37 °C. Wash for 10 min in PBS and continue with step 12.
3. Inactivate proteinase K with 5-min glycine, 2 mg/mL in PBS at RT (50 mg/mL glycine stock in distilled $H_2O$ filter sterilized and kept at 4 °C) followed by a 5-min wash in PBS.
4. Postfix for 10 min in 4% (w/v) formaldehyde in PBS, pH 7.4 and wash 2× 5 min in PBS.
5. Dehydrate in an ethanol series and leave to dry.

## 5.4. RNA *in situ* hybridization and posthybridization washes

### 5.4.1 Solutions

$10 \times$ salts: 3 $M$ NaCl, 0.1 $M$ Tris–HCl (pH 6.8), 0.1 $M$ NaPO$_4$ buffer (pH 6.8), 50 m$M$ EDTA.

Hybridization buffer (800 µL, stored at $-20\,°C$): 100 µL $10 \times$ salts; 400 µL deionized formamide, deionized (Sigma F9037); 200 µL 50% dextran sulfate; 20 µL $50 \times$ Denhardts (Sigma D2532); 80 µL H$_2$O.

Hybridization solution (40 µL/slide, for a 20 mm × 20 mm area): 4 µL formamide, deionized (Sigma F9037); 1 µL tRNA 50 µg/µL (Sigma type XXI, made up in water); 1–2 µL probe (ca. 200 ng); x µL H$_2$O to a final volume of 8 µL; 32 µL hybridization buffer.

1. Denature the hybridization solution at 80 °C for 2 min, cool in ice for 5 min.
2. Evenly distribute the solution over the sections, add a homemade coverslip cut out of a plastic bag and hybridize at 37–50 °C overnight in a humid chamber. Higher temperatures increase the stringency of the hybridization.
3. Wash slides in a Coplin jar for 3 min 2 × SSC at 50 °C to wash off coverslips in a shaking waterbath
4. 15 min wash in 2 × SSC/50% formamide (Sigma F7508) at 50 °C in a shaking waterbath
5. 15 min wash in 1 × SSC/50% formamide at 50 °C in a shaking waterbath
6. 5 min wash in 2 × SSC at room temperature on shaker
7. 2 × 5 min wash in 4 × SSC/0.2% Tween 20 at room temperature

## 5.5. DNA *in situ* hybridization and posthybridization washes

Hybridization solution (40 µL/slide, for a 20 mm × 20 mm area): 20 µL formamide, deionized (Sigma F9037); 8 µL 50% dextran sulfate (Sigma D-8906, pass through a 20-µm filter); 4 µL 20 × SSC; 0.5 µL 10% SDS; 1 µL salmon sperm DNA (Sigma D1626), 5 µg/µL, autoclaved; 5 µL probe (200 ng); 1.5 µL H$_2$O.

1. Denature the hybridization solution for 5 min at 95 °C, and cool in ice for 5 min.
2. Evenly distribute the solution over the sections, add a homemade coverslip cut out of a plastic bag, and hybridize overnight in a modified Thermocycler (Omnislide, Hybaid, Ashford, UK) using the following conditions: 75 °C for 8 min, 50 °C for 1 min, 45 °C for 90 s, 40 °C for 2 min, 38 °C for 5 min, and 37 °C for 16 h.

3. Wash the slides in a Coplin jar for 3 min in $2 \times$ SSC in a waterbath at 42 °C to remove the coverslips.
4. $2 \times 5$ min wash in 20% formamide (Sigma F7508) in $0.1 \times$ SSC at 42 °C.
5. $2 \times 5$ min wash in $2 \times$ SSC at 42 °C.
6. $2 \times 5$ min wash in $2 \times$ SSC at room temperature on the shaker.
7. $2 \times 5$ min wash in $4 \times$ SSC/0.2% Tween 20 at room temperature.

## 5.6. Immunodetection

1. Block the slides for 10 min in 5% (w/v) BSA/$4 \times$ SSC/0.2% Tween 20, 100 μL/slide under plastic coverslips as above in a humid chamber.
2. Remove coverslips and tap solution off on paper.
3. Incubate the slides in monoclonal antidigoxin mouse antibody (Sigma D-8156) 1:5000 and rabbit antidinitrophenyl antibody (Life Technologies A6430) 1:1000 in 5% BSA/$4 \times$ SSC/0.2% Tween 20, 100 μL/slide, under plastic coverslips, 1 h at 37 °C in a humid chamber in the dark.
4. Wash the slides $3 \times 5$ min in $4 \times$ SSC/0.2% Tween 20 in a jar wrapped in aluminum foil on the shaker.
5. Incubate the slides in Alexa Fluor 488 goat antimouse antibody (Invitrogen A11017) 1:300 and ExtrAvidin-Cy3 (Sigma E2511) 1:300 and antirabbit Alexa Fluor 647 antibody (Life Technologies A21246) 1:600 in 5% BSA/$4 \times$ SSC/0.2% Tween 20, 100 μL/slide, under plastic coverslips, 1 h at 37 °C in a humid chamber in the dark.
6. Wash the slides $3 \times 5$ min in $4 \times$ SSC/0.2% Tween 20 in a jar wrapped in aluminum foil on the shaker at RT.
7. Counterstain in 100 μL 1 μg/mL DAPI (Sigma 32670) with plastic coverslips in the dark for 3–4 min.
8. Quick wash in $4 \times$ SSC/0.2% Tween 20, tap excess solution off.
9. Mount slides with 15 μL of Vectashield (Vector Laboratories H-1000) using coverslips no. 1.5. Mop up excess with filter paper and seal with nail polish.

## 5.7. Imaging

As the gene probes are usually short, high sensitivity of detection coupled with deconvolution is important. The most sensitive microscope for tissue sections or chromosome/fiber spreads with weak signal and relatively high background in sections is a wide-field fluorescence microscope with a cooled CCD camera or possibly an EMCCD camera. For tissue sections, the new-generation GaAsP detectors on scanning confocal microscopes might also be sensitive enough. Deconvolution of scanning confocal microscope data is probably not necessary but is highly recommended for wide-field fluorescence microscopy of tissue sections. Good deconvolution packages are Autoquant (MediaCybernetics) and Huygens (Scientific Volume Imaging).

# 6. FUTURE PROSPECTS

Over the past 10 years or so, dramatic advances have been made in the discovery of plant secondary metabolic gene clusters (for review, see Chu et al., 2011). Other published examples of candidate gene clusters for as-yet-uncharacterized metabolic pathways include reports from *A. thaliana* (Ehlting et al., 2008; Field et al., 2011; Field & Osbourn, 2008), rice (Sakamoto et al., 2004), tomato (Falara et al., 2011), and cucumber (Huang et al., 2009). Many more secondary metabolic clusters are likely to emerge as the body of sequence information available continues to grow, accelerated by high-throughput sequencing. The application of systematic genome mining approaches in combination with functional analysis of candidate clusters offers enormous potential for the discovery of new pathways, enzymes, and chemistries.

## ACKNOWLEDGMENTS

We gratefully acknowledge the support of Biotechnology and Biological Sciences Research Council (BBSRC) responsive mode grant BB/C504435/1, the BBSRC-funded Institute Strategic Programme Grant, "Understanding and Exploiting Plant and Microbial Metabolism," Engineering and Physical Sciences Research Council (EPSRC) grant EP/H019154/1 to A. O., ESF-NSRF (Heracleitus II) grant to K. K. P., NNSF (31170278) grant of China to X. Q., and Centre National de la Recherche Scientifique funding to B F.

## REFERENCES

Allouis, S., Moore, G., Bellec, A., Sharp, R., Faivre, R. P., Mortimer, K., et al. (2003). Construction and characterisation of a hexaploid wheat BAC library from the reference germplasm 'Chinese Spring'. *Cereal Research Communications, 31*, 331–338.

Bak, S., Beisson, F., Bishop, G., Hamberger, B., Höfer, R., Paquette, S., et al. (2011). Cytochromes p450. *Arabidopsis Book, 9*, e0144 [Epub 2011 Oct 6].

Chen, S., Xiang, L., Guo, X., & Li, Q. (2011). An introduction to the medicinal plant genome project. *Frontiers of Medicine, 5*, 178–184.

Chu, H.-Y., Wegel, E., & Osbourn, A. (2011). From hormones to secondary metabolism: The emergence of secondary metabolic gene clusters in plants. *The Plant Journal, 66*, 66.

Cyr, A., Wilderman, P. R., Determan, M., & Peters, R. J. (2007). A modular approach for facile biosynthesis of labdane-related diterpenes. *Journal of the American Chemical Society, 129*, 6684–6685.

D'Auria, J. C., & Gershenzon, J. (2005). The secondary metabolism of *Arabidopsis thaliana*: Growing like a weed. *Current Opinion in Plant Biology, 8*, 308–316.

de Jong, J. H., Fransz, P., & Zabel, P. (1999). High resolution FISH in plants—Techniques and applications. *Trends in Plant Science, 4*, 258–263.

Dixon, R. A., Achnine, L., Deavours, B. E., & Naoumkina, M. (2006). Metabolomics and gene identification in plant natural product pathways. *Biotechnology in Agriculture and Forestry, 57*, 243–259.

Ehlting, J., Sauveplane, V., Olry, A., Ginglinger, J. F., Provart, N. J., & Werck-Reichhart, D. (2008). An extensive (co-)expression analysis tool for the cytochrome P450 superfamily in Arabidopsis thaliana. *BMC Plant Biology, 8*, 47.

Falara, V., Akhtar, T. A., Nguyen, T. T., Spyropoulou, E. A., Bleeker, P. M., Schauvinhold, I., et al. (2011). The tomato terpene synthase gene family. *Plant Physiology, 157*, 770–789.

Field, B., Fiston-Lavier, A. S., Kemen, A., Geisler, K., Quesneville, H., & Osbourn, A. E. (2011). Formation of plant metabolic gene clusters within dynamic chromosomal regions. *Proceedings of the National Academy of Sciences of the United States of America, 108*, 16116–16121.

Field, B., & Osbourn, A. E. (2008). Metabolic diversification—Independent assembly of operon-like gene clusters in different plants. *Science, 320*, 543–547.

Frey, M., Chomet, P., Glawischnig, E., Stettner, C., Grun, S., Winklmair, A., et al. (1997). Analysis of a chemical plant defense mechanism in grasses. *Science, 277*, 696–699.

Frey, M., Huber, K., Woong, J. P., Sicker, D., Lindberg, P., Meeley, R. B., et al. (2003). A 2-oxoglutarate-dependent dioxygenase is integrated in DIMBOA-biosynthesis. *Phytochemistry, 62*, 371–376.

Frey, M., Kliem, R., Saedler, H., & Gierl, A. (1995). Expression of a cytochrome P450 gene family in maize. *Molecular & General Genetics, 246*, 100–109.

Frey, M., Schullehner, K., Dick, R., Fiesselmann, A., & Gierl, A. (2009). Benzoxazinoid biosynthesis, a model for evolution of secondary metabolic pathways in plants. *Phytochemistry, 70*, 1645–1651.

Haralampidis, K., Bryan, G., Qi, X., Papadopoulou, K., Bakht, S., Melton, R., et al. (2001). A new class of oxidosqualene cyclases directs synthesis of antimicrobial phytoprotectants in monocots. *Proceedings of the National Academy of Sciences of the United States of America, 98*, 13431–13436.

Huang, S., Li, R., Zhang, Z., Li, L., Gu, X., Fan, W., et al. (2009). The genome of the cucumber, *Cucumis sativus* L. *Nature Genetics, 41*, 1275–1281.

Jonczyk, R., Schmidt, H., Osterrieder, A., Fiesselmann, A., Schullehner, K., Haslbeck, M., et al. (2008). Elucidation of the final reactions of DIMBOA-glucoside biosynthesis in maize: Characterization of Bx6 and Bx7. *Plant Physiology, 146*, 1053–1063.

Kushiro, T., Shibuya, M., & Ebizuka, Y. (1998). Beta-amyrin synthase—Cloning of oxidosqualene cyclase that catalyzes the formation of the most popular triterpene among higher plants. *European Journal of Biochemistry, 256*, 238–244.

Mao, G., Buschmann, H., Doonan, J., & Lloyd, C. W. (2006). The role of MAP65-1 in microtubule bundling during Zinnia tracheary element formation. *Journal of Cell Science, 119*, 753–758.

McCallum, C. M., Comai, L., Greene, E. A., & Henikoff, S. (2000). Targeted screening for induced mutations. *Nature Biotechnology, 18*, 455–457.

Mugford, S. T., Qi, X., Bakht, S., Hill, L., Wegel, E., Hughes, R. K., et al. (2009). A serine carboxypeptidase-like acyltransferase is required for synthesis of antimicrobial compounds and disease resistance in oats. *The Plant Cell, 21*, 2473–2484.

Mylona, P., Owatworakit, A., Papadopoulou, K., Jenner, H., Qin, B., Findlay, K., et al. (2008). *Sad3* and *Sad4* are required for saponin biosynthesis and root development in oat. *The Plant Cell, 20*, 201–212.

Osbourn, A., & Field, B. (2009). Operons. *Cellular and Molecular Life Sciences*, *66*, 37555–37575.

Osbourn, A. (2010). Secondary metabolic gene clusters: Evolutionary toolkits for chemical innovation. *Trends in Genetics*, *26*, 449–457.

Papadopoulou, K., Melton, R., Leggett, M., Daniels, M. J., & Osbourn, A. E. (1999). Compromised disease resistance in saponin deficient plants. *Proceedings of the National Academy of Sciences of the United States of America*, *96*, 12923–12928.

Prasad, B., Garg, A., Takwani, H., & Singh, S. (2011). Metabolite identification by liquid chromatography-mass spectrometry. *Trends in Analytical Chemistry*, *30*, 360–387.

Qi, X., Bakht, S., Leggett, M., Maxwell, C., Melton, R., & Osbourn, A. (2004). A gene cluster for secondary metabolism in oat: Implications for the evolution of metabolic diversity in plants. *Proceedings of the National Academy of Sciences of the United States of America*, *101*, 8233–8238.

Qi, X., Bakht, S., Qin, B., Leggett, M., Hemmings, A., Mellon, F., et al. (2006). A different function for a member of an ancient and highly conserved cytochrome P450 family: From essential sterol to plant defense. *Proceedings of the National Academy of Sciences of the United States of America*, *103*, 18848–18853.

Qin, B., Eagles, J., Mellon, F. A., Mylona, P., Peña-Rodriguez, L., & Osbourn, A. E. (2010). High throughput screening of mutants of oat that are defective in triterpene synthesis. *Phytochemistry*, *71*, 1245–1252.

Sainsbury, F., Thuenemann, E. C., & Lomonossoff, G. P. (2009). pEAQ: Versatile expression vectors for easy and quick transient expression of heterologous proteins in plants. *Plant Biotechnology Journal*, *7*, 682–693.

Sakamoto, T., Miura, K., Itoh, H., Tatsumi, T., Ueguchi-Tanaka, M., Ishiyama, K., et al. (2004). An overview of gibberellin metabolism enzyme genes and their related mutants in rice. *Plant Physiology*, *134*, 1642–1653.

Shimura, K., Okada, A., Okada, K., Jikumaru, Y., Ko, K. W., Toyomasu, T., et al. (2007). Identification of a biosynthetic gene cluster in rice for momilactones. *The Journal of Biological Chemistry*, *282*, 34013–34018.

Stam, P. (1993). Construction of integrated genetic linkage maps by means of a new computer package: JOINMAP. *The Plant Journal*, *3*, 739–744.

Stein, S. E. (1999). An integrated method for spectrum extraction and compound identification from GC/MS data. *Journal of American Society of Mass Spectrometry*, *10*, 770–781.

Swaminathan, S., Morrone, D., Wang, Q., Fulton, D. B., & Peters, R. J. (2009). CYP76M7 is an ent-cassadiene C11 alpha-hydroxylase defining a second multifunctional diterpenoid biosynthetic gene cluster in rice. *The Plant Cell*, *21*, 3315–3325.

Takos, A. M., Knudsen, C., Lai, D., Kannangara, R., Mikkelsen, L., Motawia, M. S., et al. (2011). Genomic clustering of cyanogenic glucoside biosynthetic genes aids their identification in *Lotus japonicus* and suggests the repeated evolution of this chemical defence pathway. *The Plant Journal*, *68*, 273–286.

von Rad, U., Hüttl, R., Lottspeich, F., Gierl, A., & Frey, M. (2001). Two glucosyltransferases are involved in detoxification of benzoxazinoids in maize. *The Plant Journal*, *28*, 633–642.

Wang, Q., Hillwig, M. L., & Peters, R. J. (2011). CYP99A3: Functional identification of a diterpene oxidase from the momilactone biosynthetic gene cluster in rice. *The Plant Journal*, *65*, 87–95.

Wegel, E., Koumproglou, R., Shaw, P., & Osbourn, A. (2009). Cell type-specific chromatin decondensation of a metabolic gene cluster in oats. *The Plant Cell*, *21*, 3926–3936.

Wilderman, P. R., Xu, M., Jin, Y., Coates, R. M., & Peters, R. J. (2004). Identification of Syn -imara-7,15-diene synthase reveals functional clustering of terpene synthases involved in rice phytoalexin/allelochemical biosynthesis. *Plant Physiology*, *135*, 2098–2105.

Wu, Y., Hillwig, M. L., Wang, Q., & Peters, R. J. (2011). Parsing a multifunctional biosynthetic gene cluster from rice: Biochemical characterization of CYP71Z6 & 7. *FEBS Letters*, *585*, 3446–3451.

Yonekura-Sakakibara, K., & Saito, K. (2009). Functional genomics for plant natural product synthesis. *Natural Product Reports*, *26*, 1466–1487.

Yu, G. X., & Wise, R. P. (2000). An anchored AFLP- and retrotransposon-based map of diploid *Avena*. *Genome*, *43*, 736–749.

CHAPTER SEVEN

# Genomic Approaches for Interrogating the Biochemistry of Medicinal Plant Species

**Elsa Góngora-Castillo\*, Greg Fedewa[†], Yunsoo Yeo[‡], Joe Chappell[‡], Dean DellaPenna[†], C. Robin Buell\*,[1]**

\*Department of Plant Biology, Michigan State University, East Lansing, Michigan, USA
[†]Department of Biochemistry & Molecular Biology, Michigan State University, East Lansing, Michigan, USA
[‡]Department of Plant and Soil Sciences, University of Kentucky, Lexington, Kentucky, USA
[1]Corresponding author: e-mail address: buell@msu.edu

## Contents

*Methods in Enzymology*, Volume 517
ISSN 0076-6879
http://dx.doi.org/10.1016/B978-0-12-404634-4.00007-3

## Abstract

Development of next-generation sequencing, coupled with the advancement of computational methods, has allowed researchers to access the transcriptomes of recalcitrant genomes such as those of medicinal plant species. Through the sequencing of even a few cDNA libraries, a broad representation of the transcriptome of any medicinal plant species can be obtained, providing a robust resource for gene discovery and downstream biochemical pathway discovery. When coupled to estimation of expression abundances in specific tissues from a developmental series, biotic stress, abiotic stress, or elicitor challenge, informative coexpression and differential expression estimates on a whole transcriptome level can be obtained to identify candidates for function discovery.

## 1. INTRODUCTION

The genomes of some plant species remain one of the final frontiers in genomics due to technical challenges in obtaining quality sequences of the genomes of polyploid, highly repetitive, and heterozygous species. For example, the genome of bread wheat, a hexaploid, is estimated at 16 Gb (Arumuganathan & Earle, 1991) and is composed of >85% repetitive sequences (Paux et al., 2006). Thus, while next-generation sequencing (NGS) platforms (Shendure & Ji, 2008) can readily generate sequences for large repetitive polyploid genomes, assembly and reconstitution of the 21 chromosomes in bread wheat using whole-genome shotgun sequencing is not feasible with current sequencing and computational methods. For other species, the high degree of heterozygosity presents a barrier to robust assembly of the gene space because of uneven degrees of similarity among the haplotypes, even in a diploid genome (The Potato Genome Sequencing Consortium, 2011). Multiple approaches can be employed to overcome these technical barriers and obtain the genome sequences of these recalcitrant species. These include reducing the inherent heterozygosity by inbreeding for several generations (Jaillon et al., 2007), sequencing a progenitor species of the polyploid (Shulaev et al., 2011), generation of unique genetic material with reduced genome complexity (The Potato Genome Sequencing Consortium, 2011), and physical separation of chromosomes and targeted sequencing of a single or a few chromosomes (Paux et al., 2008). All these efforts require considerable investment in germplasm, genetic, genomic, or molecular resources and are generally only feasible in species with deep knowledge and access to germplasm, genetic resources, and reproductive biology and in which there is a scientific community well vested in obtaining the genome sequence. However, for species with limited germplasm, knowledge, funding, or community resources, development of alternative germplasm or genomic resources is impractical. A robust, inexpensive, and rapid method for researchers to access the gene space of recalcitrant plant genomes is that of transcriptome sequencing, termed RNA-sequencing or RNA-seq (Wang, Gerstein, & Snyder, 2009).

*De novo* sequencing and assembly of transcriptomes have advanced substantially since the first publication of expressed sequence tags in 1993, which were single-pass reads of cDNA clones (Adams, Soares, Kerlavage, Fields, & Venter, 1993). Because of improvements in sequencing technology from Sanger platforms to high-throughput massively parallel sequencing platforms such as Roche/454 (Margulies et al., 2005), Illumina (Bentley et al., 2008), and SoLiD$^{TM}$ (McKernan et al., 2009), coupled with the development of efficient, robust transcript assembly algorithms (Schulz et al., 2012; Robertson et al., 2010), researchers can now readily access the breadth, depth, and expression abundances of a transcriptome from any tissue(s) of any organism with relative ease and low cost. For medicinal plant species, often there is limited knowledge of the underlying genome with respect to size, ploidy, heterozygosity, and repetitive sequence content. Further, medicinal plant species are often from taxonomic groups lacking a quality reference genome from even a single species, further limiting estimations of genome structure and content. Finally, medicinal plant research communities are limited in size and typically are highly focused on biochemical pathways and/or on downstream pharmacological properties in animal and human systems. Thus, medicinal plant species that lack the appropriate resources for robust genome sequencing are prime targets for *de novo* transcriptome sequencing and assembly which can be coupled expression abundance estimates to reveal critical genes involved in biochemical pathways of interest. This chapter describes sample selection and preparation, sequencing recommendations, assembly methods, and annotation methods for *de novo* assembly and annotation of plant transcriptomes.

## 2. PLANT MATERIAL SELECTION AND QUALITY

### 2.1. Germplasm selection

To optimize assembly and data interpretation, RNA should be obtained from a single, established genotype if at all possible. If a population is to be sampled, the initial *de novo* assembly of the transcriptome should be made from sequences from a single individual if at all possible, which, in most cases, will not interfere with transcript abundance estimates from other individuals within the population.

## 2.2. Plant health

The extracted RNA and transcript sequences from infected plant material will reflect the extent of bacterial, fungal, and insect contamination as computational approaches are limited in their ability to remove contaminating transcripts from the final assembly. Thus, any plant material to be used for RNA isolation should be free of microbial, insect, and other contaminating organisms. A special emphasis should be placed on growing plants in pasteurized or sterile soil media to reduce the contamination of RNA with fungal and oomycete pathogens. Plants should be visually inspected for insects and chemical control efforts utilized to eliminate or minimize infestation of aerial plant tissues by insects.

## 2.3. Tissue selection

1. For a robust *de novo* assembly, 4–5 diverse tissues that provide a broad representation of plant tissues and the underlying transcriptome should be selected for sequencing (e.g., roots, leaves, flowers, callus, fruit). These are referred to as the core tissues/libraries.
2. The core tissues should be augmented with multiple tissues/organs or treatments that reflect a range of conditions in which the metabolite of interest is present at variable levels (including absent) to provide contrasting conditions for coexpression analyses with transcripts and metabolites.

## 3. ISOLATION OF RNA

1. Excise tissue samples of interest, weigh, snap-freeze in liquid nitrogen, and place at $-80\ ^{\circ}$C until RNA extraction. Once excised, tissue should be handled rapidly and placed into liquid nitrogen to limit degradation of the RNA.
2. Grind 1–5 gm of the plant tissue into a fine powder in liquid nitrogen using mortars and pestles precooled with liquid nitrogen. Place in a sterile plastic 50-ml tube cooled with liquid nitrogen and return to the $-80\ ^{\circ}$C freezer until processing further. Only one sample should be processed at a time, and every precaution should be taken to keep samples from thawing.

3. Total RNA can be routinely isolated using the Qiagen RNeasy Plant RNA isolation kit (Catalog # 74903, Qiagen, Valencia, CA). However, for recalcitrant tissues (tissues yielding low levels of RNA or yielding RNA contaminated by carbohydrates or phenolics as evident by aberrant spectrophotometric ratios [ABS 260 nm/280 nm $<1.7$ or $>1.85$; ABS $300/260 > 0.1$]), the Spectrum Plant Total RNA isolation kit (Catalog # STRN10) from Sigma (St. Louis, MO) is recommended. When the Spectrum isolation kit is used, volumes of reagents should be according to the manufacturer's recommendations and include all the precautions noted below.

4. Efforts should be taken to minimize nucleases at all steps of the procedure by wearing gloves and using RNAase-free plastic ware. Mortars and pestles should be thoroughly washed, sprayed with RNAse ZAP (Catalog # AM9780, Ambion/Life Technologies, Carlsbad, CA), incubated 10 min at room temperature, rewashed, and then dried at room temperature or in a drying oven; if not used immediately, mortars and pestles should be stored wrapped in aluminum foil.

5. For RNA extraction using the Qiagen RNAeasy Plant RNA kit, retrieve a plant sample from the $-80\,^\circ$C freezer in a liquid nitrogen bath and weigh up to 100 mg (maximum) into a prechilled weigh boat. Transfer the sample to a prechilled 2.0-ml microcentrifuge tube, immediately add 450 µl of RLT buffer containing 4.5 µl of β-mercaptoethanol, and vortex vigorously for 1 min according to the manufacturer's recommendations. Place the sample in a preheated $56\,^\circ$C heat block for 3 min to aid tissue disruption. Centrifuge for 3 min at $>10,000 \times g$, transfer the lysate to a QIAshredder spin column, and centrifuge for 2 min at $>10,000 \times g$. Add one-half volume of 100% ethanol to the filtrate, mix by inverting, transfer to an RNeasy spin column, and then centrifuge at $>10,000 \times g$ for 15 s. Discard the filtrate and add 80 µl of DNase I solution directly to the RNeasy membrane column. Prepare the DNase I solution by gently mixing 10 µl of DNase I (RNase-free DNase, Catalog # 79254, Qiagen, Valencia, CA) with 70 µl of RDD buffer provided in the Qiagen RNeasy RNA isolation kit. Incubate the column at room temperature for 15 min before adding 350 µl of RW1 buffer (also provided in the kit), centrifuge for 15 s at $>10,000 \times g$, and discard the filtrate. Wash the column twice with 500 µl of the RPE buffer provided in the kit, and centrifuge for 15 s at $>10,000 \times g$. Finally, elute the RNA from the column by adding 50 µl of RNase-free water, incubating for 10 min at room temperature, then centrifuging at

$>10,000 \times g$ for 1 min. A second water elution should be performed to qualify the completeness of the first elution as determined by spectrophotometric assessment.

6. RNA yields should be calculated from absorbance readings at 260 nm using a conventional or Nanodrop (Thermo Scientific, Wilmington, DE) spectrophotometer ($Abs_{260}$ $1.0 = 44$ µg/ml). Quality should be assessed by the absorbance ratio 260/280 ($\geq 1.7$) and 300/260 ($\leq 0.1$) and evaluation by electrophoretic separation with an Agilent 2100 BioAnalyzer (RIN values $\geq 8.0$) or equivalent.

## 4. GENERATION OF NEXT-GENERATION WHOLE TRANSCRIPTOME SEQUENCES

In this step, two types of libraries will be constructed and sequenced. The core library will contain RNA from multiple tissues, sequenced in the paired end (PE) mode, and used to generate the initial *de novo* assembly. Additional libraries need to be constructed from single tissues/treatments and sequenced in the single end (SE) mode for transcript abundance estimates.

### 4.1. Removal of contaminating DNA in the RNA samples

Prior to library production, contaminating DNA should be removed from the RNA samples.

1. The TURBO DNA-free kit (Catalog # AM1907, Ambion/Life Technologies, Carlsbad, CA) can be used to remove DNA from 12–15 µg of RNA in a 50-µl sample.
2. RNA should be checked again for quantity and quality using the same methods as for RNA isolation described above.

### 4.2. Construction of cDNA libraries for next-generation sequencing

1. Pooling RNA from multiple samples.

Typically, RNAs from as divergent tissues as possible (e.g., root, leaf, flower, stem, callus, fruit, or seedlings) will provide the broadest representation of the transcriptome. For optimal assembly of a reference transcriptome, PE sequences greatly enhance the quality of the assembly. For the reference transcriptome, individual libraries should be made from the core tissues. Alternatively, RNA from multiple core tissues can be pooled and sequenced as a single library. For pooling, equal quantities of TURBO DNA-free RNA isolated from multiple tissues, from a single individual plant if possible, should be pooled to provide at least 20 μg of total RNA for library production. The advantage of keeping the core tissue libraries separate is that the reads from each library can also be used for expression abundance estimates, whereas once the RNA samples are mixed, expression abundances are uninformative.

For noncore tissues, libraries should be made from single tissues/treatments and sequenced from a single end.

**2.** cDNA library construction.

cDNA libraries for sequencing on the Illumina platform should be made using the TruSeq RNA Sample Preparation kit (e.g., Catalog # RS-122-2001 (Set A) or RS-122-2002 (Set B), Illumina, San Diego, CA).

## 4.3. Sequencing of cDNA libraries

**1.** Sequence the core cDNA library(s) using either the Illumina or SoLiD platforms. PE reads ($>30$ $M$ each library, $\geq 100$ bp) from $\sim 300$ to 400-bp fragments should be generated.

**2.** Sequence each of the single-tissue libraries using either the Illumina or the SoLiD platforms. SE reads ($>20$ $M$ each library, 36–50 bp) from $\sim 300$ to 400-bp fragments should be generated.

## 5. ASSEMBLY OF A REFERENCE TRANSCRIPTOME

In this step, the PE reads from the core library(ies) will be used to generate a *de novo* reference assembly of the transcriptome; SE reads from individual tissues will then be mapped to the initial assembly to identify novel reads in the single libraries, and a final, annotated assembly will be generated.

## 5.1. Quality assessment of sequences and cleaning of reads

NGS transcriptome assemblers such as Velvet/Oases (Schulz, Zerbino, Vingron, & Birney, 2012; Zerbino & Birney, 2008) are not quality aware. Therefore, short reads need to be preprocessed and checked for quality before assembling. The FASTX toolkit (Blankenberg et al., 2010; http://hannonlab.cshl.edu/fastx_toolkit/) is a collection of command-line programs designed for FASTA or FASTQ short-read sequence files. To use these tools, first check the FASTX toolkit usage information (http://hannonlab.cshl.edu/fastx_toolkit/) and view the basic help page by typing –h for each one of the programs. The preprocessing pipeline includes assessment of the quality of reads, removal of adapters (optional), trimming, and filtering of low-quality sequences.

1. To measure the quality of your short reads, run the `fastx_quality_stats` program by typing the following commands in a terminal window:

```
/path/to/fastx-toolkit/bin/fastx_quality_stats –i
/path/to/myFASTQfile –o myStatsOutputFile.txt
```

The output is a tabular file containing the following information for each nucleotide in the cycle (ALL/A/C/G/T/N):

count = Number of bases found in this column.

min = Lowest quality score value found in this column.

max = Highest quality score value found in this column.

sum = Sum of quality score values for this column.

mean = Mean quality score value for this column.

Q1 = First quartile quality score.

med = Median quality score.

Q3 = Third quartile quality score.

IQR = Interquartile range (Q3–Q1).

lW = "Left-whisker" value (for box plots).

rW = "Right-whisker" value (for box plots).

2. To visualize the quality score in a box plot, type the following command in a terminal window:

```
/path/to/fastx-toolkit/bin/fastq_quality_boxplot_graph.sh –i
MyStatsOutputFile.txt –o MyQualityBoxplot.jpg
```

*Tip*: To visualize the image, use imaging software.

Two sample sequencing runs are shown in Fig. 7.1. Based on base quality scores (Ewing & Green, 1998; Ewing, Hillier, Wendl, & Green, 1998), the quality for each base is shown on the Y axis. A minimum quality of 20 is recommended. It is more likely to observe homogeneous quality in "short" reads (36–50 bp) than in longer reads (>50 bp) due to a decrease in the quality of the sequence in each sequencing cycle.

3. (Optional) To remove low-quality bases from the 3′ end, use `fastx_trimmer` from the FASTX toolkit. The key parameters are the –i flag to provide the input file, the –o flag to provide the output file, the –t flag to define the number of bases trimming from the 3′ end, and the –m flag to define the minimum read length to retain. By using –v, it will print a short summary of the process. The following example is based on the box plot of 120-bp reads in Fig. 7.1.

```
/path/to/fastx-toolkit/bin/fastx_trimmer –v –i
/path/to/MySequenceFile.fastq –o MyTrimmedSequence.fastq –t 40 –m 75
```

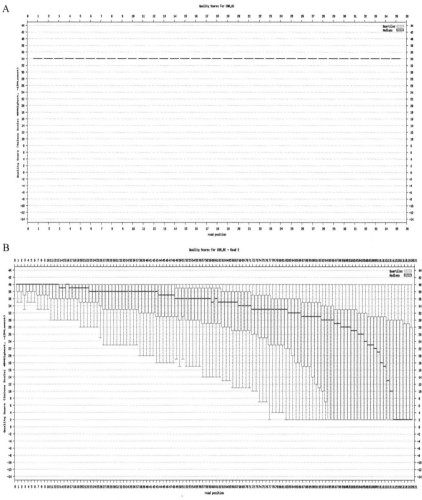

**Figure 7.1** The quality of base calls as a function of cycle length. The base quality is plotted on the *y*-axis and the cycle number on the *x*-axis. (A) Plot showing 36-bp reads with a quality score of 34 for each base as there is no variation in the quality of the reads. (B) Plot showing 120-bp reads with a high variability in quality (range of 40 to 2). (For color version of this figure, the reader is referred to the online version of this chapter.)

**4.** (Optional) In some cases, the library has been barcoded or has adapters present. If so, the fastx_clipper program from the FASTX toolkit can be used to remove the barcodes or adapters. The key parameters are the –a flag that specifies the adapter sequence, the -l flag which specifies the minimum length to retain after clipping the adapter, and the –i and –o flags that specify the input and output file, respectively. Using the –v flag will display a summary of the process.

*Tip*: If the sequences were not trimmed, use the raw sequence file in this step.

```
/path/to/fastx-toolkit/bin/fastx_clipper –v –a ATGC –l 36 –i
myTrimmedSequence.fastq –o myClippedSequence.fastq
```

5. To remove low-complexity sequences, run the `fastx_artifacts_filter` program from the FASTX toolkit by typing the following commands in the terminal window:

    *Tip*: If the sequences were not trimmed or clipped, use the raw sequence file in this step.

```
/path/to/fastx-toolkit/bin/fastx_artifacts_filter –v –i
myClippedSequence.fastq –o myFilteredSequences.fastq
```

6. To remove the low-quality sequences from the sequence files, run the `fastq_quality_trimmer` from the FASTX toolkit. The key parameters are the –t flag that specifies the minimum quality to retain (a minimum quality of 20 is recommended), the –l flag that specifies the minimum read length to retain, and the –i and –o flags to specify the input and output files. Note that the input file to run this tool is the output file from the previous step.

```
/path/to/fastx-toolkit/bin/fastq_quality_trimmer –v –t 20 –l 75 –i
MyFilteredSequences.fastq –o MyHighQualitySequences.fastq
```

7. To convert FASTQ to FASTA type, type the following command in the terminal window:

```
/path/to/fastx-toolkit/bin/fastq_to_fasta –v –i
MyHighQualitySequences.fastq –o MyHighQualitySequences.fasta
```

    *Tip*: Some FASTX programs use FASTQ format only, for instance, `fastq_quality_trimmer`. Tools named as "fastx" can use either FASTA or FASTQ format.

    *Tip*: For PE reads, the cleaning pipeline needs to be applied to each sequence in the pair.

    *Tip*: The estimated time to run each FASTX command will be between 5 and 20 min depending of the file size.

## 5.2. *De novo* assembly of reads

The assembly algorithms for NGS are based on the mathematical concept of a graph, which is a set of vertices or nodes that can be connected by edges (MacLean, Jones, & Studholme, 2009). The Velvet assembler (Zerbino & Birney, 2008) uses a refinement of this approach called the de Brujin graph in which the edges are *k*-mers. In a de Brujin graph, all of the reads are broken into $k$-mers, and the path represents a series of overlapping $k$-mers that overlap by a length of $k-1$ (MacLean et al., 2009; Zerbino & Birney, 2008). For *de novo* transcriptome assembly, Velvet uses the Oases module (Schulz et al., 2012). It takes the preliminary assembly produced by Velvet and clusters the contigs into small groups called loci. Then, if information is available, it will use the read sequence and the pairing information to infer transcript isoforms (http://www.ebi.ac.uk/~zerbino/oases/). Velvet runs `velveth` and `velvetg`. Velveth takes the sequences and produces a hash table; it produces two files, `Sequences` and `Roadmaps`. Velvetg is the core of Velvet where the de Brujin graph is built. To use the Velvet/Oases package, first check the manual (http://www.ebi.ac.uk/~zerbino/velvet/).

1. A *de novo* assembly of PE read libraries is performed to create the reference transcriptome. To assemble PE reads with Velvet, merge the PE sequences into a single file.

   a. After preprocessing and cleaning, the PE sequences from the pairs will be uneven, that is, one end may be retained and the other trimmed and eliminated. Therefore, it will be necessary to run `select_paired.pl` script. This perl script is part of the Velvet package and takes sequences in the FASTA format (convert sequences, if needed, by typing command line in step 7 of Section 5.1). To run this program, type the following commands in the terminal window:

   ```
   /path/to/Velvet/contrib/select_paired/select_paired.pl
   MyHighQualitySequence_read_1.fasta mySortedForwardReads.fasta
   MyHighQualitySequence_read_2.fasta mySortedReverseReads.fasta
   mySingletonsReads.fasta
   ```

   b. Prepare the sequences by merging the forward and reverse reads files into a single file using the perl programs included in the velvet package: `shuffleSequences_fasta.pl` (or `shuffleSequences_fastq.pl`). An example using a FASTA format is provided:

   ```
   /path/to/velvet/shuffleSequences_fasta.pl
   mySortedForwardReads.fasta mySortedReverseReads.fasta
   myMergedReads.fasta
   ```

   c. To assemble using Velvet and Oases, it is necessary to run three different commands: velveth, velvetg, and oases. The main arguments for velveth are the `-fasta` flag that specifies the file format, the `-shortPaired` flag that specifies the merged PE short-reads file, and the `-short` flag that specifies the SE short-read file. Velvetg uses the `-read_trkg` flag necessary to run oases. The key arguments for oases are `-min_trans` flag that specifies the minimum contig length to retain (by default is 100) and the `–ins_lenght` that specifies the insert size between the pairs of the PE library. The following example uses a standard 31 *k*-mer length:

```
/path/to/velvet/velveth myOutputDirectory 31 –fasta
–shortPaired myMergedReads.fasta –short mySingletonsReads.fasta
/path/to/velvet/velvetg myOutputDirectory –read_trkg yes
/path/to/oases/oases myOutputDirectory –min_trans_lgth 250
–ins_lenght 200
```

*Tip*: Be aware that the insert size is different for each PE library.

*Tip*: A *k*-mer of 31 is a standard *k*-mer length. It is recommended to determine the optimal *k*-mer by performing several assemblies with different *k*-mers. The hash length *k* must follow these three rules (Zerbino, 2008):

- It must be an odd number, to avoid palindromes. If you put in an even number, Velvet will just deincrement it and proceed.
- It must be below or equal to the MAXKMERHASH length, which is a compilation parameter.
- It is less than the read length; otherwise you will not observe any overlap between the reads.

*Tip*: Velveth produces two output files: Roadmaps and Sequences; velvetg produces five output files: contig.fa, stats.txt, PreGraph, Graph2, LastGraph, and Log; oases produces three output files: contig-ordering, splicing-events, and transcripts.fa

*Tip*: The assembly may take 1–3 h to run depending on the file size. The Velvet/Oases algorithms require large amounts of RAM memory, and it is recommended that you have at least 60 GB of RAM.

## 5.3. Quality of the *de novo* assembly

To measure the quality of the assembly, examine the size of the assembled transcripts by calculating the N50 length, average contig size, and transcript size distribution; an N50 length around 1 kb is recommended. The completeness of the assembly is obtained by calculating the percentage of reads that map back to the assembly (described in Section 5.4) and assigning functional annotation (described in Section 5.6). Typical or good results are indicated by 65–80% of reads mapping to the reference transcriptome; less than 50% mapped reads may indicate an incomplete reference assembly and less than 20% mapped reads may indicate sample contamination

1. The N50 is defined as the length of the smallest contig such that at least 50% of the bases can be found in a contig of at least this length. Execute the code in Supplementary file 1 (https://www.elsevierdirect.com/companions/9780124046344) by typing:

```
/path/to/getN50.pl –f transcripts.fa
```

## 5.4. Mapping of reads from single-tissue libraries

1. For downstream analyses, a reference set of transcripts needs to be identified from the assembly. During the assembly process, Oases will use read pair and sequence overlap information and generate isoforms which represent true alternative isoforms, alleles, close paralogs, close homologs, or close homeologs (chromosomes in an allopolyploid) depending on the extent of sequence similarity among these transcripts. For most contigs, only a single isoform is generated, and this can be used to construct the reference transcriptome. However, for those contigs with multiple isoforms, as a matter of ease and to maximally reflect the transcript, the longest transcript of the contig is retained as the representative transcript. The representative transcripts are then stitched together into a single, artificial pseudomolecule for downstream read mapping. To create the pseudomolecule, use the `pseudomolecule.pl` code (Supplementary file 2, https://www.elsevierdirect.com/companions/9780124046344). It will generate three output files: a FASTA file containing the pseudomolecule, a gff3 file, and a gtf3 containing the representative transcripts. Execute the code in Supplementary file 2 by typing:

```
/path/to/pseudomolecule.pl –f transcripts.fa –s species –o
myReferenceSequence
```

   *Tip*: View the basic help page by typing –h for each one of the programs above.

2. After preprocessing the reads from the tissue/condition-specific SE libraries (Section 5.1), map the reads to the reference pseudomolecule using the TopHat software (Trapnell, Pachter, & Salzberg, 2009), a quality-aware short read aligner for RNA-seq data which is built upon the ultrafast short-read mapping program Bowtie (Langmead, Trapnell, Pop, & Salzberg, 2009). The aligner uses the quality of the sequences to perform the alignment; therefore, it is highly recommended to use the FASTQ format for the SE reads. To map the SE reads to the reference created in the previous step, perform the following steps:

**a.** Create an index of the reference sequence by typing the following command in the terminal window:

```
/path/to/bowtie/bowtie-build  -f myReferenceSequence.fasta
myReferenceSequence
```

**b.** Map the SE reads to the reference pseudomolecule. The main arguments to run TopHat are the –o flag that specifies the name of the output directory, the –solexa1.3-quals flag that specifies the quality scores are encoded in Phred-scale base-64 for FASTQ files from Illumina GA version 1.3 or later, and the –G flag that specifies the gtf or gff3 file containing the gene models (representative transcripts).

```
/path/to/tophat/tophat –o myOutputDirectory –solexa1.3-quals
–G myReferenceSequence.gtf /path/to/myReferenceSequence /path/
to/myCleanedSeReads.fastq
```

*Tip*: The Bowtie and SamTools programs (Li et al., 2009) are necessary to run TopHat. To install and use these programs, check the manuals (http://tophat.cbcb.umd.edu/manual.html; http://bowtie-bio.sourceforge.net/manual.shtml; http://samtools.sourceforge.net/samtools.shtml).

*Tip*: The results of the alignment are stored in a BAM file that is a binary file. The SamTools program is used to manipulate this file. The latest version of TopHat (1.4.0) reports the unmapped reads in the output directory as unmapped_left.fq.z; these reads will be used to improve the assembly.

## 5.5. Improved *de novo* assembly

**1.** To improve the reference transcriptome, assemble the unmapped SE reads from the previous step (unmapped_left.fq.z) with the PE reads from above (Section 5.2). To run Velvet and Oases, use the following commands:

```
/path/to/velvet/velveth myOutputDirectory 31 –fasta –shortPaired
myMergedReads.fasta  –short mySingletons.fasta UnmappedReads.
fasta
/path/to/velvet/velvetg myOutputDirectory –read_trkg yes
/path/to/oases/oases myOutputDirectory –min_trans_lgth 250
–ins_lenght 200
```

*Tip*: Before using the unmapped SE reads for the second assembly, convert from FASTQ format to FASTA (if needed) by typing into the command line the commands in step 7 of Section 5.1.

*Tip*: Select random SE reads to use in the assembly if there are too many sequences present, and you are memory limited. You can use the "awk" command for this matter by typing:

```
awk 'NR < n' mySingletons.fasta (or UnmappedReads.fasta) >
myRandomSEreads.fasta
```
"n" = number of lines to print out.

## 5.6. Quality assessment and functional annotation of the final assembly

1. After assembling SE and PE reads and constructing the final assembly, filter out low-complexity sequences and sequences with gaps equal to or larger than two-thirds of the sequence length by executing the code in the Supplementary file 3 (https://www.elsevierdirect.com/companions/9780124046344) by typing:

```
/path/to/lowcomplexity.pl –f /path/to/transcripts.fa –o
myCleanedAssembly.fasta
```

   *Tip*: The script above requires Bioperl modules. To download and install Bioperl, check the manual (http://www.bioperl.org/wiki/Main_Page).

2. Measure the quality of the assembly as listed in Section 5.3.

3. To annotate the sequences for function, use BLAST (Basic Local Alignment Tool; Altschul et al., 1997) and HMMPFAM (Hidden Markov Models—Pfam; Punta et al., 2012).

It is recommended that different directories for each one of these programs be used by typing the following commands in the terminal window:

```
mkdir myBlastWorkDirectory
mkdir myPfamWorkDirectory
```

   *Tip*: To annotate using BLAST, any collection of publicly available annotated sequences can be used as reference. The Arabidopsis proteome (http://www.arabidopsis.org/) and/or Uniref100 (http://www.uniprot.org/help/uniref; Suzek, Huang, McGarvey, Mazumder, & Wu, 2007) are recommended.

   *Tip*: To download, install, and use BLAST, check the manual (http://www.ncbi.nlm.nih.gov/books/NBK1762/).

a. To execute BLAST, it is necessary to create a database of your reference sequences (Arabidopsis proteome, Uniref100, etc.). To create a database, use the `makeblastdb` command. The main parameters are the `–in` flag that specifies the input file in FASTA format, the `–dbtype` flag that specifies the molecule type (`nucl/prot`), and the `–parse_seqid` flag that enables retrieval of sequences based upon sequence identifiers. Create the database by typing the following commands in the terminal window (the

Arabidopsis proteome is used as an example). Make sure that you are working in the BLAST working directory:

```
/path/to/blast/makeblastdb –in arabidopsis_proteome.fasta
–parse_seqid –dbtype prot
```

b. Following the example above, run BLAST using the `blastx` search application that will translate the query sequences. The main parameters are the `–db` flag that specifies the database name, the `–query` flag that specifies the name of the sequence file, the `–out` flag that specifies the name of the file to write the application output, the `–evalue` flag that specifies the expectation value threshold for saving hits, the `–num_alignments` flag that specifies the number of alignments to show in the BLAST output, and the `–num_descriptions` flag that specifies the number of one-line descriptions shown in the BLAST output. To run blastx, use the following command line:

```
/path/to/blast/blastx –db arabidopsis_proteome.fasta –query
myCleanedAssembly.fasta –out myOutputBlast.txt –evalue 10e-10
–num_descriptions 20 –num_alignments 20
```

c. Pfam families are divided into two categories: Pfam-A entries consist of high-quality manually curated families and Pfam-B families, which are automatically generated from the ProDom database and represented by a single alignment (Finn et al., 2010). To annotate the assembled sequences using the Pfam domains (Punta et al., 2012), it is necessary to download the database from

```
ftp://ftp.sanger.ac.uk/pub/databases/Pfam/current_release/.
```

*Tip*: To download, install, and use HMM3, check the manual (http://hmmer.janelia.org/). The following command line is compatible with HMM3 (Finn, Clements, & Eddy, 2011) and Pfam 24 or later (ftp://ftp.sanger.ac.uk/pub/databases/Pfam/releases/).

d. To search protein sequences against the Pfam HMM library, type the following command in the terminal window (make sure you are working in the working Pfam directory):

```
/path/to/PfamScan/pfam_scan.pl –fasta myCleanedAssembly.fasta
–dir.
```

*Tip*: Running these programs will take hours to several days depending on the database size and available RAM.

## 6. ASSESSMENT OF EXPRESSION ABUNDANCES

In this step, expression abundances will be determined using the reads from the single-tissue libraries.

## 6.1. Mapping of reads to final *de novo* assembly

1. Create a pseudomolecule defining the representative transcripts as shown in step 1 of Section 5.4
2. Map the SE reads to the pseudomolecule of the final *de novo* assembly as shown in step 2 of Section 5.4

## 6.2. Quantification of the transcript abundances

1. To estimate the transcript abundances in the transcriptome, use the Cufflinks program (Trapnell et al., 2010). Cufflinks accepts aligned RNA-Seq reads and assembles the alignments into a parsimonious set of transcripts, and then estimates the relative abundances of these transcripts based on how many reads support each one. Properly normalized, the RNA-Seq fragment counts can be used as measures of relative abundance of transcripts. Cufflinks measures transcript abundances in Fragments Per Kilobase of exon model per million fragments mapped (FPKM). In this case, the transcripts represent the exon models. Cufflinks uses the following parameters: the –G flag that specifies the gtf or gff3 file containing the gene models and the –o flag that specifies the name of the output directory. The last argument that Cufflinks will take is the alignment file in BAM format.

   ```
   /path/to/cufflinks/cufflinks –G myReferenceSequence.gtf –o
   myOutputDir accepted_hits.bam
   ```

   *Tip*: To download, install, and use Cufflinks, check the manual (http://cufflinks.cbcb.umd.edu/index.html).

   *Tip*: The estimated time to run Cufflinks is between 5 and 30 min for each BAM file.

   *Tip*: The genes.expr file contains the transcript abundances, and the expression values are reported in FPKM.

## 6.3. Quality assessment of the expression matrix

1. To visualize this file in the terminal window, use one of the following commands:

   ```
   less genes.expr
   more genes.expr
   ```

2. The genes.expr file is a tabular file that can be easily exported to Excel. If expression values were estimated for several SE libraries, it is easier to visualize them in a table as shown in Table 7.1.

Table 7.1 Example of a matrix table containing FPKM values derived from several RNA-Seq experiments

| Transcript ID | Flower | Mature seed | Primary tap root | Sterile seedling |
|---|---|---|---|---|
| Aba_locus_10000_iso_5_len_2002_ver_2 | 53.3573 | 16.8643 | 60.4732 | 88.7859 |
| Aba_locus_10001_iso_1_len_768_ver_2 | 1.28434 | 0.991732 | 17.2949 | 0 |
| Aba_locus_10003_iso_4_len_1024_ver_2 | 4.76948 | 0 | 3.62686 | 27.7211 |
| Aba_locus_10004_iso_3_len_1395_ver_2 | 148.113 | 38.1735 | 83.8227 | 61.5356 |
| Aba_locus_100067_iso_1_len_323_ver_2 | 0 | 0 | 2.55721 | 2.97049 |
| Aba_locus_10007_iso_1_len_1475_ver_2 | 12.3071 | 11.9107 | 8.68283 | 8.25411 |
| Aba_locus_10008_iso_8_len_2119_ver_2 | 71.3359 | 55.5804 | 104.715 | 69.1209 |

*Tip*: Within a single library, if the FPKM values for most of the genes are zero, it could be an indicator that the library is contaminated and should be investigated further.

## 7. SUMMARY

Robust, inexpensive methods for transcriptome sequencing and assembly are available. Meaningful approximations of gene function can be made using open-source software that includes well-supported biochemical function annotations. When coupled to determination of expression abundances across a developmental or treatment series, inferences in biological processes can be made. Current limitations of *de novo* transcriptome sequencing include lack of resolution of isoforms, alleles, paralogs, homologs, and homeologs within a specific transcript assembly, presence of partial transcripts and chimeras, and lack of full representation of all transcripts. However, for the majority of transcripts in many medicinal plant species, access to a transcriptome assembly and expression abundances is a rapid, efficient, inexpensive method to access genes relevant to biochemical pathways of interest.

## 8. USEFUL LINKS

1. FASTX Toolkit
      http://hannonlab.cshl.edu/fastx_toolkit/
2. Velvet/Oases Package
      http://www.ebi.ac.uk/~zerbino/oases/
      http://www.ebi.ac.uk/~zerbino/velvet/
3. Bowtie, Tophat, & Cufflinks
      http://bowtie-bio.sourceforge.net/manual.shtml
      http://tophat.cbcb.umd.edu/manual.html
      http://cufflinks.cbcb.umd.edu/index.html
4. Samtools
      http://samtools.sourceforge.net/samtools.shtml
5. Protein Sequences & Protein Domain Databases
      http://www.arabidopsis.org/
      http://www.uniprot.org/help/uniref
      ftp://ftp.sanger.ac.uk/pub/databases/Pfam/current_release/
6. BLAST & HMMER
      http://www.ncbi.nlm.nih.gov/books/NBK1762/
      http://hmmer.janelia.org/

## ACKNOWLEDGMENTS

Funding for medicinal plant transcriptome work was provided by a grant to J. C., D. D. P., and C. R. B. from the National Institute of General Medical Sciences (1RC2GM092521) and from the Michigan State University GREEEN program to C. R. B. (GR11-181).

## REFERENCES

Adams, M. D., Soares, M. B., Kerlavage, A. R., Fields, C., & Venter, J. C. (1993). Rapid cDNA sequencing (expressed sequence tags) from a directionally cloned human infant brain cDNA library. *Nature Genetics, 4*, 373–380.

Altschul, S. F., Madden, T. L., Schaffer, A. A., Zhang, J., Zhang, Z., Miller, W., et al. (1997). Gapped BLAST and PSI-BLAST: A new generation of protein database search programs. *Nucleic Acids Research, 25*, 3389–3402.

Arumuganathan, K., & Earle, E. (1991). Nuclear DNA content of some important plant species. *Plant Molecular Biology Reporter, 9*, 208–218.

Bentley, D. R., Balasubramanian, S., Swerdlow, H. P., Smith, G. P., Milton, J., Brown, C. G., et al. (2008). Accurate whole human genome sequencing using reversible terminator chemistry. *Nature, 456*, 53–59.

Blankenberg, D., Gordon, A., Von Kuster, G., Coraor, N., Taylor, J., & Nekrutenko, A. (2010). Manipulation of FASTQ data with Galaxy. *Bioinformatics, 26*, 1783–1785.

Ewing, B., & Green, P. (1998). Base-calling of automated sequencer traces using phred. II. Error probabilities. *Genome Research, 8*, 186–194.

Ewing, B., Hillier, L., Wendl, M. C., & Green, P. (1998). Base-calling of automated sequencer traces using phred. I. Accuracy assessment. *Genome Research, 8*, 175–185.

Finn, R. D., Mistry, J., Tate, J., Coggill, P., Heger, A., Pollington, J. E., et al. (2010). The Pfam protein families database. *Nucleic Acids Research, 38*, D211–D222.

Jaillon, O., Aury, J. M., Noel, B., Policriti, A., Clepet, C., Casagrande, A., et al. (2007). The grapevine genome sequence suggests ancestral hexaploidization in major angiosperm phyla. *Nature, 449*, 463–467.

Langmead, B., Trapnell, C., Pop, M., & Salzberg, S. L. (2009). Ultrafast and memory-efficient alignment of short DNA sequences to the human genome. *Genome Biology, 10*, R25.

Li, H., Handsaker, B., Wysoker, A., Fennell, T., Ruan, J., Homer, N., et al. (2009). The Sequence Alignment/Map format and SAMtools. *Bioinformatics, 25*, 2078–2079.

MacLean, D., Jones, J. D., & Studholme, D. J. (2009). Application of 'next-generation' sequencing technologies to microbial genetics. *Nature Reviews. Microbiology, 7*, 287–296.

Margulies, M., Egholm, M., Altman, W. E., Attiya, S., Bader, J. S., Bemben, L. A., et al. (2005). Genome sequencing in microfabricated high-density picolitre reactors. *Nature, 437*, 376–380.

McKernan, K. J., Peckham, H. E., Costa, G. L., McLaughlin, S. F., Fu, Y., Tsung, E. F., et al. (2009). Sequence and structural variation in a human genome uncovered by short-read, massively parallel ligation sequencing using two-base encoding. *Genome Research, 19*, 1527–1541.

Paux, E., Roger, D., Badaeva, E., Gay, G., Bernard, M., Sourdille, P., et al. (2006). Characterizing the composition and evolution of homoeologous genomes in hexaploid wheat through BAC-end sequencing on chromosome 3B. *The Plant Journal, 48*, 463–474.

Paux, E., Sourdille, P., Salse, J., Saintenac, C., Choulet, F., Leroy, P., et al. (2008). A physical map of the 1-gigabase bread wheat chromosome 3B. *Science, 322*, 101–104.

Punta, M., Coggill, P. C., Eberhardt, R. Y., Mistry, J., Tate, J., Boursnell, C., et al. (2012). The Pfam protein families database. *Nucleic Acids Research, 40*, D290–D301.

Robertson, G., Schein, J., Chiu, R., Corbett, R., Field, M., Jackman, S. D., et al. (2010). De novo assembly and analysis of RNA-seq data. *Nature Methods*, 7, 909–912.

Schulz, M. H., Zerbino, D. R., Vingron, M., & Birney, E. (2012). Oases: Robust de novo RNA-seq assembly across the dynamic range of expression levels. *Bioinformatics*, 28, 1086–1092.

Shendure, J., & Ji, H. (2008). Next-generation DNA sequencing. *Nature Biotechnology*, 26, 1135–1145.

Shulaev, V., Sargent, D. J., Crowhurst, R. N., Mockler, T. C., Folkerts, O., Delcher, A. L., et al. (2011). The genome of woodland strawberry (*Fragaria vesca*). *Nature Genetics*, 43, 109–116.

Suzek, B. E., Huang, H., McGarvey, P., Mazumder, R., & Wu, C. H. (2007). UniRef: Comprehensive and non-redundant UniProt reference clusters. *Bioinformatics*, 23, 1282–1288.

The Potato Genome Sequencing Consortium, (2011). Genome sequence and analysis of the tuber crop potato. *Nature*, 475, 189–195.

Trapnell, C., Pachter, L., & Salzberg, S. L. (2009). TopHat: Discovering splice junctions with RNA-Seq. *Bioinformatics*, 25, 1105–1111.

Trapnell, C., Williams, B. A., Pertea, G., Mortazavi, A., Kwan, G., van Baren, M. J., et al. (2010). Transcript assembly and quantification by RNA-Seq reveals unannotated transcripts and isoform switching during cell differentiation. *Nature Biotechnology*, 28, 511–515.

Wang, Z., Gerstein, M., & Snyder, M. (2009). RNA-Seq: A revolutionary tool for transcriptomics. *Nature Reviews. Genetics*, 10, 57–63.

Zerbino, D. (2008). *Velvet Manual*. (http://www.ebi.ac.uk/zerbino/velvet).

Zerbino, D. R., & Birney, E. (2008). Velvet: Algorithms for de novo short read assembly using de Bruijn graphs. *Genome Research*, 18, 821–829.

CHAPTER EIGHT

# Phylogenetic Approaches to Natural Product Structure Prediction

## Nadine Ziemert, Paul R. Jensen

Center for Marine Biotechnology and Biomedicine, Scripps Institution of Oceanography, University of California San Diego, La Jolla, California, USA

## Contents

## Abstract

Phylogenetics is the study of the evolutionary relatedness among groups of organisms. Molecular phylogenetics uses sequence data to infer these relationships for both organisms and the genes they maintain. With the large amount of publicly available sequence data, phylogenetic inference has become increasingly important in all fields of biology. In the case of natural product research, phylogenetic relationships are proving to be highly informative in terms of delineating the architecture and function of the genes involved in secondary metabolite biosynthesis. Polyketide synthases and nonribosomal peptide synthetases provide model examples in which individual domain phylogenies display different predictive capacities, resolving features ranging from substrate specificity to structural motifs associated with the final metabolic product. This chapter provides examples in

*Methods in Enzymology*, Volume 517
ISSN 0076-6879
http://dx.doi.org/10.1016/B978-0-12-404634-4.00008-5

which phylogeny has proven effective in terms of predicting functional or structural aspects of secondary metabolism. The basics of how to build a reliable phylogenetic tree are explained along with information about programs and tools that can be used for this purpose. Furthermore, it introduces the Natural Product Domain Seeker, a recently developed Web tool that employs phylogenetic logic to classify ketosynthase and condensation domains based on established enzyme architecture and biochemical function.

# 1. INTRODUCTION

## 1.1. A short introduction to phylogeny

All life on earth is united by a shared evolutionary history. Phylogenetics is the study of that history based on the principles of common ancestry and descent. In the premolecular age, organismal phylogenies were generally created based on morphological character states. With the advent of DNA sequencing, molecular phylogenetics has become the standard for inferring evolutionary relationships. In general, molecular methods are considered far superior since the actions of evolution are ultimately reflected in genetic sequences. The analysis of DNA and protein sequences also provides unprecedented opportunities to infer gene phylogenies, which in many cases may not be congruent with the phylogenies of the organisms in which the genes reside. These incongruences can be due to different rates of gene evolution and, more dramatically, to the process of horizontal gene transfer (HGT), which is now widely recognized as a major force driving bacterial evolution (Ochman, Lerat, & Daubin, 2005).

With the enormous advances being made in next generation sequencing technologies, the analysis of DNA and amino acid sequence data, loosely defined as bioinformatics, has become increasingly important in all fields of biology (Mak, 2010). In natural product research, bioinformatic tools have been developed for a variety of applications including the *in silico* analysis of secondary metabolite biosynthetic gene clusters and the small molecules they produce. Online tools such as the nonribosomal peptide synthetases (NRPS)/polyketide synthases (PKS) database (Yadav, Gokhale, & Mohanty, 2009), NP searcher (Li, Ung, Zajkowski, Garneau-Tsodikova, & Sherman, 2009), and antiSMASH (Medema et al., 2011) have made biosynthetic gene analysis highly accessible. Many of these tools have been reviewed (Bachmann & Ravel, 2009) and will not be discussed in detail here. In general, they are based on the identification of DNA and amino acid sequence similarities and the assumption that these similarities imply similar

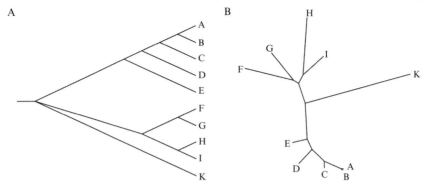

**Figure 8.1** Phylogenetic trees. Molecular phylogenetic analyses are usually displayed in the form of trees. Examples include (A) a rooted rectangular tree or (B) an unrooted radial tree. Both maximum likelihood trees were generated using MEGA.

function. An additional approach is to put sequences into an evolutionary context using phylogenetic methods. The advantage of this approach is that similar sequences can have a diversity of functions that can be resolved based on evolutionary relationships (Eisen, 1998).

Phylogenetic analyses are usually displayed graphically in so-called phylogenetic trees, where each branch of the tree represents one organism or gene (Fig. 8.1). Contemporary phylogenetic concepts were first developed in the 1960s and 1970s (O'Malley & Koonin, 2011) and, with the introduction of DNA sequence data, revolutionized our understanding of microbial evolution and systematics (Woese, 1987). In natural products chemistry, "species trees" based on phylogenetic markers have mainly been used to provide a more accurate identification of the source organism and, in some cases, to draw correlations between taxonomy and secondary metabolite production (Engene et al., 2011; Jensen, 2010; Larsen, Smedsgaard, Nielsen, Hansen, & Frisvad, 2005). During the past decade, the applications of molecular phylogeny have grown exponentially. Phylogeny is now routinely used to improve functional predictions, and "phylogenomics" has been adopted to trace the history of functional change (Eisen, 1998; Eisen & Fraser, 2003). The increased use of phylogenetics in natural product research has provided remarkable new insight into the evolution of the extraordinarily large and complex genes and gene pathways responsible for secondary metabolite biosynthesis.

This chapter provides a short overview of the applications of phylogenetics in natural product research. The aims are to demonstrate the tremendous predictive powers of these methods in terms of identifying common biosynthetic capabilities and new biosynthetic paradigms. It is not intended to be a comprehensive review of phylogenetic methods or theory as provided elsewhere(Salemi & Vandamme, 2003; Schmitt & Barker, 2009). Instead, the goals are to provide a brief introduction on how to build and interpret a reliable phylogenetic tree. This is followed by a discussion of select bioinformatic tools with a focus on the Natural Product Domain Seeker (NaPDoS), which can be used to classify biosynthetic genes based on their phylogenetic relationships.

## 1.2. The biosynthetic logic of secondary metabolism

Two of the most common enzyme families associated with natural product biosynthesis are PKSs and NRPSs. These genes are responsible for the biosynthesis of the majority of bioactive microbial metabolites identified today. Polyketide and nonribosomal peptide biosynthetic pathways are multienzyme complexes that sequentially construct natural products in an assembly line process from carboxylic acid and amino acid building blocks, respectively (Hertweck, 2009; Marahiel, Stachelhaus, & Mootz, 1997). They consist of multiple domains that are responsible for the activation, thiolation (T), condensation (C), and modification of the individual monomers that are incorporated into the final product. In certain PKS classes and most NRPSs, these domains occur in multimodular architectures, resulting in single genes that can exceed 40 kb, making them among the largest bacterial genes known. The evolutionary history of these domains and modules can be highly complex, revealing rapid rates of evolution through recombination, gene duplication, and HGT (Jenke-Kodama & Dittmann, 2005; Jenke-Kodama, Sandmann, Muller, & Dittmann, 2005).

### 1.2.1 Polyketide synthases

Polyketides are polymers of acetate and other simple carboxylic acids. Despite the simplicity of these building blocks, they display remarkable levels of structural diversity due to the combinatorial nature of the assembly line process and frequent postassembly modifications (Fischbach & Walsh, 2006). Many well-known antibiotics including erythromycin and tetracycline are polyketides, as are the dinoflagellate polyethers, which are among the largest secondary metabolites known (Kellmann, Stuken, Orr, Svendsen, & Jakobsen, 2010). PKSs are highly diverse and widespread having been detected in bacteria, fungi, plants, and various eukaryotic genomes, however they are best known as bacterial secondary metabolites. Their sporadic taxonomic distributions and known propensity for HGT makes their evolutionary histories especially interesting (Jenke-Kodama et al., 2005).

PKS genes are generally too large and complex for meaningful phylogenetic analysis; however, individual domain phylogenies are remarkably informative. While T domains are generally too short for analysis, the elongation or ketosynthase (KS) domains have proven highly predictive of pathway associations and enzyme architecture (Jenke-Kodama et al., 2005; Moffitt & Neilan, 2003; Nguyen et al., 2008; Ridley, Lee, & Khosla, 2008). On the other hand, the substrate activating or acyltransferase (AT) domains clade based on substrate specificity and can be used to predict the incorporation of malonyl- or methylmalonyl-CoA into the growing polyketide chain. Of the three optional reductive domains, ketoreductase phylogeny can be used to predict the stereochemistry of the resulting hydroxyl group (Jenke-Kodama, Börner, & Dittmann, 2006). Finally, the phylogeny of thioesterase domains, which cleave the polyketide product from the carrier protein, can be used to predict if this product will be linear or cyclic. KS domains are the most conserved and form an essential part of each PKS gene cluster. These domains have been used to fingerprint PKS genes from individual strains (Edlund, Loesgen, Fenical, & Jensen, 2011) and environmental DNA (Wawrik et al., 2007). KS phylogeny has even been used to predict secondary metabolite diversity (Foerstner, Doerks, Creevey, Doerks, & Bork, 2008; Metsa-Ketela et al., 1999), structures (Freel, Nam, Fenical, & Jensen, 2011; Gontang, Gaudencio, Fenical, & Jensen, 2010), and the evolutionary processes that generate new structural diversity (Freel et al., 2011)

PKS genes are broadly divided into three types (PKSI-III) (Shen, 2003). These types are clearly resolved in a KS-based phylogenetic tree (Fig. 8.2) and reveal the close evolutionary history they share with fatty acid synthases (Jenke-Kodama et al., 2005). Type I PKSs are the most diverse and generally encode all catalytic domains on a single protein that acts iteratively or in a modular fashion. Iterative acting type I PKSs in fungi evolved independently from the iterative type I PKSs observed in bacteria (Kroken, Glass, Taylor, Yoder, & Turgeon, 2003) and can be further divided into reductive and nonreductive clades (Yadav et al., 2009). Remarkably, KS phylogeny can be used to identify at least eight well-supported type I PKS clades, each of which represents a distinct enzyme architecture or biochemical function (Ziemert et al., 2012). One of these clades comprises the iterative acting type I PKSs that are responsible for the biosynthesis of enediynes. This is one of the most biologically active classes of natural products yet to be discovered and includes the potent anticancer agent calicheamicin. More detailed

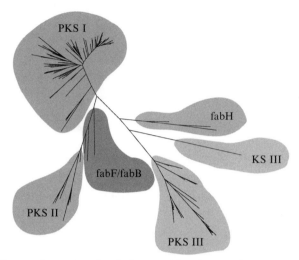

**Figure 8.2** Ketosynthase domain phylogeny. The three PKS types (I–III) are clearly resolved in this KS phylogenetic tree as is their close relationship to various FAS (*fab*) genes. Sequences classified as KS III form a distinct lineage that is involved in the initiation of aromatic polyketide biosynthesis. This maximum likelihood tree was generated with PhyML using a manually curated alignment generated with muscle. (For color version of this figure, the reader is referred to the online version of this chapter.)

phylogenetic analyses of this clade distinguish between genes that produce 9- or 10-membered core enediyne ring structures (Liu et al., 2003).

Type I KS domain phylogeny reveals another well-supported clade comprising modular PKSs that lack integrated AT domains. In these "trans-AT" PKSs, the AT catalytic activity is generally complemented by a freestanding enzyme (Nguyen et al., 2008). Trans-AT PKSs evolved by extensive HGT and maintain considerably greater modular diversity than the cis-AT group. Whereas the close cladding of cis-AT KS domains can be used to predict the production of similar compounds (Gontang et al., 2010), trans-AT KS phylogeny can be used to predict substrate specificity (Nguyen et al., 2008). This was a surprising finding, given that substrate specificity can be inferred from AT domain phylogeny in cis-AT PKSs.

Iterative acting type II PKSs encode each catalytic site on a distinct protein. Typical type II PKSs encode two distinct KS domains: KSα, which catalyzes the condensation reaction, and KSβ, also known as the chain length factor, which determines the number of iterative condensation steps that occur. These type II KS subclasses form two distinct phylogenetic lineages within the larger type II KS clade. Finer level phylogenetic relationships within the KSα clade correspond to the structural classes of the metabolites produced and can be used to distinguish between spore pigments, antracyclines, tetracyclines, and angucyclines, among others (Metsä-Ketelä et al., 2002).

The KS phylogenetic tree reveals another distinct clade that has been called KS III (Fig. 8.2). These sequences are most closely related to FabH proteins, which are involved in fatty acid biosynthesis. KS III domains are involved in the initiation of aromatic polyketide biosynthesis and can incorporate unusual PKS starter units (Xu, Schenk, & Hertweck, 2007). Recently, a new type of KS III domain was discovered in the cervimycin biosynthetic pathway (Bretschneider et al., 2011). Phylogenetic analysis of the CerJ KS domain positioned it between the known KS III domains and ATs, suggesting it may have a new biochemical function. It was subsequently demonstrated that this KS domain is not involved in a typical Claisen condensation reaction but instead transfers activated malonyl units onto a sugar residue. A final KS clade comprises sequences derived from type III PKSs (Moore & Hopke, 2001). This family of multifunctional enzymes includes chalcone and stilbene synthases and was originally considered to be specific for plants before bacterial homologues were discovered (Moore et al., 2002).

### 1.2.2 Nonribosomal peptide synthetases

NRPSs are multimodular enzymes that are structurally similar to type I modular PKSs. Like PKSs, they are generally found clustered in operons that include genes associated with transport, resistance, posttranslational modification, and other functions required for the effective use of the natural product. NRPSs produce small peptides by condensing activated amino acids onto a growing peptide chain that is bound as a thioester to the enzyme (Fischbach & Walsh, 2006). NRPS genes have only been detected in prokaryotes and fungi (Bushley & Turgeon, 2010), where they are responsible for the biosynthesis of a variety of well-known bioactive compounds including penicillin and vancomycin. The minimal domain requirements of a typical NRPS module consist of an adenylation (A) domain that is responsible for substrate specificity and activation, a T domain that covalently tethers the substrate to the enzyme via a thioester bond, and a C domain that catalyzes peptide bond formation between the substrate and the growing peptide chain. Peptide modifying domains responsible for amino acid methylation or cyclization are sometimes observed and create additional structural diversity.

NRPS domain phylogenies are complex and reflect different evolutionary paradigms. C and A domains are the largest and most conserved and have been shown to evolve independently in the same pathway (Fewer et al., 2007).

A major bioinformatic breakthrough was made with the discovery that the amino acids lining the A domain binding pocket are highly predictive of the amino acid substrate that is incorporated into the growing peptide (Stachelhaus, Mootz, & Marahiel, 1999). This discovery made it possible to use bioinformatics to predict the amino acid sequences of NRPS-derived peptides (Challis, Ravel, & Townsend, 2000). While A domain amino acid specificity can be resolved phyogenetically when limited to the eight amino acids in the binding pocket, these signatures are obscured when the larger (180–200 aa) domain sequences are considered (Challis et al., 2000). Nonetheless, A domain phylogenies have proven highly informative in that they tend to reveal clades that correspond to the gene in which they reside, making structural predictions of unknown NRPSs possible when compared to experimentally characterized biosynthetic pathways (Cramer et al., 2006). In addition, A domains associated with the biosynthesis of hybrid PKS/NRPS genes or with the incorporation of N-methylated amino acids and dioxypiperazines can be resolved (Cramer et al., 2006). A recent A domain phylogenomic study in fungi revealed two major clades representing the more ancient mono/bimodular NRPSs and the more recently evolved multimodular NRPSs (Bushley & Turgeon, 2010). These authors suggested that the rapid evolution of multimodular NRPS A domains reflect niche-specific adaptations.

C domain phylogeny clearly reflects the stereochemistry of the amino acids that are added to the growing peptide chain or other functional features of the enzyme. Six characteristic clades have been identified (Rausch, Hoof, Weber, Wohlleben, & Huson, 2007). These include LCL domains, which catalyze peptide bond formation between two L-amino acids, DCL domains, which condense an L-amino acid to a growing peptide ending with a D-amino acid, and starter C domains, which acylate the first amino acid with a β-hydroxy-carboxylic acid. In addition, cyclization domains catalyze both peptide bond formation and the subsequent cyclization of cysteine, serine, or threonine residues; epimerization (E) domains switch the chirality of the last amino acid in the growing peptide generally from L to D; and dual E/C domains catalyze both E and C reactions.

### 1.2.3 More examples

Phylogenetics is yielding useful information in the analysis of virtually all classes of biosynthetic enzymes. For example, terpenes are assembled from five-carbon isoprene units, which can subsequently be attached to other compound classes via prenyltransferases (PTases) (Heide, 2009). PTases have been divided into three major classes: isoprenyl pyrophosphate synthases

(IPPSs), protein PTases, and aromatic PTases. The evolutionary relationships of aromatic PTases containing a PT-barrel fold suggest that fungal and bacterial enzymes share a common ancestry (Bonitz, Alva, Saleh, Lupas, & Heide, 2011). The DMATS family of fungal indole PTases catalyzes, among others, the prenylation of ergot alkaloids. Although no significant sequence similarity is detected between the fungal and bacterial enzymes, a sensitive analytical method called HHsearch (Soding, Biegert, & Lupas, 2005) revealed clear homology (Bonitz et al., 2011). No common ancestry could be detected between membrane-bound PTases, which are mostly involved in primary metabolism, and the soluble PT-barrel containing PTases associated with secondary metabolite biosynthesis, suggesting the evolution of multiple prenylation mechanisms in nature.

Ribosomally produced peptides (RPs) represent a class of secondary metabolites that is receiving increased attention. Bacteriocins represent one well-studied group of RPs and include the microcins of *Escherichia coli* and the lantibiotics of Gram-positive bacteria (Jack & Jung, 2000). Most bacteriocins contain a characteristic N-terminal leader sequence that is cleaved concomitant with translocation across the membrane (Michiels, Dirix, Vanderleyden, & Xi, 2001). Phylogenetic analysis of the peptidase domain revealed a clear distinction between Gram-positive and Gram-negative bacteria and a clade comprising cyanobacteria (Dirix et al., 2004). The colicins represent a family of RPs that can be divided into two different evolutionary lineages based on their mode of action (Riley & Wertz, 2002). Other RPs include the cyanobactins, which are widespread among cyanobacteria (Leikoski, Fewer, & Sivonen, 2009; Schmidt et al., 2005; Sudek, Haygood, Youssef, & Schmidt, 2006; Ziemert et al., 2008). Recent phylogenetic analyses could distinguish four different cyanobactin clades that can be linked to structural features of the compounds. In addition, a phylogenetic model was created to predict the products of orphan RP gene clusters (Donia & Schmidt, 2011). As shown in the examples above, phylogeny is increasingly being used to make effective predictions of secondary metabolite gene function.

## 2. WORKING WITH SEQUENCE DATA

### 2.1. Assembling the dataset

The general steps required for a phylogenetic analysis are outlined in Fig. 8.3. The first step is to find sequences that are homologous to the gene of interest. This is a crucial but often undervalued part of the analysis. Distinguishing homologs, that is, sequences that share a common ancestry, from sequences

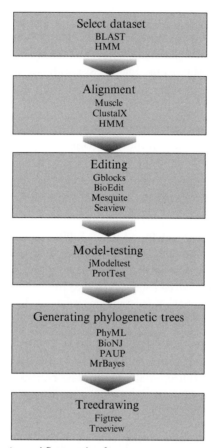

**Figure 8.3** Phylogenetic workflow and software.

that share a random level of similarity is challenging but can be overcome by setting a conservative similarity threshold. Although there are exceptions, sequence similarities should generally be higher than 25% for proteins and 60% for DNA to produce meaningful phylogenies. It can also be helpful to select one or more sequences to function as out-groups. These should be homologous sequences that are more distantly related to all other sequences in the analysis than they are to each other. Out-groups are used to root the tree and help infer the direction of evolution. However, it can be difficult to find an appropriate out-group, as it implies the evolutionary context of the gene of interest is known. Midpoint rooting or creating an unrooted tree makes the selection of an out-group unnecessary.

The easiest way to find sequences of interest is to perform a database search. Public sequence databases such as the National Center for Biotechnology

Information (NCBI) allow keyword and sequence similarity searches. The most popular search algorithm is the Basic Local Alignment Search Tool (BLAST) (Table 8.1), which can accommodate nucleotide or protein sequences and identifies local regions of similarity and their statistical significance

**Table 8.1** Select bioinformatic programs

| Application | Program | Source |
|---|---|---|
| Similarity searches | BLAST | http://blast.ncbi.nlm.nih.gov/Blast.cgi |
| | HMMER | http://hmmer.janelia.org/ |
| Multiple alignments | ClustalX | http://www.clustal.org/ |
| | Muscle | http://www.drive5.com/muscle/ |
| Alignment editing | BioEdit | http://www.mbio.ncsu.edu/bioedit/bioedit.html |
| | Mesquite | http://mesquiteproject.org/mesquite/mesquite.html |
| Model-testing | jMODELTEST | http://darwin.uvigo.es/software/jmodeltest.html |
| | PROTTEST | http://darwin.uvigo.es/software/prottest.html |
| Generating trees | PAUP* | http://paup.csit.fsu.edu/ |
| | Phylip | http://evolution.genetics.washington.edu/phylip.html |
| | BioNJ | http://www.atgc-montpellier.fr/bionj/ |
| | TREE-PUZZLE | http://www.tree-puzzle.de/ |
| | PhyML | http://www.atgc-montpellier.fr/phyml/ |
| | MrBayes | http://mrbayes.sourceforge.net/ |
| Tree display | Figtree | http://tree.bio.ed.ac.uk/software/figtree/ |
| | Treeview | http://taxonomy.zoology.gla.ac.uk/rod/treeview.html |
| Multipurpose | MEGA | http://www.megasoftware.net/ |
| | Seaview | http://pbil.univ-lyon1.fr/software/seaview.html |
| | Geneious | http://www.geneious.com/ |

(Altschul et al., 1997). The BLAST tool provides a table of significant alignment hits that can be downloaded and used for further analyses. For protein searches, BLAST also offers the option of a position-specific iterative BLAST (PSI-BLAST) that creates a more sensitive profile for weak but biologically relevant sequence similarities (Altschul et al., 1997). For a more detailed review of how to use BLAST to find homologous sequences, see Ladunga (2002).

In general, protein sequence similarity searches are more sensitive and therefore preferred to nucleic acid searches. However, if the protein of interest contains different functional domains, as in type I PKS and NRPSs, a comparison of the complete protein may not be very informative relative to independent domain analyses. A slightly different but potentially more sensitive approach to homolog searching is to use a Hidden-Markov-Model (HMM). HMMs are probabilistic models used to create sensitive protein family profiles that can be used to screen genomes or databases for homologous sequences (Finn, Clements, & Eddy, 2011).

## 2.2. Creating alignments

Before running a phylogenetic analysis, it is important to make sure that homologous sites are compared. This is accomplished by creating an alignment in which each sequence is assigned a separate row and homologous positions in different sequences aligned in columns. Generating an accurate alignment is easier when the sequences are similar and becomes more difficult when diverse or repetitive sequences are analyzed. A variety of software packages are available to perform multiple alignments including ClustalX (Thompson, Gibson, Plewniak, Jeanmougin, & Higgins, 1997) and Muscle (Edgar, 2004; Table 8.1). ClustalX belongs to the older class of programs in which sequences are progressively aligned starting with the most similar sequences. Newer programs such as Muscle work iteratively and are considered to be more accurate since they reoptimize the initial alignment. An additional approach uses HMMs (Section 2.1) to generate alignments (Finn et al., 2011). The HMMER software can be used for both sequence alignment and the detection of sequence similarity. As with BLAST searches, amino acid sequence alignments are generally easier to generate and less ambiguous than nucleic acid alignments.

Depending on the alignment program, there might be various options and parameters to select. One important option is to choose a protein or DNA weight matrix. These are empirically based models of how likely it is that one amino acid or nucleotide changes into another. Another option is the gap penalty, which regulates the number of gaps that are allowed in the alignment. It is important to explore these options and test what impact they have on the alignment, as it is the foundation of all analyses that follow.

## 2.3. Editing the alignment

Once an alignment has been created, manual curation is highly recommended to maximize accuracy and avoid artifacts. Truncated sequences should be deleted and longer sequences shortened so that all are equal in length. Highly variable regions can be masked as they may not be phylogenetically informative. Likewise, gaps increase the risk of misalignment, which can result in inaccurate trees. However, variable regions can provide important phylogenetic information so they are best interpreted on a case-by-case basis. For proteins, knowledge about active sites and structure can be taken into consideration when editing the alignment. If it is not clear whether regions are important for the analysis, it is recommended to test different alignments by generating preliminary trees. There are also automated methods such as AltAVisT (Morgenstern, Goel, Sczyrba, & Dress, 2003) and gblocks (Talavera & Castresana, 2007) that perform alignment sensitivity tests and eliminate poorly aligned and divergent regions. However, these methods should never replace a careful manual inspection of the alignment. Free software programs that can be used to edit alignments include Bioedit, Mesquite (Maddison & Maddison, 2009), and Seaview (Gouy, Guindon, & Gascuel, 2010; Table 8.1). These programs can also be used to convert the alignments into the different formats needed for phylogenetic analysis.

## 2.4. Model tests

Generating a phylogenetic tree with maximum likelihood (ML) or Bayesian methods is based on statistical models. Although it is important to test different parameters to determine the robustness of a tree, it is also important to identify which model best fits the data. One popular program is ProtTest (Abascal, Zardoya, & Posada, 2005; Table 8.1), which calculates likelihood values using different models and estimates the optimal parameters for the subsequent tree calculation. Models of nucleotide substitution can be calculated with the jmodeltest software (Posada, 2008). Model testing and alignment editing are not essential steps in generating phylogenetic trees, but both are recommended to improve accuracy and branch support.

## 2.5. Generating phylogenetic trees

It is important to keep in mind that the "true" tree cannot be identified and that phylogeny is a statistical estimation of the most likely evolutionary relationships of the sequences. This is why it is called phylogenetic inference and why it is important to use more than one method to test the consistency of the results and the robustness of the trees. There are four major methods to generate phylogenetic trees from amino acid or nucleotide sequences. The fastest method for most alignments is neighbor-joining (NJ) (Saitou & Nei, 1987). It is the most commonly used distance-based method and calculates a distance matrix for all pairs of sequences in the alignment. It then builds a tree based on the minimum-evolution criterion and the distance relationships. Since it is relatively fast, NJ is widely used to produce preliminary trees and as a starting point for other model-based methods. However, other methods should always be used to support the results. A variety of software packages are available that provide NJ analysis such as BioNJ (Gascuel, 1997), PAUP★ (http://paup.csit.fsu.edu/), MEGA (Tamura et al., 2011), and PHYLIP (Felsenstein, 2005; Table 8.1).

A method that also uses the minimum-evolution criterion is maximum parsimony (MP). However, MP and the following methods introduced here differ fundamentally from distance methods in that they calculate the optimal tree from a diversity of possible trees. Among these "tree searching" methods, Parsimony is known to be the most intuitive because it detects the tree that requires the fewest number of changes in the data. However, with larger datasets, the number of possibilities increases exponentially as do the computational demands. Furthermore, MP often calculates multiple trees that are equally parsimonious and therefore a comparison with other treeing methods is recommended. Commonly used software to generate MP trees is PAUP★, but packages such as PHYLIP (Felsenstein, 2005) can also be used.

Statistical methods based on specific models of evolution include ML and Bayesian analyses. ML calculates the probability of a tree, given certain parameters, and produces a tree with the highest likelihood score. Bayesian approaches are similar in that likelihood scores are calculated; however, instead of looking for one tree, the best set of trees is calculated. Posterior probabilities are then calculated using the Markov chain Monte Carlo algorithm, which results in a collection of trees that can be summarized in a consensus tree (Larget & Simon, 1999). Both treeing methods demand more computational power than MP and distance methods but are thought to be more accurate. ML methods are implemented in the programs TREE-PUZZLE (Schmidt, Strimmer, Vingron, & von Haeseler, 2002) and PhyML (Guindon & Gascuel, 2003), which was especially developed to deal with larger datasets. The Bayesian method is implemented in a program called MrBayes (Huelsenbeck & Ronquist, 2001).

Independent of which phylogenetic method is used, it is important to estimate the reliability of a given tree. The most common statistical method applied to phylogenetic trees is bootstrapping, which randomly samples with replacement the columns in the alignment and generates new trees using the same parameters. Bootstrap values represent the percentage of trees that possess each specific node. While bootstrap values can be statistically biased, values >75% are generally considered significant. Bayesian methods have the advantage that they provide posterior probabilities that identify the percent each clade occurs among all trees sampled. ML methods also offer the option to perform the Approximate Likelihood Ratio Test, which is derived from the likelihood score of each branch that is calculated during the tree search. These methods have the advantage that they require almost no additional computational time.

Finally, trees need to be visualized. Free and easy to use programs include Treeview and Figtree (Table 8.1). The type of tree generated depends on the data and objectives; however, published trees should display a scale bar and some method of statistical support. For more detailed information about phylogenetic analyses, we refer to other sources (Hall, 2007; Salemi & Vandamme, 2003).

## 2.6. Bioinformatic programs

Many of the specialized software packages described above perform one step in the phylogenetic analysis. Alternative packages perform multiple steps and include a user-friendly graphical interface. Free examples include MEGA (Tamura et al., 2011) and Seaview (Gouy et al., 2010), which generate both sequence alignments and phylogenetic trees (Table 8.1). Geneious is a more general bioinformatic software package that includes alignment algorithms and phylogenetic analyses (Drummond et al., 2011); however, it must be purchased. A useful program that allows complete phylogenetic analyses on a Web server is the phylogeny.fr platform (http://www.phylogeny.fr/). This program was developed to produce robust trees even by those with no experience in phylogeny (Dereeper et al., 2008). It also offers useful options for more experienced users and does not require software to be downloaded.

However, this program is not applicable for larger datasets, no model testing is available, and bootstrapping is limited to 100 replicates. Nonetheless, the pipeline is perfect to test datasets, generate preliminary trees, and compare different phylogenetic methods.

## 3. NaPDoS

### 3.1. Scope of NaDoS

NaPDoS (http://napdos.ucsd.edu/) is a recently released, Web-based bioinformatic tool that uses phylogenetic information to predict the class and, in some cases, structure of the natural products produced by bacterial PKS and NRPS genes. It can detect and extract KS and C domains from DNA and amino acid sequences derived from PCR products, genes, whole or draft genomes, and metagenomic data. NaPDoS classifies these sequences based on the phylogenetic relationships of more than 200 KS and C reference sequences. This Web-tool provides a rapid method to evaluate the biosynthetic richness and novelty of individual bacterial strains, communities, or environments and offers a rational guide to identify known secondary metabolites (dereplicate) and facilitate the discovery of new compounds and mechanistic biochemistry.

### 3.2. How NaPDoS works

The bioinformatic pipeline employed by NaPDoS includes HMM and BLAST searches and is constructed to be fast and flexible. NaPDoS first detects and excises KS or C domains from the query sequences. In a second step, these sequences are BLASTed against a reference database of experimentally characterized KS and C domains and assigned an initial classification that defines enzyme architecture or biochemical function. The third step generates a profile alignment by incorporating the sequences into a carefully curated reference alignment generated from all known biochemical classes of KS and C domains. This alignment is then used to create a phylogenetic tree, which is manually interpreted to establish a final classification for each sequence. Trimmed and aligned sequences can then be downloaded for subsequent analysis.

The NaPDoS Web site includes a detailed tutorial. A graphical interface indicates where to upload query sequences and a clickable SEEK button to run the analyses. Advanced options for BLAST and HMM search parameters are available, but the default settings should work well for most data.

A preliminary output table provides the coordinates for any KS or C domains detected and their top BLAST hits, e-values, and alignment lengths. Information describing the biosynthetic pathways associated with the top BLAST hits is provided for comparative purposes.

It should be emphasized that the initial BLAST-based classifications provided by NaPDoS are preliminary and may not reflect the phylogenetic position of the query sequences, especially in cases where the sequence similarities are low. In general, KS or C domains derived from the same pathway often share $\geq 90\%$ amino acid sequence identity. In cases where a query sequence shares this level of identity with a reference sequence, it can be predicted that the pathway from which the sequence was derived has a high probability of producing compounds in the same structural class, as has been demonstrated previously (Edlund et al., 2011; Gontang et al., 2010). For domains that share $< 90\%$ identity to the top NaPDoS match, an NCBI BLAST search is highly recommended as the NaPDoS database is not comprehensive. If the results of this search do not yield a top match that shares $\geq 90\%$ identity, then it should be anticipated that the pathway has not been experimentally characterized and that the product may be new.

To generate final KS or C domain classifications, they should be inserted into the NaPDoS reference alignment along with select NCBI BLAST matches. The trimmed alignment can then be used by NaPDoS to build a ML tree in which the query sequences are indicated in red, or a Newick file, which can be opened with a user-chosen tree-viewing program. This tree can then be manually interpreted to determine the phylogenetic relationship of the query sequences relative to the NaPDoS classification system. In cases where a query sequence does not clade with any of the reference sequences, it may be associated with a new biochemical mechanism or enzyme architecture. For example, a group of C domains that clades outside of the eight functional types identified in NaPDoS appears to be associated with the condensation and subsequent dehydration of serine to dehydroalanine (Ziemert et al., 2012).

## 4. CONCLUSIONS AND FUTURE DIRECTIONS

Increased access to DNA sequencing has created a need for new bioinformatic tools that can be used to analyze and interpret the large volumes of sequence data that are now publically available. In the case of natural products research, these tools are increasingly being used to facilitate the discovery process. Phylogenetics provides a platform to generate biosynthetic hypotheses that can facilitate the discovery of new biochemistry, as functional differences are almost always reflected in phylogenetic trees. Tools such as NaPDoS can help provide a logical guide to the identification of organisms or environments that present the greatest potential for natural product discovery. These predictive capabilities will continue to increase as more biosynthetic pathways are characterized. Sequence-based approaches are providing a new paradigm that promises to increase the rate and efficiency with which natural products are discovered and insight into the evolutionary processes that have generated the extraordinary levels of structural diversity observed among secondary metabolites.

## ACKNOWLEDGMENTS

We thank Sheila Podell, Eric Allen, and Kevin Penn for their contributions to creating the NaPDoS bioinformatic pipeline. P. R. J. acknowledges financial support from the National Institutes of Health (grant 1R01GM086261-O1). N. Z. acknowledges financial support from the German Research Foundation (DFG 1325/1-1).

## REFERENCES

Abascal, F., Zardoya, R., & Posada, D. (2005). ProtTest: Selection of best-fit models of protein evolution. *Bioinformatics*, *21*, 2104–2105.

Altschul, S. F., Madden, T. L., Schaffer, A. A., Zhang, J., Zhang, Z., Miller, W., et al. (1997). Gapped BLAST and PSI-BLAST: A new generation of protein database search programs. *Nucleic Acids Research*, *25*, 3389–3402.

Bachmann, B. O., & Ravel, J. (2009). Methods for in silico prediction of microbial polyketide and nonribosomal peptide biosynthetic pathways from DNA sequence data. *Methods in Enzymology*, *458*, 181–217 chap. 8.

Bonitz, T., Alva, V., Saleh, O., Lupas, A. N., & Heide, L. (2011). Evolutionary relationships of microbial aromatic prenyltransferases. *PLoS One*, *6*, e27336.

Bretschneider, T., Zocher, G., Unger, M., Scherlach, K., Stehle, T., & Hertweck, C. (2011). A ketosynthase homolog uses malonyl units to form esters in cervimycin biosynthesis. *Nature Chemical Biology*, *8*, 154–161.

Bushley, K. E., & Turgeon, B. G. (2010). Phylogenomics reveals subfamilies of fungal nonribosomal peptide synthetases and their evolutionary relationships. *BMC Evolutionary Biology*, *10*, 26.

Challis, G. L., Ravel, J., & Townsend, C. A. (2000). Predictive, structure-based model of amino acid recognition by nonribosomal peptide synthetase adenylation domains. *Chemical Biology*, *7*, 211–224.

Cramer, R. A., Jr., Stajich, J. E., Yamanaka, Y., Dietrich, F. S., Steinbach, W. J., & Perfect, J. R. (2006). Phylogenomic analysis of non-ribosomal peptide synthetases in the genus Aspergillus. *Gene*, *383*, 24–32.

Dereeper, A., Guignon, V., Blanc, G., Audic, S., Buffet, S., Chevenet, F., et al. (2008). Phylogeny.fr: Robust phylogenetic analysis for the non-specialist. *Nucleic Acids Research*, *36*, W465–W469.

Dirix, G., Monsieurs, P., Dombrecht, B., Daniels, R., Marchal, K., Vanderleyden, J., et al. (2004). Peptide signal molecules and bacteriocins in Gram-negative bacteria: A genome-wide in silico screening for peptides containing a double-glycine leader sequence and their cognate transporters. *Peptides, 25*, 1425–1440.

Donia, M. S., & Schmidt, E. W. (2011). Linking chemistry and genetics in the growing cyanobactin natural products family. *Chemical Biology, 18*, 508–519.

Drummond, A., Ashton, B., Buxton, S., Cheung, M., Cooper, A., Duran, C., et al. (2011). *Geneious v5.4*. http://www.geneious.com/.

Edgar, R. C. (2004). MUSCLE: A multiple sequence alignment method with reduced time and space complexity. *BMC Bioinformatics, 5*, 113.

Edlund, A., Loesgen, S., Fenical, W., & Jensen, P. R. (2011). Geographic distribution of secondary metabolite genes in the marine actinomycete *Salinispora arenicola*. *Applied and Environmental Microbiology, 77*, 5916–5925.

Eisen, J. A. (1998). Phylogenomics: Improving functional predictions for uncharacterized genes by evolutionary analysis. *Genome Research, 8*, 163–167.

Eisen, J. A., & Fraser, C. M. (2003). Phylogenomics: Intersection of evolution and genomics. *Science, 300*, 1706–1707.

Engene, N., Choi, H., Esquenazi, E., Byrum, T., Villa, F. A., Cao, Z., et al. (2011). Phylogeny-guided isolation of ethyl tumonoate A from the marine cyanobacterium cf. *Oscillatoria margaritifera*. *Journal of Natural Products, 74*, 1737–1743.

Felsenstein, J. (2005). *PHYLIP (Phylogeny Inference Package) version 3.6*. Seattle: Department of Genome Sciences, University of Washington Distributed by the author.

Fewer, D. P., Rouhiainen, L., Jokela, J., Wahlsten, M., Laakso, K., Wang, H., et al. (2007). Recurrent adenylation domain replacement in the microcystin synthetase gene cluster. *BMC Evolutionary Biology, 7*, 183.

Finn, R. D., Clements, J., & Eddy, S. R. (2011). HMMER web server: Interactive sequence similarity searching. *Nucleic Acids Research, 39*, W29–W37.

Fischbach, M. A., & Walsh, C. T. (2006). Assembly-line enzymology for polyketide and nonribosomal Peptide antibiotics: Logic, machinery, and mechanisms. *Chemical Reviews, 106*, 3468–3496.

Foerstner, K. U., Doerks, T., Creevey, C. J., Doerks, A., & Bork, P. (2008). A computational screen for type I polyketide synthases in metagenomics shotgun data. *PLoS One, 3*, e3515.

Freel, K. C., Nam, S. J., Fenical, W., & Jensen, P. R. (2011). Evolution of secondary metabolite genes in three closely related marine actinomycete species. *Applied and Environmental Microbiology, 77*, 7261–7270.

Gascuel, O. (1997). BIONJ: An improved version of the NJ algorithm based on a simple model of sequence data. *Molecular Biology and Evolution, 14*, 685–695.

Gontang, E. A., Gaudencio, S. P., Fenical, W., & Jensen, P. R. (2010). Sequence-based analysis of secondary-metabolite biosynthesis in marine actinobacteria. *Applied and Environmental Microbiology, 76*, 2487–2499.

Gouy, M., Guindon, S., & Gascuel, O. (2010). SeaView version 4: A multiplatform graphical user interface for sequence alignment and phylogenetic tree building. *Molecular Biology and Evolution, 27*, 221–224.

Guindon, S., & Gascuel, O. (2003). A simple, fast, and accurate algorithm to estimate large phylogenies by maximum likelihood. *Systematic Biology, 52*, 696–704.

Hall, B. (2007). *Phylogenetic trees made easy: A how-to manual*. Sinuaer Associates, Sunderland, MA: 3rd ed.

Heide, L. (2009). Prenyl transfer to aromatic substrates: Genetics and enzymology. *Current Opinion in Chemical Biology, 13*, 171–179.

Hertweck, C. (2009). The biosynthetic logic of polyketide diversity. *Angewandte Chemie (International Ed. in English), 48*, 4688–4716.

Huelsenbeck, J. P., & Ronquist, F. (2001). MRBAYES: Bayesian inference of phylogenetic trees. *Bioinformatics, 17*, 754–755.

Jack, R. W., & Jung, G. (2000). Lantibiotics and microcins: Polypeptides with unusual chemical diversity. *Current Opinion in Chemical Biology*, *4*, 310–317.

Jenke-Kodama, H., Börner, T., & Dittmann, E. (2006). Natural biocombinatorics in the polyketide synthase genes of the actinobacterium Streptomyces avermitilis. *PLoS Computational Biology*, *2*, e132.

Jenke-Kodama, H., & Dittmann, E. (2005). Combinatorial polyketide biosynthesis at higher stage. *Molecular Systems Biology*, *1* (2005), 0025.

Jenke-Kodama, H., Sandmann, A., Muller, R., & Dittmann, E. (2005). Evolutionary implications of bacterial polyketide synthases. *Molecular Biology and Evolution*, *22*, 2027–2039.

Jensen, P. R. (2010). Linking species concepts to natural product discovery in the post-genomic era. *Journal of Industrial Microbiology and Biotechnology*, *37*, 219–224.

Kellmann, R., Stuken, A., Orr, R. J., Svendsen, H. M., & Jakobsen, K. S. (2010). Biosynthesis and molecular genetics of polyketides in marine dinoflagellates. *Marine Drugs*, *8*, 1011–1048.

Kroken, S., Glass, N. L., Taylor, J. W., Yoder, O. C., & Turgeon, B. G. (2003). Phylogenomic analysis of type I polyketide synthase genes in pathogenic and saprobic ascomycetes. *Proceedings of the National Academy of Sciences of the United States of America*, *100*, 15670–15675.

Ladunga, I. (2002). Finding homologs to nucleotide sequences using network BLAST searches. *Current Protocols in Bioinformatics*, *26*, 3.3.1–3.3.26.

Larget, B., & Simon, D. L. (1999). Markov chain Monte Carlo algorithms for the Bayesian analysis of phylogenetic trees. *Molecular Biology and Evolution*, *16*, 750–759.

Larsen, T. O., Smedsgaard, J., Nielsen, K. F., Hansen, M. E., & Frisvad, J. C. (2005). Phenotypic taxonomy and metabolite profiling in microbial drug discovery. *Natural Product Reports*, *22*, 672–695.

Leikoski, N., Fewer, D. P., & Sivonen, K. (2009). Widespread occurrence and lateral transfer of the cyanobactin biosynthesis gene cluster in cyanobacteria. *Applied and Environmental Microbiology*, *75*, 853–857.

Li, M. H., Ung, P. M., Zajkowski, J., Garneau-Tsodikova, S., & Sherman, D. H. (2009). Automated genome mining for natural products. *BMC Bioinformatics*, *10*, 185.

Liu, W., Ahlert, J., Gao, Q., Wendt-Pienkowski, E., Shen, B., & Thorson, J. S. (2003). Rapid PCR amplification of minimal enediyne polyketide synthase cassettes leads to a predictive familial classification model. *Proceedings of the National Academy of Sciences of the United States of America*, *100*, 11959–11963.

Maddison, W. P., & Maddison, D. R. (2009). *Mesquite: A modular system for evolutionary analysis*. Version 2.71. http://mesquiteproject.org. 18 March 2012.

Mak, H. C. (2010). Trends in computational biology-2010. *Nature Biotechnology*, *29*, 45–49.

Marahiel, M. A., Stachelhaus, T., & Mootz, H. D. (1997). Modular peptide synthetases involved in nonribosomal peptide synthesis. *Chemical Reviews*, *97*, 2651–2674.

Medema, M. H., Blin, K., Cimermancic, P., de Jager, V., Zakrzewski, P., Fischbach, M. A., et al. (2011). antiSMASH: Rapid identification, annotation and analysis of secondary metabolite biosynthesis gene clusters in bacterial and fungal genome sequences. *Nucleic Acids Research*, *39*, W339–W346.

Metsa-Ketela, M., Halo, L., Munukka, E., Hakala, J., Mantsala, P., & Ylihonko, K. (2002). Molecular evolution of aromatic polyketides and comparative sequence analysis of polyketide ketosynthase and 16S ribosomal DNA genes from various streptomyces species. *Applied and Environmental Microbiology*, *68*, 4472–4479.

Metsa-Ketela, M., Salo, V., Halo, L., Hautala, A., Hakala, J., Mantsala, P., et al. (1999). An efficient approach for screening minimal PKS genes from Streptomyces. *FEMS Microbiology Letters*, *180*, 1–6.

Michiels, J., Dirix, G., Vanderleyden, J., & Xi, C. (2001). Processing and export of peptide pheromones and bacteriocins in Gram-negative bacteria. *Trends in Microbiology*, *9*, 164–168.

Moffitt, M. C., & Neilan, B. A. (2003). Evolutionary affiliations within the superfamily of ketosynthases reflect complex pathway associations. *Journal of Molecular Evolution*, *56*, 446–457.

Moore, B. S., Hertweck, C., Hopke, J. N., Izumikawa, M., Kalaitzis, J. A., Nilsen, G., et al. (2002). Plant-like biosynthetic pathways in bacteria: From benzoic acid to chalcone. *Journal of Natural Products*, *65*, 1956–1962.

Moore, B. S., & Hopke, J. N. (2001). Discovery of a new bacterial polyketide biosynthetic pathway. *ChembioChem*, *2*, 35–38.

Morgenstern, B., Goel, S., Sczyrba, A., & Dress, A. (2003). AltAVisT: Comparing alternative multiple sequence alignments. *Bioinformatics*, *19*, 425–426.

Nguyen, T., Ishida, K., Jenke-Kodama, H., Dittmann, E., Gurgui, C., Hochmuth, T., et al. (2008). Exploiting the mosaic structure of trans-acyltransferase polyketide synthases for natural product discovery and pathway dissection. *Nature Biotechnology*, *26*, 225–233.

Ochman, H., Lerat, E., & Daubin, V. (2005). Examining bacterial species under the specter of gene transfer and exchange. *Proceedings of the National Academy of Sciences of the United States of America*, *102* (Suppl 1), 6595–6599.

O'Malley, M. A., & Koonin, E. V. (2011). How stands the Tree of Life a century and a half after The Origin? *Biology Direct*, *6*, 32.

Posada, D. (2008). jModelTest: Phylogenetic model averaging. *Molecular Biology and Evolution*, *25*, 1253–1256.

Rausch, C., Hoof, I., Weber, T., Wohlleben, W., & Huson, D. H. (2007). Phylogenetic analysis of condensation domains in NRPS sheds light on their functional evolution. *BMC Evolutionary Biology*, *7*, 78.

Ridley, C. P., Lee, H. Y., & Khosla, C. (2008). Evolution of polyketide synthases in bacteria. *Proceedings of the National Academy of Sciences of the United States of America*, *105*, 4595–4600.

Riley, M. A., & Wertz, J. E. (2002). Bacteriocins: Evolution, ecology, and application. *Annual Review of Microbiology*, *56*, 117–137.

Saitou, N., & Nei, M. (1987). The neighbor-joining method: a new method for reconstructing phylogenetic trees. *Molecular Biology and Evolution*, *4*, 406–425.

Salemi, M., & Vandamme, A.-M. (2003). *The phylogenetic handbook: A practical approach to DNA and protein phylogeny*. Sinuaer Associates, Sunderland, MA: Cambridge University Press.

Schmidt, E. W., Nelson, J. T., Rasko, D. A., Sudek, S., Eisen, J. A., Haygood, M. G., et al. (2005). Patellamide A and C biosynthesis by a microcin-like pathway in Prochloron didemni, the cyanobacterial symbiont of Lissoclinum patella. *Proceedings of the National Academy of Sciences of the United States of America*, *102*, 7315–7320.

Schmidt, H. A., Strimmer, K., Vingron, M., & von Haeseler, A. (2002). TREE-PUZZLE: Maximum likelihood phylogenetic analysis using quartets and parallel computing. *Bioinformatics*, *18*, 502–504.

Schmitt, I., & Barker, F. K. (2009). Phylogenetic methods in natural product research. *Natural Product Reports*, *26*, 1585–1602.

Shen, B. (2003). Polyketide biosynthesis beyond the type I, II and III polyketide synthase paradigms. *Current Opinion in Chemical Biology*, *7*, 285–295.

Soding, J., Biegert, A., & Lupas, A. N. (2005). The HHpred interactive server for protein homology detection and structure prediction. *Nucleic Acids Research*, *33*, W244–W248.

Stachelhaus, T., Mootz, H. D., & Marahiel, M. A. (1999). The specificity-conferring code of adenylation domains in nonribosomal peptide synthetases. *Chemical Biology*, *6*, 493–505.

Sudek, S., Haygood, M. G., Youssef, D. T., & Schmidt, E. W. (2006). Structure of trichamide, a cyclic peptide from the bloom-forming cyanobacterium Trichodesmium erythraeum, predicted from the genome sequence. *Applied and Environmental Microbiology*, *72*, 4382–4387.

Talavera, G., & Castresana, J. (2007). Improvement of phylogenies after removing divergent and ambiguously aligned blocks from protein sequence alignments. *Systematic Biology*, *56*, 564–577.

Tamura, K., Peterson, D., Peterson, N., Stecher, G., Nei, M., & Kumar, S. (2011). MEGA5: Molecular evolutionary genetics analysis using maximum likelihood, evolutionary distance, and maximum parsimony methods. *Molecular Biology and Evolution*, *28*, 2731–2739.

Thompson, J. D., Gibson, T. J., Plewniak, F., Jeanmougin, F., & Higgins, D. G. (1997). The CLUSTAL_X windows interface: Flexible strategies for multiple sequence alignment aided by quality analysis tools. *Nucleic Acids Research*, *25*, 4876–4882.

Wawrik, B., Kutliev, D., Abdivasievna, U. A., Kukor, J. J., Zylstra, G. J., & Kerkhof, L. (2007). Biogeography of actinomycete communities and type II polyketide synthase genes in soils collected in New Jersey and Central Asia. *Applied and Environmental Microbiology*, *73*, 2982–2989.

Woese, C. R. (1987). Bacterial evolution. *Microbiological Reviews*, *51*, 221–271.

Xu, Z., Schenk, A., & Hertweck, C. (2007). Molecular analysis of the benastatin biosynthetic pathway and genetic engineering of altered fatty acid-polyketide hybrids. *Journal of the American Chemical Society*, *129*, 6022–6030.

Yadav, G., Gokhale, R. S., & Mohanty, D. (2009). Towards prediction of metabolic products of polyketide synthases: An in silico analysis. *PLoS Computational Biology*, *5*, e1000351.

Ziemert, N., Ishida, K., Quillardet, P., Bouchier, C., Hertweck, C., & de Marsac, N. T. (2008). Microcyclamide biosynthesis in two strains of Microcystis aeruginosa: From structure to genes and vice versa. *Applied and Environmental Microbiology*, *74*, 1791–1797.

Ziemert, N., Podell, S., Penn, K., Badger, J. H., Allen, E. E., & Jensen, P. R. (2012). The natural product domain seeker NaPDoS: A phylogeny based bioinformatic tool to classify secondary metabolite gene diversity. *PLoS One*, *7(3)*, e34064.

# Heterologous Expression of Pathways

CHAPTER NINE

# Using a Virus-Derived System to Manipulate Plant Natural Product Biosynthetic Pathways

Frank Sainsbury*, Pooja Saxena†, Katrin Geisler‡, Anne Osbourn‡, George P. Lomonossoff†,1

*Département de Phytologie, Pavillon des Services, Université Laval, Québec, QC, Canada
†Department of Biological Chemistry, John Innes Centre, Norwich, United Kingdom
‡Department of Metabolic Biology, John Innes Centre, Norwich, United Kingdom
1Corresponding author: e-mail address: george.lomonossoff@jic.ac.uk

## Contents

## Abstract

A series of vectors (the pEAQ series) based on cowpea mosaic virus has been developed which allows the rapid transient expression of high levels of foreign protein in plants without the need for viral replication. The plasmids are small binary vectors, which are introduced into plant leaves by agroinfiltration. They are modular in design and allow the insertion of multiple coding sequences on the same segment of T-DNA. These properties make the pEAQ vectors particularly suitable for use in situations, such as the investigation and manipulation of metabolic pathways, where the coexpression of multiple proteins within a cell is required.

*Methods in Enzymology*, Volume 517
ISSN 0076-6879
http://dx.doi.org/10.1016/B978-0-12-404634-4.00009-7

## 1. INTRODUCTION

The past 20 years have seen the development of many RNA viruses as vectors for the transient expression of foreign peptides and polypeptides in plants (Porta & Lomonossoff, 2002; Scholthof, Scholthof, & Jackson, 1996). The advantages of using viruses, as opposed to stable genetic transformation, for such expression include the facts that (i) viral genomes are small and therefore easy to manipulate, (ii) infection of plants with modified viruses is much simpler and quicker than the regeneration of stably transformed lines, and (iii) a sequence inserted into a virus vector will be highly amplified during viral replication. Initially, plant virus–based vectors were based on replication-competent full-size virus genomes, with the gene to be expressed being added to the full complement of viral genes. Inevitably, there were a number of disadvantages to this approach: there are size constraints on the sequences which can be inserted while retaining virus viability, the inserted sequence is susceptible to "genetic drift" during virus replication, and there are bio-containment concerns over the use of vectors based on fully competent viruses as these retain their ability to spread in the environment. As a result, in the past decade, attention has turned toward the development of plant virus–based expression systems based on defective versions of viral RNAs, which alleviate some or all of these disadvantages (Cañizares, Liu, Perrin, Tsakiris, & Lomonossoff, 2006; Gleba, Marillonnet, & Klimyuk, 2004). These studies have resulted in the creation of systems in which the ability of the virus to spread both within the plant and in the environment is curtailed and has culminated in the development of a system in which the need for replication to achieve high-level expression has been eliminated.

## 2. DELETED VECTORS BASED ON COWPEA MOSAIC VIRUS

Among the plant viruses which have been developed into an expression system is cowpea mosaic virus (CPMV; Sainsbury, Cañizares, & Lomonossoff, 2010). This virus infects a number of legume species and grows to particularly high titers in its natural host, cowpea (*Vigna unguiculata*); it also infects the commonly used experimental host, *Nicotiana benthamiana*. The genome of CPMV consists of two separately encapsidated positive-strand RNA molecules of 5889 (RNA-1) and 3481 (RNA-2) nucleotides. The RNAs each contain a single open reading frame (ORF) and are expressed through the synthesis and subsequent processing of precursor polyproteins. RNA-1 encodes proteins involved in the replication of viral RNAs and polyprotein processing while RNA-2, which is entirely dependent on the proteins encoded by RNA-1 for its replication, encodes the movement protein and the two coat proteins, large (L) and small (S) that are essential for cell-to-cell movement and systemic spread. The development of CPMV-based expression systems has focused entirely on modifying the sequence of RNA-2; replication functions, when required, are provided by coinoculating the RNA-2 constructs with RNA-1 (Fig. 9.1A).

A particular attraction of CPMV as a virus-based vector is the fact that it is naturally bipartite. This means that two RNA molecules have to be replicated within the same cell, implying that virus exclusion will not occur. Virus exclusion is the phenomenon whereby the presence of a replicating RNA within a cell effectively excludes the replication of a second construct based on the same virus. Thus, systems based on monopartite, replication-competent viruses, such as tobacco mosaic virus, are essentially limited to the production of a single protein within a given cell unless a second, different noncompeting virus is used for the expression of the second protein (Giritch et al., 2006). The utility of CPMV for the coexpression of multiple proteins was first demonstrated using full-length versions of RNA-2 harboring either the yellow or cyan fluorescent protein (YFP or CFP) or the heavy and light chains of an IgG. In each case, the two RNA-2-based constructs were inoculated in conjunction with RNA-1 to provide the replication functions (Sainsbury, Lavoie, D'Aoust, Vezina, & Lomonossoff, 2008). However, although successful coexpression was demonstrated in the inoculated tissue, segregation of the different RNA-2-based constructs occurred on systemic movement, leading to cells expressing either YFP or CFP but rarely both. Since, for practical purposes, this limits the expression of multiple genes to the inoculated tissue, it was rationalized that there was little to be lost in terms of levels of expression if those features necessary for the spread of the virus were removed from RNA-2. This would have the advantage of creating vectors which are unable to spread in the environment and are therefore biocontained.

## 2.1. Replication-competent deleted vectors

An expression system based on a defective form of CPMV RNA-2 (Fig. 9.1B) was created using the observation that the sequences necessary for replication of RNA-2 by the RNA-1-encoded replication complex lie exclusively at the 5$'$ and 3$'$ ends of the RNA (Rohll, Holness,

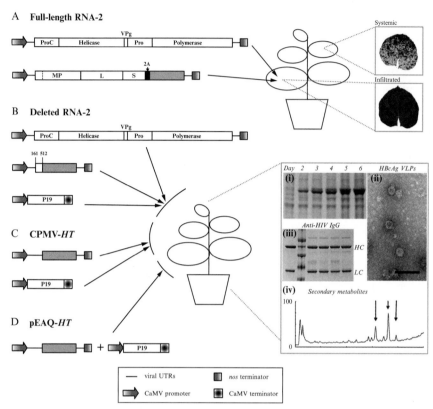

**Figure 9.1** Development of CPMV-based expression vectors. (A) Full-length vectors based on the entire genome of CPMV where RNA-1 provides replication functions for RNA-2 molecules modified to contain the gene of interest (gray) following the FMDV 2A catalytic peptide. Also shown is an example of an expression pattern following infiltration and systemic movement (black tissue). (B) Deleted RNA-2 vectors where the entire coding sequence of RNA-2 is replaced by the gene of interest. RNA-1 provides replication functions and P19 supplies silencing suppressor functions in lieu of the small coat protein. (C) CPMV-*HT* vectors based on the modified UTRs of RNA-2 where P19 provides a silencing suppressor function. (D) Schematic representation of pEAQ-*HT* where all elements for CPMV-*HT* are present on a single T-DNA. (i) Time-course of GFP expression from CPMV-*HT*, (ii) TEM image of crude extracts showing assembled HBcAg particles following expression from CPMV-*HT*, (iii) purified anti-HIV antibody following expression from deleted RNA-2 and CPMV-*HT*, (iv) example high performance liquid chromatography (HPLC) trace representing metabolic engineering enabled by CPMV-based vectors.

Lomonossoff, & Maule, 1993). This allows most of the RNA-2 ORF to be deleted without affecting the ability of RNA-2 to be replicated. However, while the essential 3′-terminal sequences lie exclusively within the 3′-UTR, the essential 5′ sequence extends beyond the first in-frame AUG (position 161) as far as the second in-frame AUG at position 512; deletion or frameshift mutations introduced between the two AUG codons abolish replication, as does elimination of the AUG at 161 (Holness, Lomonossoff, Evans, & Maule, 1989; Rohll et al., 1993). This means that initiation of translation of the foreign gene must be driven by the second in-frame AUG at position 512 if the ability of the construct to be replicated by RNA-1 is to be retained (Cañizares et al., 2006). To create a vector in which the heterologous coding sequence can be precisely fused to AUG 512, site-directed mutagenesis of a full-length copy of RNA-2 in an *Escherichia coli* plasmid was used to introduce a *Bsp*HI site (TCATGA) around AUG 512 and a *Stu*I site (AGGCCT) after UAA 3299, the termination codon for the RNA-2-encoded polyprotein; in addition, two *Bsp*HI sites from the vector backbone were removed (Liu & Lomonossoff, 2002). The resulting vector, termed pM81B-S2NT-1, allows the whole of the RNA-2 ORF downstream of AUG 512 to be excised by digestion with *Bsp*HI and *Stu*I and replaced with any sequence with *Bsp*HI and *Stu*I (blunt)-compatible ends. The use of the *Bsp*HI site is important as it preserves the AUG at 512 and this initiator is used to drive translation of the inserted gene. To express the foreign gene in plants, the pM81B-S2NT-1-derived plasmids are digested with *Asc*I and *Pac*I and the fragment containing the foreign sequence flanked by the CaMV 35S promoter and *nos* terminator, as well as the CPMV RNA-2-derived UTRs, is transferred to similarly digested pBINPLUS and the resulting plasmids are finally introduced by transformation into *Agrobacterium tumefaciens*. For expression, *A. tumefaciens* suspensions harboring the desired RNA-2-based constructs are coinfiltrated into *N. benthamiana* leaves with plasmids encoding a full-length copy of RNA-1 and a suppressor of gene silencing, usually P19 from Tomato bushy stunt virus (TBSV; Fig. 9.1B). This approach has been used successfully to express assembled particles of Hepatitis B core antigen (HBcAg; Mechtcheriakova, Eldarov, Beales, Skryabin, & Lomonossoff, 2008) and to express a fully assembled, functional IgG (Sainsbury, Sack, et al., 2010). Initially, it was believed that replication of the deleted RNA-2 by RNA-1 was essential for high levels of protein expression. However, the presence of a strong suppressor of silencing, such as P19, increases the stability of mRNA to such a degree that amplification of RNA levels by replication results in little further increase in expression levels (Sainsbury, Sack, et al., 2010). As a result, the deleted RNA-2-based vectors (delRNA-2 vectors) described above, though replication competent, can be deployed for expression of proteins in plants in the absence of RNA-1. Details of the construction and method of use of these replication-competent vectors are described in detail in Sainsbury, Liu, and Lomonossoff (2009).

The vectors described above were originally created to express proteins, such as vaccine candidates and antibodies, which would be subsequently purified and characterized. However, Mugford et al. (2009) showed that the vectors could also be used to express active enzymes within *N. benthamiana* leaves with the aim of manipulating plant metabolism and analyzing enzyme function rather than obtaining the protein itself. In these experiments, the expression of a serine carboxypeptidase-like acyltransferase from oat leaves (*Sad 7*, an enzyme from the avenicin biosynthetic pathway; see Chapter 6) was shown to be able to catalyze the transfer of both *N*-methyl anthraniloyl- and benzoyl-groups to des-acyl avenacins when expressed in *N. benthamiana*. These studies paved the way for the use of virus-based vectors for the manipulation of metabolic pathways in plants. The use of replication-competent CPMV RNA-2 based vectors for the expression of active enzymes in plants is described in detail in Chapter 14.

## 2.2. The CPMV-*HT* expression system

A major drawback with the replication-competent RNA-2 system is the need to precisely fuse the sequence to be expressed to AUG 512. This makes the cloning strategy a somewhat cumbersome, two-step procedure (see Sainsbury, Liu, & Lomonossoff, 2009). Further, coinfiltration with a separate construct containing a suppressor of silencing is necessary to obtain high levels of expression. The observation that replication is not essential for high-level protein expression led Sainsbury and Lomonossoff (2008) to examine whether it is possible to simplify the structure of the expression plasmids, particularly around the 5'-UTR, to facilitate the insertion of foreign sequences. These studies showed that elimination of both AUG 161 and an upstream out-of-frame AUG at position 115, though abolishing replication, actually substantially increased expression levels of a variety of proteins. This effect was caused by the modified 5'-UTR rendering the mRNAs "hyper-translatable" and expression systems using them have been termed CPMV-*HT*. The levels of foreign protein produced using the *HT* leaders far exceeded those achieved from replicating full-length or delRNA-2 vectors in transient expression studies. Achieving high-level expression without replication has a number of substantial advantages: the problems associated with genetic drift and virus exclusion are eliminated and there is no obvious limitation on the size or complexity of the sequences that can be expressed. Thus, CPMV-*HT* is ideally suited to the simultaneous expression of multiple proteins within the same cell. A number of pharmaceutically relevant proteins have been expressed to high level using the CPMV-*HT* system (Fig. 9.1C and D). They include approximately 1 g/kg of agroinfiltrated tissue of assembled HBcAg particles (Sainsbury & Lomonossoff, 2008), and up to 0.4 g/kg of the human antihuman immunodeficiency virus IgG, 2G12 (Sainsbury, Sack, et al., 2010). However, in its initial form the CPMV-*HT* expression system still required a two-step cloning procedure and coinfiltration with a suppressor of silencing (Fig. 9.1C).

## 2.3. The pEAQ series of vectors

To refine the CPMV-*HT* expression system, its various components (the CPMV-*HT* expression cassette and the P19 sequence) were placed on the T-DNA region of a binary vector for Agrobacterium-mediated delivery to plant cells (Fig. 9.1D). This was achieved by constructing a new series of binary vectors, the pEAQ series (Table 9.1; Fig. 9.2A). These vectors are less than half the size of the original pBINPLUS plasmid while retaining all the essential components for efficient transient expression and stable transformation (Sainsbury, Thuenemann, & Lomonossoff, 2009). In addition to a variety of T-DNA configurations, each pEAQ vector contains a polylinker of unique restriction sites that, while designed for use with the modular cloning of the CPMV-*HT* cassette as discussed below, may also be used to insert any expression cassette or sequence (Fig. 9.2A). The removal of reading frame dependence, essential for replication but evidently adversely affecting translation, permits the use of a one-step cloning procedure either through the use of a multiple cloning site for restriction enzyme-based cloning or via the GATEWAY® system of recombination-based cloning. Further expanding their usefulness, these expression vectors also contain the appropriate sequences for N- or C-terminal His-tagging of proteins of interest, allowing for straightforward purification from plant tissue (Tables 9.2 and 9.3). The use of such pEAQ-based CPMV-*HT* vectors has so far enabled high-level expression and purification via a His-tag of a number of heterologous proteins, including human gastric lipase (Vardakou, Sainsbury, Rigby, Mulholland, & Lomonossoff, 2012) and a rice chitinase (Miyamoto et al., 2012).

Though expression of a single protein during transient expression is useful in several cases, there are many instances where coexpression of more

**Table 9.1** The pEAQ vector series

| Plasmid name | GenBank accession | Features |
|---|---|---|
| pEAQ-*HT* | GQ497234 | Designed for easy and quick cloning of a gene of interest into the CPMV-*HT* system. Its T-DNA comprises:<br><br>• the CPMV-*HT* expression cassette with a polylinker (Table 9.2) to insert the gene of interest;<br>• the suppressor of gene silencing P19;<br>• neomycin phosphotransferase II (*nptII*) to confer resistance to kanamycin. |
| pEAQexpress | GQ497230 | Designed for cloning multiple CPMV-*HT* expression cassettes in the same vector for transient expression only. Its T-DNA comprises:<br><br>• a multiple cloning site for insertion of multiple expression cassettes digested using enzymes *Pac*I and *Asc*I;<br>• the suppressor of gene silencing P19. |
| pEAQselectK | GQ497231 | Designed for expression from a CPMV-*HT* expression cassette in the absence of a suppressor of silencing. Its T-DNA comprises:<br><br>• a multiple cloning site for insertion of the expression cassettes;<br>• neomycin phosphotransferase II (*nptII*) to confer resistance to kanamycin. |
| pEAQspecialK | GQ497232 | Designed for expression from a CPMV-*HT* expression cassette in the presence of a suppressor of silencing. Its T-DNA comprises:<br><br>• a multiple cloning site for insertion of the expression cassette;<br>• the suppressor of gene silencing P19;<br>• neomycin phosphotransferase II (*nptII*) to confer resistance to kanamycin. |
| pEAQspecialKm | GQ497233 | Best suited for stable expression of proteins in whole plants. Its T-DNA comprises:<br><br>• a multiple cloning site for insertion of the expression cassette; |

**Table 9.1**  The pEAQ vector series—cont'd

| Plasmid name | GenBank accession | Features |
|---|---|---|
| | | • the modified suppressor of gene silencing P19/R43W (Saxena et al., 2011);<br>• neomycin phosphotransferase II (*nptII*) to confer resistance to kanamycin. |
| pEAQ-*HT*-DEST1 | GQ497235 | Designed for easy and quick cloning using the gateway® system to express wild-type protein in plants. Its T-DNA comprises:<br><br>• the CPMV-*HT* expression cassette with attR sites to introduce the gene of interest from the entry clone via recombination;<br>• the suppressor of gene silencing P19;<br>• neomycin phosphotransferase II (*nptII*) that confers resistance to kanamycin. |
| pEAQ-*HT*-DEST2 | GQ497236 | Designed for easy and quick cloning using the gateway® system to express N-terminally His-tagged protein in plants. Its T-DNA comprises:<br><br>• the CPMV-*HT* expression cassette with attR sites to introduce the gene of interest from the entry clone via recombination;<br>• the suppressor of gene silencing P19;<br>• neomycin phosphotransferase II (*nptII*) that confers resistance to kanamycin. |
| pEAQ-*HT*-DEST3 | GQ497237 | Designed for easy and quick cloning using the gateway® system to express C-terminally His-tagged protein in plants. Its T-DNA comprises:<br><br>• the CPMV-*HT* expression cassette with attR sites to introduce the gene of interest from the entry clone via recombination;<br>• the suppressor of gene silencing P19;<br>• Neomycin phosphotransferase II (*nptII*) that confers resistance to kanamycin. |

than one protein is required. The simplest way of achieving this is to coinfiltrate Agrobacterium suspensions each containing a plasmid designed to express a single protein. Although this is a highly flexible approach, to ensure efficient coexpression of constructs within the same cell, each

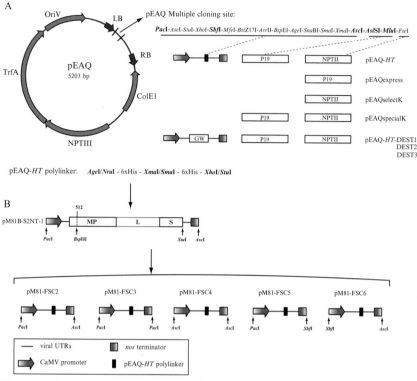

**Figure 9.2** Schematic representation of the pEAQ vector system. The series of vectors enables direct cloning into expression vectors of the assembly of multiple expression cassettes via a modular series of cloning vectors. (A) Representation of the minimal plasmid backbone of the pEAQ series and various configurations of the T-DNA designed for different applications. GW, GATEWAY. (B) Representation of the modular series of cloning vectors each containing the pEAQ-*HT* polylinker and designed for CPMV-*HT* cassette insertion into the pEAQ multiple cloning site.

**Table 9.2** Restriction enzyme pairs to use for restriction enzyme-based cloning into pEAQ-*HT* depending on whether or not a His-tag is required

| | | |
|---|---|---|
| For expression of wild-type protein | *Nru*I or *Age*I at the 5′ end | *Xho*I or *Stu*I at the 3′ end |
| For expression of protein with a His-tag at the N-terminus | *Xma*I or *Sma*I at the 5′ end | *Xho*I or *Stu*I at the 3′ end |
| For expression of protein with a His-tag at the C-terminus | *Nru*I or *Age*I at the 5′ end | *Xma*I or *Sma*I at the 3′ end |

Each position contains the option of using a restriction enzyme that leaves an overhang or a blunt end.

**Table 9.3** Appropriate pEAQ-HT-DEST vector to use depending on whether or not a His-tag is required

| For expression of wild-type protein | Do the LR reaction of your entry clone with pEAQ-*HT*-DEST1 |
| --- | --- |
| For expression of protein with a His-tag at the N-terminus | Do the LR reaction of your entry clone with pEAQ-*HT*-DEST2 |
| For expression of protein with a His-tag at the C-terminus | Do the LR reaction of your entry clone with pEAQ-*HT*-DEST3 |

bacterial suspension has to be present at a high concentration, making the inoculum dense and difficult to infiltrate. Even then, it cannot be guaranteed that every cell will receive all the infiltrated constructs (Montague et al., 2011). This problem can be addressed by expressing multiple CPMV-*HT* expression cassettes from the same T-DNA, which is facilitated by the fact that an essential part of the design of the pEAQ vector series is its modular nature (Fig. 9.2).

The CPMV-*HT* cassette, consisting of the gene of interest flanked by the 35S promoter and the modified RNA-2 5′-UTR on one side and the 3′-UTR and *nos* terminator on the other, is also present on a series of cloning vectors based on pM81B-S2NT-1, named pM81-FSC2 through to pM81-FSC6 (Fig. 9.2B). Each cassette contains the same polylinker found in pEAQ-*HT* (Table 9.2; Fig. 9.2B) replacing the entire coding region of RNA-2 and enabling N- or C-terminal His-tagging of proteins to be expressed. They are flanked by restriction sites that allow for the insertion of multiple cassettes into the pEAQ polylinker (Fig. 9.2). For example, up to five cassettes may be inserted sequentially into the basic transient expression vector, pEAQexpress. Alternatively, two CPMV-*HT* cassettes can be simultaneously inserted through multipart ligations into *Pac*I/*Asc*I-digested pEAQ vectors. This approach has been used to assemble an antibody-expression construct in pEAQspecialK using pM81-FSC3 and pM81-FSC4 containing the heavy and light chain of an IgG molecule, respectively (Frank Sainsbury, Pooja Saxena & George Lomonossoff, unpublished data).

The benefit of expressing genes from the same T-DNA rather than coinfiltrating separate constructs has been illustrated for the production of assembled antibodies, where coexpression of the heavy and light chains in the same cell is required (Sainsbury, Thuenemann, & Lomonossoff, 2009) and for the production of empty (RNA-free) particles of CPMV, which requires the coexpression of the coat protein precursor and the enzyme necessary for its processing (Montague et al., 2011).

The properties of the CPMV-*HT* system make it ideal for use in situations where the simultaneous expression of multiple proteins within the same cell is required. Thus, the pEAQ series of vectors have proved extremely useful not only in situations where large amounts of a protein are required (e.g., Matić, Rinaldi, Masenga, & Noris, 2012; Saunders, Sainsbury, & Lomonossoff, 2009; Vardakou et al., 2012) but also in situations where manipulation of metabolism, through the coexpression of several enzymes, is required. This has been illustrated in experiments in which an oxidosqualene cyclase and a CYP450 from oat were expressed in *N. benthamiana* leaves, either separately or in combination. Expression of the oxidosqualene cyclase alone resulted in the *N. benthamiana* leaves producing β-amyrin (Fig. 9.3), a metabolite not normally produced by this species. Infiltration with a pEAQexpress-based construct containing the sequence of both the oxidosqualene cyclase and the CYP450 resulted in reduction of the signal for β-amyrin and the appearance of a new peak representing a novel compound, the structure of which is currently being examined. Thus, the pEAQ vectors can be used to express metabolic enzymes with the aim of understanding their function, the analysis of intermediates produced by their action or, ultimately, to produce novel compounds in plants.

## 3. EXPRESSION OF ENZYMES IN PLANTS USING pEAQ VECTORS

This section constitutes a practical guide for using the pEAQ vector series to express foreign proteins, including active enzymes in plants.

## 3.1. Materials
### 3.1.1 Media, buffers, and solutions
- Luria–Bertani (LB) medium: 10 g/L bacto-tryptone, 10/L NaCl, and 5 g/L yeast extract, pH 7.0.
- LB agar: as LB with 10 g/L agar added.
- SOC: 20 g/L bacto-tryptone, 5 g/L yeast extract, 0.58 g/L NaCl, 0.19 g/L KCl, 2.03 g/L $MgCl_2$, 2.46 g/L magnesium sulfate 7–hydrate, 3.6 g glucose.
- MMA: 10 m$M$ MES (2-[*N*-morpholino]ethanesulfonic acid; Sigma-Aldrich), pH 5.6, 10 mM $MgCl_2$, 100 μ$M$ acetosyringone (Sigma-Aldrich).
- 1 × TBE: 10.8 g/L Tris–HCl, 5.5 g/L boric acid, 2 m$M$ EDTA.

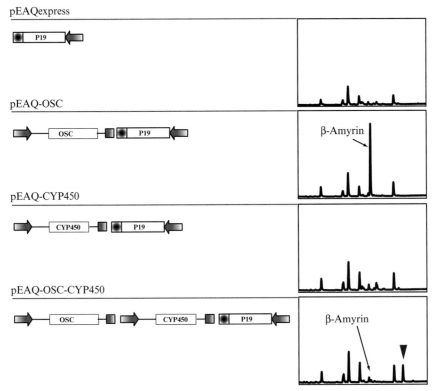

**Figure 9.3** Use of CPMV-*HT*-based vectors for metabolic engineering. *N. benthamiana* leaves were infiltrated with either: the empty vector pEAQexpress, which contains just the P19 silencing suppressor expression cassette, Fig. 9.2; pEAQ-OSC, which contains the sequence of an oat oxidosqualene cyclase (OSC) in a CPMV-*HT* cassette; pEAQ-CYP450, which contains a cytochrome P450 (CYP450) in a CPMV-*HT* cassette; pEAQ-OSC-CYP450, which contains the sequences of both OSC and CYP450 on the same T-DNA. The position of β-amyrin is indicated in the HPLC traces on the right. The arrowhead indicates the presence of a novel compound resulting from the expression of both OSC and CYP450 from a single construct.

### 3.1.2 Bacterial strains

- One Shot[®] TOP10 chemically competent *E. coli* (Invitrogen) is used for propagation of recombinant plasmids.
- One Shot[®] ccdB survival 2T1R chemically competent *E. coli* (Invitrogen) is used for propagation of pEAQ-DEST plasmids before recombination with entry clones.
- *A. tumefaciens* strain LBA4404 (Hoekema, Hirsch, Hooykaas, & Schilperoort, 1983) is used for plant transformations.

### 3.1.3 Plants

N. benthamiana plants are grown in glasshouses maintained at 25 °C with supplemental lighting to provide 16 h of daylight throughout the year. Plants are watered daily.

## 3.2. Creating expression plasmids

### 3.2.1 Choice of expression vector

Using the features described in Table 9.1, choose the most appropriate pEAQ vector for the expression of your gene of interest.

### 3.2.2 Restriction enzyme-based cloning

1. Generate the insert with appropriate restrictions sites at both ends using PCR.
2. Digest both the insert and plasmid pEAQ-*HT* with appropriate restriction enzymes (Table 9.2) by mixing components including the appropriate buffer according to the enzyme manufacturer's recommendation. For expression using pEAQ vectors other than pEAQ-*HT*, the insert should first be cloned in the polylinker of pEAQ-*HT* or one of the FSC cloning vectors (Fig. 9.2B). Then, the entire expression cassette can moved to other pEAQ vectors using the appropriate restriction enzymes.
3. Dephosphorylate linearized vector with alkaline phosphatase following the manufacturer's instructions. This should be done after heat inactivation of the restriction enzymes.
4. Resolve digests on a 1% agarose gel and purify the vector and inserts using QIAQuick gel extraction kit.
5. Combine the vector and insert (in molar ratio 1:3) in ligase buffer with T4 DNA ligase and incubate according to manufacturer's recommendations.
6. Transform competent *E. coli* and plate onto LB agar plates with kanamycin (50 μg/mL) selection.
7. Colonies may be screened by PCR or restriction analysis. Positive clones are grown overnight and plasmids are extracted for sequencing (to confirm insertion) and Agrobacterium transformation.

### 3.2.3 Cloning by GATEWAY recombination

1. Generate the insert with bacteriophage lambda attachment B (*attB*) sites at both ends using PCR.
2. Using BP clonase II, transfer the PCR fragment to a GATEWAY donor vector via directional recombination. The resultant plasmid is the entry clone.

3. Using LR clonase II, transfer the gene of interest from the entry vector to the appropriate pEAQ-*HT*-DEST vector (see Table 9.3).

4. Transform competent *E. coli* and plate onto LB agar plates with kanamycin (50 µg/mL) selection.

5. Colonies may be screened by PCR or restriction analysis. Positive clones are grown overnight, and plasmids are extracted for sequencing (to confirm insertion) and Agrobacterium transformation.

### 3.2.4 Transformation of Agrobacterium

Once the expression construct is generated, it is introduced into *A. tumefaciens* strain LBA4404 or AGL1 using electroporation. After electroporation at 2.5 kV, cells are recovered at 28 °C for 1 h and plated on LB agar containing rifampicin 50 µg/mL (for LBA4404) and kanamycin 50 µg/mL (for carried plasmid). Transformed colonies are visible on plates in 2–3 days.

## 3.3. Agroinfiltration of *N. benthamiana*

### 3.3.1 Preparation of Agrobacterium suspensions

1. Prepare 10–100 mL LB with appropriate antibiotics for the Agrobacterium strain (rifampicin 50 µg/mL for LBA4404) and carried plasmid (kanamycin 50 µg/mL for pEAQ vectors). The volume of the culture depends on the scale of your experiment. Generally, a 10-mL culture is enough to infiltrate 4–5 leaves (ca. 5 g fresh-weight tissue).

2. Inoculate the liquid culture by picking a single colony from a plate. Grow the culture at 28 °C in a shaking incubator until the optical density (OD) at 600 nm is $\geq 2$. Typically, inoculate the culture in the afternoon and grow overnight.

3. Spin cells at $4000 \times g$ for 10 min at room temperature to pellet them and discard the supernatant.

4. Resuspend cells gently in the required volume of MMA to make a solution of final $OD_{600} = 0.4$. For coexpression of two constructs, prepare solutions of individual $OD_{600} = 0.8$ which when mixed 1:1 will result in a final $OD_{600} = 0.4$ for each construct.

5. Leave the solutions at room temperature for 0.5–3 h.

### 3.3.2 Infiltration of leaves with Agrobacterium suspensions

There are two different methods for introducing Agrobacterium suspensions into leaves. The method of choice largely depends on the scale of the experiment.

- Syringe-infiltration (small-scale expression; 1–10 plants)

To infiltrate leaves, nick the leaf surface with a sterile needle or a sterile pipette tip. Aspirate infiltration solution into a sterile 1-mL plastic syringe (take care to avoid bubbles) and, place the syringe over the leaf wound while keeping a finger behind the leaf for support. Gently press the solution into the intercellular space. For best results, 3–4-week-old *N. benthamiana* plants are used and the youngest fully expanded leaves are chosen for infiltration.

• Vacuum-infiltration (large-scale expression; 10–100 plants)

Cover the base of a 3–4-week-old plant such that the soil is retained in the pot during the infiltration procedure. Invert the plant into a beaker containing the Agrobacterium suspension and place the beaker in the center of the vacuum desiccator unit. Ensure that all the leaves are submerged in the Agrobacterium suspension. Close and seal the desiccator and apply vacuum for 60 s at negative pressure of 25 in. Hg (170 mbar). Release the vacuum gently and return the infiltrated plant to the growth room. More than one plant can be infiltrated at the same time depending on the size of the desiccator unit.

## 3.4. Monitoring expression levels in *N. benthamiana* leaves

The precise method for monitoring expression will depend on the nature of the experiment. Harvesting of *N. benthamiana* leaves is typically done at 3–10 days postinfiltration (dpi). Tissue can be processed fresh or snap-frozen and stored at −80 °C. A time-course should be done to assess optimum expression as expression levels might vary depending on the nature of your protein. In the case of green fluorescent protein (GFP), expression is visible *in vivo* under UV illumination. Therefore, it is sensible to include inoculation with a GFP-expressing construct (such as pEAQ-*HT*-GFP) as a control in each experiment. Using the *HT* system, expression of GFP in *N. benthamiana* becomes visible after 2 days, reaches its peak at 5 dpi, and remains at this level until day 7, after which the leaves start showing necrotic symptoms.

## 4. CONCLUSIONS AND PERSPECTIVES

The use of the pEAQ vector series to manipulate plant natural product biosynthetic pathways is currently in its infancy. However, the data obtained to date suggest that this approach is likely to be widely applicable. The ease of use of the vectors, coupled with the speed of expression, means that many possible combinations of enzymes can be rapidly screened. The modular nature of the vectors means that a mix-and-match approach, in which enzymes from different origins are combined, is feasible, raising the possibility of the synthesis of new-to-nature compounds.

Aside from allowing extremely high-level transient protein expression, the pEAQ series has also been adapted for stable genetic transformation. Although the vectors can be used directly to produce transgenic cell cultures (Sun et al., 2011), expression of the wild-type P19 suppressor of silencing is highly detrimental to the regeneration of fertile plants. This can be overcome by deploying versions of the pEAQ series harboring a mutant version of P19 with reduced suppression activity (Saxena et al., 2011). Using this approach, it is possible to produce homozygous plants expressing high levels of the protein(s) of interest. Thus, transient expression can initially be used to rapidly investigate and/or manipulate a biosynthetic pathway followed by stable transformation to produce lines of plants with the appropriate characteristics.

## ACKNOWLEDGMENTS

This work was supported by a Danish FOBI International PhD studentship (K.G.), and the UK Biotechnological and Biological Sciences Research Council (BBSRC) Institute Strategic Programme Grant "Understanding and Exploiting Plant and Microbial Secondary Metabolism" (BB/J004596/1) and the John Innes Foundation (F.S., P.S. G.P.L., A.O.).

## REFERENCES

Cañizares, M. C., Liu, L., Perrin, Y., Tsakiris, E., & Lomonossoff, G. P. (2006). A bipartite system for the constitutive and inducible expression of high levels of foreign proteins in plants. *Plant Biotechnology Journal, 4,* 183–193.

Giritch, A., Marillonnet, S., Engler, C., van Eldik, G., Botterman, J., Klimyuk, V., et al. (2006). Rapid high-yield expression of full-size IgG antibodies in plants coinfected with noncompeting viral vectors. *Proceedings of the National Academy of Sciences of the United States of America, 103,* 14701–14706.

Gleba, Y., Marillonnet, S., & Klimyuk, V. (2004). Engineering viral expression vectors for plants: The 'full virus' and the 'deconstructed virus' strategies. *Current Opinion in Plant Biology, 7,* 182–188.

Hoekema, A., Hirsch, P. R., Hooykaas, P. J. J., & Schilperoort, R. A. (1983). A binary plant vector strategy based on separation of vir- and T-region of the *Agrobacteria tumefaciens* Ti-plasmid. *Nature, 303,* 179–180.

Holness, C. L., Lomonossoff, G. P., Evans, D., & Maule, A. J. (1989). Identification of the initiation codons for translation of cowpea mosaic virus middle component RNA using site directed mutagenesis of an infectious cDNA clone. *Virology, 172,* 311–320.

Liu, L., & Lomonossoff, G. P. (2002). Agroinfection as a rapid method for propagating cowpea mosaic virus-based constructs. *Journal of Virological Methods, 105,* 343–348.

Matić, S., Rinaldi, R., Masenga, V., & Noris, E. (2012). Efficient production of chimeric human papillomavirus 16 L1 protein bearing the M2e influenza epitope in *Nicotiana benthamiana* plants. *BMC Biotechnology, 11,* 106.

Mechtcheriakova, I. A., Eldarov, M. A., Beales, L., Skryabin, K. G., & Lomonossoff, G. P. (2008). Production of hepatitis B virus core particles in plants using a CPMV-based vector. *Voprosy Virusologii, 53,* 15–20.

Miyamoto, K., Shimizu, T., Lin, F., Sainsbury, F., Thuenemann, E., Lomonossoff, G., et al. (2012). Identification of an E-box motif responsible for the expression of jasmonic acid-induced chitinase gene OsChia4a in rice. *Journal of Plant Physiology, 169,* 621–627.

Montague, N. P., Thuenemann, E. C., Saxena, P., Saunders, K., Lenzi, P., & Lomonossoff, G. P. (2011). Recent advances of cowpea mosaic virus-based particle technology. *Human Vaccines*, 7, 383–390.

Mugford, S. T., Qi, X., Bakht, S., Hill, L., Wegel, E., Hughes, R. K., et al. (2009). A serine carboxypeptidase-like acyltransferase is required for synthesis of antimicrobial compounds and disease resistance in oats. *The Plant Cell*, 21, 2473–2484.

Porta, C., & Lomonossoff, G. P. (2002). Viruses as vectors for the expression of foreign sequences in plants. *Biotechnology and Genetic Engineering Reviews*, 19, 245–291.

Rohll, J. B., Holness, C. L., Lomonossoff, G. P., & Maule, A. J. (1993). 3′ Terminal nucleotide sequences important for the accumulation of cowpea mosaic virus M-RNA. *Virology*, 193, 672–679.

Sainsbury, F., Cañizares, M. C., & Lomonossoff, G. P. (2010). Cowpea mosaic virus: The plant virus-based biotechnology workhorse. *Annual Review of Phytopathology*, 48, 437–455.

Sainsbury, F., Lavoie, P.-O., D'Aoust, M.-A., Vezina, L.-P., & Lomonossoff, G. P. (2008). Expression of multiple proteins using full-length and deleted versions of *Cowpea Mosaic Virus* RNA-2. *Plant Biotechnology Journal*, 6, 82–92.

Sainsbury, F., Liu, L., & Lomonossoff, G. P. (2009). Cowpea mosaic virus-based systems for the expression of antigens and antibodies in plants. In L. Faye (Ed.), *Methods in molecular biology*, Vol. 483, (pp. 25–39). Humana Press Inc. New York, USA.

Sainsbury, F., & Lomonossoff, G. P. (2008). Extremely high-level and rapid transient protein production in plants without the use of viral replication. *Plant Physiology*, 148, 1212–1218.

Sainsbury, F., Sack, M., Stadlman, J., Quendler, H., Fischer, R., & Lomonossoff, G. P. (2010). Rapid transient production in plants by replicating and non-replicating vectors yields high quality functional anti-HIV antibody. *PLoS One*, 5, e13976. http://dx.plos.org/10.1371/journal.pone.0013976.

Sainsbury, F., Thuenemann, E. C., & Lomonossoff, G. P. (2009). pEAQ: Versatile expression vectors for easy and quick transient expression of heterologous proteins in plants. *Plant Biotechnology Journal*, 7, 682–693.

Saunders, K., Sainsbury, F., & Lomonossoff, G. P. (2009). Efficient generation of *Cowpea Mosaic Virus* empty virus-like particles by the proteolytic processing of precursors in insect cells and plants. *Virology*, 393, 329–337.

Saxena, P., Hsieh, Y. C., Alvarado, V. Y., Sainsbury, F., Saunders, K., Lomonossoff, G. P., et al. (2011). Improved foreign gene expression in plants using a virus-encoded suppressor of RNA silencing modified to be developmentally harmless. *Plant Biotechnology Journal*, 9, 703–712.

Scholthof, H. B., Scholthof, K. B., & Jackson, A. O. (1996). Plant virus gene vectors for transient expression of foreign proteins in plants. *Annual Review of Phytopathology*, 34, 299–323.

Sun, Q. Y., Ding, L. W., Lomonossoff, G. P., Sun, Y. B., Luo, M., Li, C. Q., et al. (2011). Improved expression and purification of recombinant human serum albumin from transgenic tobacco suspension culture. *Journal of Biotechnology*, 155, 164–172.

Vardakou, M., Sainsbury, F., Rigby, N., Mulholland, F., & Lomonossoff, G. P. (2012). Expression of active recombinant human gastric lipase in *Nicotiana benthamiana* using the CPMV-*HT* transient expression system. *Protein Expression and Purification*, 81, 69–74.

CHAPTER TEN

# DNA Assembler: A Synthetic Biology Tool for Characterizing and Engineering Natural Product Gene Clusters

**Zengyi Shao\*, Huimin Zhao\*,†,‡,1**
\*Department of Chemical and Biomolecular Engineering, University of Illinois at Urbana-Champaign, Urbana, Illinois, USA
†Institute for Genomic Biology, University of Illinois at Urbana-Champaign, Urbana, Illinois, USA
‡Departments of Chemistry, Biochemistry, and Bioengineering, University of Illinois at Urbana-Champaign, Urbana, Illinois, USA
[1]Corresponding author: e-mail address: zhao5@illinois.edu

## Contents

## Abstract

The majority of existing antibacterial and anticancer drugs are natural products or their derivatives. However, the characterization and engineering of these compounds are often hampered by limited ability to manipulate the corresponding biosynthetic pathways. Recently, we developed a genomics-driven, synthetic biology-based method, DNA assembler, for discovery, characterization, and engineering of natural product biosynthetic pathways (Shao, Luo, & Zhao, 2011). By taking advantage of the highly efficient yeast *in vivo* homologous recombination mechanism, this method synthesizes the entire expression vector containing the target biosynthetic pathway and the genetic elements needed for DNA maintenance and replication in individual hosts in a single-

*Methods in Enzymology*, Volume 517
ISSN 0076-6879
http://dx.doi.org/10.1016/B978-0-12-404634-4.00010-3

step manner. In this chapter, we describe the general guidelines for construct design. By using two distinct biosynthetic pathways, we demonstrate that DNA assembler can perform multiple tasks, including heterologous expression, introduction of single or multiple point mutations, scar-less gene deletion, generation of product derivatives, and creation of artificial gene clusters. As such, this method offers unprecedented flexibility and versatility in pathway manipulations.

# 1. INTRODUCTION

Microorganisms and plants produce numerous secondary metabolites or natural products, which are a prolific source of therapeutic agents (Dewick, 2002; Herbert, 1989; Li & Vederas, 2009). For example, it has been estimated that 77% of antibacterial drugs and 78% of anticancer drugs are either natural products or their derivatives (Newman & Cragg, 2007). Recent advances in DNA sequencing technologies and bioinformatics indicate that sequenced genomes and metagenomes may represent a tremendously rich source for discovery of novel pathways involved in natural product biosynthesis (Challis, 2008; Zerikly & Challis, 2009). However, only a tiny fraction of these biosynthetic pathways have been characterized, and discovery and sustainable production of natural products are often hampered by our limited ability to manipulate the corresponding biosynthetic pathways.

Existing strategies for characterizing natural product gene clusters can be broadly classified into two groups including the native host gene inactivation method and the heterologous host gene expression method (Shao, Luo, & Zhao, 2011). The native host-based method involves introducing mutation (s) and deleting gene(s) directly from the chromosome and comparing the metabolite profile of the mutant with that of the wild type; the heterologous host-based method involves moving the gene cluster from the native producer to a secondary host and comparing the metabolite profile of the resulting strain with that of the background strain. However, both approaches suffer a few limitations. For the native host-based method, first, most native producers are not cultivable in the laboratory and many microorganisms grow very slowly and produce minute amount of target compounds. It has been estimated that only 1% of bacteria and 5% of fungi have been cultivated in the laboratory (Bull, Goodfellow, & Slater, 1992; Davies, 1999; Demain, 2006; Leadbetter, 2003). Second, genetic tools have not been developed for the majority of organisms, which makes manipulation of a target biosynthetic pathway inconvenient and inefficient since each host has to be examined individually. On the other hand, the heterologous host-based method suffers from difficulties in the transfer and genetic manipulation of large biosynthetic pathways ($>30$ kb), problems with promoter recognition and gene expression, the limited availability of expression hosts, and low yield (Gross, 2007). Therefore, it is highly desirable to establish novel efficient modification, transfer, and expression technologies.

Recently, we developed the DNA assembler approach to assemble multiple gene expression cassettes into biochemical pathways either on a plasmid or on a chromosome in *Saccharomyces cerevisiae* in a single-step fashion (Shao, Zhao, & Zhao, 2009) and further demonstrated it as a genomics-driven, synthetic biology-based method for discovery, characterization, and engineering of natural product biosynthetic pathways (Shao et al., 2011). This approach remedies several of the key drawbacks associated with the traditional native host-based method and the heterologous host-based method, offering a novel and powerful way to study and engineer natural product biosynthetic clusters.

## 2. GENERAL GUIDELINES FOR CONSTRUCT DESIGN

The plasmid to be assembled is composed of pathway fragments and helper fragments (Fig. 10.1A). The target gene cluster is split into several pathway fragments, each of which has a length of up to 6 kb. Pathway fragments are amplified by polymerase chain reaction (PCR) from the isolated genomic DNA of the native producer if it is cultivable, or chemically synthesized *de novo*. Three helper fragments carrying the genetic elements needed for DNA maintenance and replication in *S. cerevisiae* (assembly host), *Escherichia coli* (DNA enrichment host), and the target heterologous expression host are amplified from the corresponding vectors. Since PCR primers are designed to generate an overlap region between two adjacent fragments, cotransformation of these fragments will allow them to be assembled into a single DNA molecule in *S. cerevisiae* through homologous recombination. The isolated plasmids are subsequently introduced by transformation into *E. coli* for plasmid enrichment and verification, and the correct construct is introduced by transformation into the desired host for heterologous expression of the target biosynthetic pathway. The two example gene clusters in this chapter are both isolated from *Streptomyces* species, so *Streptomyces lividans* is chosen as the heterologous expression host due to its common usage as a heterologous host for expressing a variety of natural product gene clusters isolated from *Streptomyces* (Eliot et al., 2008; Shao et al., 2011;

A

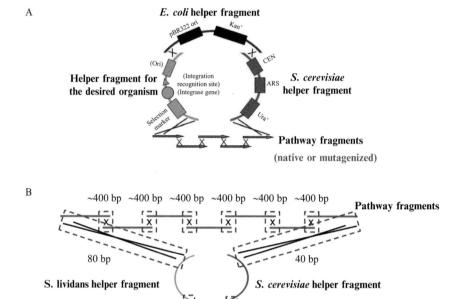

**Figure 10.1** (A) The DNA assembler-based strategy for efficient manipulation of natural product biosynthetic pathways. Various genetic modifications are introduced in the pathway fragments to be assembled. (B) The overlap lengths between adjacent fragments in assembly. *Reproduced by permission of The Royal Society of Chemistry.* (See Color Insert.)

Woodyer et al., 2006). Because the pathway fragments can be readily modified by PCR, various genetic manipulations such as site-directed mutagenesis and scar-less gene deletions and insertions can be introduced into the target pathway. In addition, individual genes can be chemically synthesized *de novo* with optimized codons, resulting in a translationally improved pathway. Such genetic manipulations are extremely difficult to accomplish by conventional methods.

In order to ensure high assembly efficiency, it is critical to design appropriate lengths of overlaps between adjacent fragments. For larger gene clusters ($>20$ kb), increasing the length of overlap is necessary for obtaining a decent efficiency. As shown in Fig. 10.1B, overlaps of $\sim 400$ bp between internal adjacent pathway fragments can be easily generated. For example, the forward primer for amplifying the second pathway fragment can be located $\sim 400$ bp upstream of the annealing position of the reverse primer for amplifying the first pathway fragment. Overlaps between other fragments are generated by adding tails to primers; thus, they cannot be very long and usually have lengths of 40–80 bp. We encountered difficulties in amplifying the *S. cerevisiae* helper fragment, and in the end, the correct product was only obtained by using the pair of primers without any tails. As a result, the *S. cerevisiae* helper fragment only overlaps with the last pathway fragment and the adjacent helper fragment by 40 bp. Overlaps of 80 bp are generated between the *S. lividans* helper fragment and its neighbors.

## 3. HETEROLOGOUS EXPRESSION AND FINE MODIFICATION OF NATURAL PRODUCT GENE CLUSTERS

To demonstrate the ease of using DNA assembler for heterologous expression and fine modification of natural product gene clusters, the aureothin biosynthetic pathway (He & Hertweck, 2003; Traitcheva, Jenke-Kodama, He, Dittmann, & Hertweck, 2007; Fig. 10.2A) is chosen as an example. Aureothin is a rare nitroaryl-substituted polyketide (Fig. 10.2B, compound **1**) from *Streptomyces thioluteus*, which exhibits antitumor, antifungal, and insecticidal activities (He & Hertweck, 2003; Traitcheva et al., 2007).

## 3.1. Heterologous expression of the aureothin biosynthetic gene cluster

### 3.1.1 DNA preparation

1. Amplification of pathway fragments: The 29-kb aureothin biosynthetic pathway is split into seven pathway fragments, each 4–6 kb long. In principle, the number of pathway fragments to be assembled should be minimized, but low yield or even failure could occur more frequently when the PCR product is relatively long (>6 kb). The pathway fragments are PCR-amplified from the genomic DNA of *S. thioluteus* using the primers listed in Table 10.1. Set up the reaction mixtures as follows: 50 μL of FailSafe PCR 2× PreMix G (EPICENTRE Biotechnologies, Madison, WI, USA), 2.5 μL of forward primer (20 pmol/μL), 2.5 μL of reverse primer (20 pmol/μL), 1 μL of template (10–50 ng of *S. thioluteus* genomic DNA), 1 μL of Phusion DNA polymerase (New England Biolabs, Beverly, MA, USA), 5 μL of dimethyl sulfoxide (DMSO), and 38 μL of ddH$_2$O in a total volume of 100 μL.

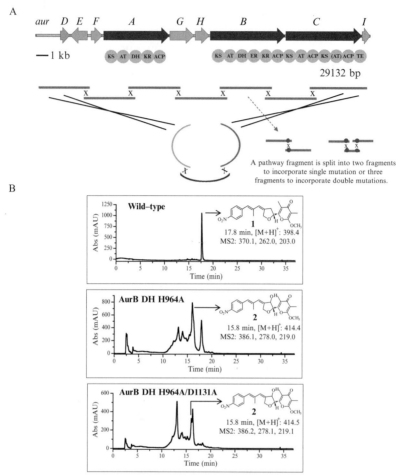

**Figure 10.2** Assembly of the wild-type aureothin biosynthetic pathway and the mutant aureothin biosynthetic pathways for creation of new derivatives. (A) The aureothin biosynthetic pathway. (B) LC–MS analysis of the *S. lividans* clones carrying the wild-type aureothin biosynthetic pathway and the mutant aureothin biosynthetic pathways. *Reproduced by permission of The Royal Society of Chemistry.* (For color version of this figure, the reader is referred to the online version of this chapter.)

2. Amplification of helper fragments: The *S. cerevisiae* helper fragment is amplified from the vector pRS416 (New England Biolabs). The *E. coli* and *S. lividans* helper fragments are amplified from pAE4, which is a *Streptomyces*–*E. coli* shuttle vector obtained from Professor William Metcalf (University of Illinois, Urbana, IL, USA; the complete sequence of the plasmid can be obtained by request from the authors). The

**Table 10.1** Primers used in assembling the aureothin gene cluster variants

| Name | Sequence |
|---|---|
| Aur-1-for | TGGTACTGCAAATACGGCATCAGTTACCGTGAGCAGATCGGATCAGCTCGTCCCGTTCGG |
| Aur-1-rev | GCTGCTCTTCTCGCATCGTC |
| Aur-2-for | CGTAGAGGAGCTCCAGCAGC |
| Aur-2-rev | CTCCTCCAGCACCTCGCAGC |
| Aur-3-for | TCCTGACCTTCGACTCGCTG |
| Aur-3-rev | CATGTTCGATCCTTCCGTTG |
| Aur-4-for | GACGGTGCACCAGCTGGTCA |
| Aur-4-rev | GTTGCCGGTCATGTGGTAGC |
| Aur-5-for | ATGACCAATGACGCGCCAAGAC |
| Aur-5-rev | CCGTCCATCAGGTCGAACGC |
| Aur-6-for | CCTACTACGGCCTGGTGGAC |
| Aur-6-rev | CATCGCCGTCATCGAGACGA |
| Aur-7-for | GCAACGAAGGACATGTCCAG |
| Aur-7-rev | TTATAGCACGTGATGAAAAGGACCCAGGTGGCACTTTTCGTCAGTCAGTCGTCCAGGCGC |
| Yeast-for | CGAAAAGTGCCACCTGGGTC |
| Yeast-rev | AATATTGTGAGTTTAGTATACATGCA |

*Continued*

**Table 10.1** Primers used in assembling the aureothin gene cluster variants—cont'd

| Name | Sequence |
|---|---|
| Strep-for | GTATTATAAGTAAATGCATGTATACTAAACTCACAATATT ATGGGCGCCGACGTGCTCA |
| Strep-rev | ATTAGCCATGGCATCACAGTATCGTGATGACATTAATTAAACGCAATCCAGTGCAAAGCTA |
| E. coli-for | AGACGGAAGAAGCTAGCTTTGCACTGGATTGCGTTAATTAATGTCATCACGATACTGTGAT |
| E. coli-rev | GCCCATGACCACCGTCGTCTCCGAACGGGACGAGCTGATCCGATCTGCTCACGGTAACTG |
| AurB-DH-H964A-for | TGTCGGCGGGACCGAGTCCTGGCTGGCCGACGCCGTCGTGCTGCGGCTCCACGCTCGTCC |
| AurB-DH-H964A-rev | GGACGAGCGGTGGAGCCGAGCACGACGGCGTCGGCCAGCCAGGACTCGGTCCGCGCCGACA |
| AurB-DH-D1131A-for | GCTACGGCGCGTCCACCCCGCCGCTCCTCGCCGCCGCACTGCACACCGCCCTCCTGAAGGAGG |
| AurB-DH-D1131A-rev | CCTCCTTCAGGAGGGCGGTGTGCAGTGCGGCGGCGAGGAGCGGGGGTGGACGCGCCGTAGC |

The mutated codons are underlined.

corresponding primers are listed in Table 10.1. Set up the reaction for amplifying the *S. cerevisiae* helper fragment as follows: 20 μL of 5× Phusion GC reaction buffer (New England Biolabs), 2.5 μL of dNTP premix containing 10 m*M* each nucleotide, 2.5 μL of forward primer (20 pmol/μL), 2.5 μL of reverse primer (20 pmol/μL), 1 μL of template pRS416 at the concentration of 50–100 ng/μL, 1 μL of Phusion DNA polymerase, and 70.5 μL of ddH$_2$O in a total volume of 100 μL. Set up the reactions for amplifying the *E. coli* and the *S. lividans* helper fragments as follows: 50 μL of FailSafe PCR 2× PreMix G, 2.5 μL of forward primer (20 pmol/μL), 2.5 μL of reverse primer (20 pmol/μL), 1 μL of template plasmid pAE4 at the concentration of 50–100 ng/μL, 1 μL of Phusion DNA polymerase, and 43 μL of ddH$_2$O in a total volume of 100 μL.

3. Amplify the *S. cerevisiae* helper fragment using the following condition: Fully denature at 98 °C for 30 s, followed by 25 cycles of 98 °C for 10 s, 45 °C for 30 s, and 72 °C for 1 min, with a final extension at 72 °C for 10 min. We encountered difficulties in amplifying the *S. cerevisiae* helper fragment and in the end the correct product was obtained by using the designated reagents and PCR program mentioned above.

4. Amplify all the other fragments using the following condition: Fully denature at 98 °C for 30 s, followed by 25 cycles of 98 °C for 10 s, 58 °C for 30 s, and 72 °C for 1–3 min, with a final extension at 72 °C for 10 min. 58 °C is used as a standard annealing temperature. In some cases, especially for amplifying certain fragments from *Streptomyces*, fine-tuning annealing temperature is necessary for obtaining the correct amplicons or improving PCR yields. Due to the high GC content of the *Streptomyces* genome (>70%), including 5% DMSO in the reaction mixture will reduce the chance of forming secondary structures in general, resulting in better amplification efficiency. However, in some cases, we did encounter lower amplification efficiencies or even failures when DMSO was used. Generally, when difficulties are encountered to amplify a certain fragment, a set of PCR conditions with various annealing temperatures and inclusion or exclusion of 5% of DMSO need to be tested.

5. As shown by the vector maps of the assembled constructs (Fig. 10.3), the minimal genetic elements needed for selecting and maintaining successful transformants include an origin of replication (*ori*) and a selection marker. The *S. cerevisiae* helper fragment is designed to contain *CEN6* and *ARS H4* as an *ori*, and *ura3* as a selection marker; the *E. coli* helper fragment is designed to contain *oriR6K* as an *ori* and *accIV* encoding the

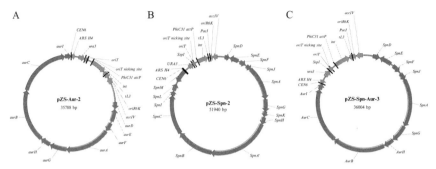

**Figure 10.3** (A) The construct pZS-Aur-2 containing the wild-type aureothin biosynthetic pathway. (B) The construct pZS-Spn-2 containing the wild-type spectinabilin biosynthetic pathway. (C) The construct pZS-Spn-Aur-3 containing the hybrid spectinabilin–aureothin biosynthetic pathway. *Reproduced by permission of The Royal Society of Chemistry.* (For color version of this figure, the reader is referred to the online version of this chapter.)

apramycin resistance gene (it can also be used as the selection marker in *Streptomyces*). The *Streptomyces* helper fragment does not have the corresponding *ori* used in *Streptomyces*; therefore, the assembled construct could not be maintained as a plasmid, and instead has to be integrated into the chromosome. It also contains *oriT* which is the conjugal transfer *ori*, *PhiC31 attP* as the phage ΦC31 recognition site, *int* encoding for the ΦC31 integrase, and *tL3* as a terminator.

6. Load the 100 µL of PCR products onto 0.7% agarose gels and perform electrophoresis at 120 V for 20 min. Gel-purify PCR products using the QIAquick Gel Extraction Kit (QIAGEN, Valencia, CA, USA). Check the concentrations of the purified products using NanoDrop.

7. Take 200–300 ng of each fragment, mix in a tube, and calculate the final volume. Add 10% (v/v) 3 *M* sodium acetate and 2% (v/v) 10 mg/mL glycogen (e.g., if there is 100 µL of mixture, add 10 µL of sodium acetate and 2 µL of glycogen) and mix well. Add 2 × (v/v) 100% ethanol (e.g., if the final volume is about 110 µL, add 220 µL ethanol) and mix well. Store the DNA mixture at −80 °C for at least 1 h. Centrifuge at 4 °C, 13,200 rpm for 20 min. Usually, the precipitated DNA can be seen at the bottom of the tube. Remove the supernatant completely (do not touch the DNA). Add 500 µL of 70% ethanol to wash the DNA pellet and centrifuge at room temperature, 13,200 rpm for 3 min. Remove the ethanol completely and air-dry the pellet for 1–2 min (do not over-dry it).

Resuspend the DNA pellet in 4 μL ddH$_2$O. Now the DNA is ready for transformation. The fragment mixture can be maintained at $-20\,°C$ for several months.

### 3.1.2 Transformation

1. Inoculate a single colony of *S. cerevisiae* strain HZ848 into 3 mL YPAD medium and grow overnight in a shaker at $30\,°C$ and 250 rpm. *S. cerevisiae* HZ848 (*MATα, ade2-1, Δura3, his3-11, 15, trp1-1, leu2-3, 112, can1-100*) is used as the host for DNA assembly. However, any *S. cerevisiae* auxotrophic strain can be used as a host if the vector carrying the corresponding selection marker is provided.

2. Measure the OD$_{600}$ of the seed culture and inoculate the appropriate amount into 50 mL of fresh YPAD medium to obtain an OD$_{600}$ of 0.2 (e.g., if the overnight culture has an OD$_{600}$ of 10, then add 1 mL into 50 mL of fresh YPAD medium). Continue growing the 50 mL culture for ~4 h to obtain an OD$_{600}$ of 0.8 (the doubling time for a *S. cerevisiae* laboratory strain is ~2 h).

3. Spin down the yeast cells at $4\,°C$, 4000 rpm for 10 min and remove the spent medium. Use 50 mL of ice-cold ddH$_2$O to wash the cells once and centrifuge again. Discard water, add 1 mL of ice-cold ddH$_2$O to resuspend the cells, and move them to a sterile Eppendorf tube. Spin down the cells using a benchtop centrifuge for 30 s at $4\,°C$, 7000 rpm. Remove water and use 1 mL of 1 $M$ ice-cold sorbitol to wash the cells once (now the cells look slightly yellow). Centrifuge again and remove the sorbitol. Resuspend the cells in 250–300 μL of chilled 1 $M$ sorbitol and distribute them into 50 μL aliquots. Now each 50 μL of cells is ready for electroporation. Unlike *E. coli*, yeast competent cells need to be freshly prepared each time.

4. Mix the 4 μL of DNA with 50 μL of yeast cells and put the mixture into a chilled electroporation cuvette. Electroporate the cells at 1.5 kV and quickly add 1 mL of prewarmed ($30\,°C$) YPAD medium to resuspend cells. For an efficient electroporation, a time constant of 5.0–5.2 ms should be obtained. Grow in a shaker at $30\,°C$, 250 rpm for 1 h. Spin down the cells in a sterile tube at 13,200 rpm for 30 s and remove the YPAD medium. Use 1 mL of room temperature sorbitol solution to wash the cells two to three times and finally resuspend the cells in 1 mL sorbitol. Spread 100 μL of resuspended cells onto SC-Ura plates. Incubate the plates at $30\,°C$ for 2–3 days until colonies appear. Normally, 200–300 colonies will be obtained.

### 3.1.3 Verification of the correctly assembled pathways

1. Assembly efficiency is defined as the percentage of correct clones among the transformants appearing on the plate. Usually, 10 colonies are randomly picked from the SC-Ura plate and each colony is inoculated into 1.5 mL of SC-Ura liquid medium. Grow at 30 °C for 1.5 days. Purify yeast plasmid DNA from each 1.5 mL of culture using the Zymoprep II kit (Zymo Research, Orange, CA, USA).

2. Mix 2 µL of isolated plasmid with 50 µL of *E. coli* BW25141 cells and put the mixture into a chilled electroporation cuvette. We used *E. coli* strain BW25141 for plasmid enrichment and verification. However, any *E. coli* strain suitable for DNA cloning, such as DH5α and JM109, can be used. Electroporate the cells at 2.5 kV, and quickly add 1 mL of SOC medium to resuspend the cells. For an efficient electroporation, a time constant of 5.0–5.2 ms should be obtained. Grow in a shaker at 37 °C, 250 rpm for 1 h. Spin down the cells, remove 800 µL of SOC medium, resuspend the pellet with the remaining 200 µL of SOC medium, and spread the cells on LB plates supplemented with 50 µg/mL apramycin (Apr). Incubate the plates at 37 °C for 16–18 h until colonies appear. The number of *E. coli* transformants obtained could vary from a few to several thousands. This is mainly due to the low quality of the isolated yeast plasmids. However, as long as colonies appear, experiments can proceed.

3. Inoculate a single colony from each plate to 5 mL of LB supplemented with 50 µg/mL apramycin, and grow at 37 °C for 12–16 h. Purify *E. coli* plasmids from each 5 mL of culture using the QIAgen Miniprep kit (QIAGEN). Check the plasmid concentrations by NanoDrop.

4. Verify the correctly assembled pathway through two separate restriction digests for each construct. In order to verify the correctly assembled constructs by restriction digestion, a set of digestions consisting of one or two enzymes that cut the expected construct multiple times is chosen. Usually, two to three sets of digestions need to be set up in order to ensure the correct assembly. For a plasmid with a size of 20–30 kb, find one or two enzymes which cut the DNA molecule ~10 times. Try to avoid using enzyme digestions that will result in multiple fragments with similar sizes. For example, set up the following two digestions to verify the aureothin constructs: (a) *Sac*I digestion at 37 °C for 3 h: 1.5 µL of 10 × NEB buffer, 0.15 µL of 100 × BSA, 300 ng of plasmid, and 5 units of *Sac*I. Add ddH$_2$O to a final volume of 15 µL. Expected bands for the assembled aureothin gene cluster: 163, 696, 752, 839, 1207, 1965, 2370, 2956, 3074, 3612, 3693, 5250, and 9211 bp. (b) *Asc*I digestion at 37 °C for

3 h: 1.5 μL of 10 × NEB buffer, 0.15 μL of 100 × BSA, 300 ng of plasmid, and 5 units of *Asc*I. Add ddH$_2$O to a final volume of 15 μL. Expected bands for the assembled aureothin gene cluster: 487, 1890, 2927, 3154, 3716, 5419, 5433, 5810, and 6952 bp. In addition to restriction digestion, the correctly assembled constructs can be confirmed by DNA sequencing. Usually, for a gene cluster of ∼30 kb, an assembly efficiency of 60–80% is expected.

### 3.1.4 Conjugation and heterologous expression of the aureothin pathway variants in S. lividans

1. Mix 50–100 ng of each verified plasmid with 50 μL of *E. coli* WM6026 cells and put the mixture into a chilled electroporation cuvette. *E. coli* strain WM6026 (obtained from Professor William Metcalf) serves as the donor strain for transferring plasmids into *S. lividans* (the details of strain construction can be obtained by request from the authors). It is an auxotrophic strain whose growth relies on exogenously supplemented 2,6-diaminopimelic acid (DAP).

2. Electroporate the cells at 2.5 kV and quickly add 1 mL of SOC medium and 5 μL of 38 mg/mL DAP to resuspend cells. Grow in a shaker at 37 °C, 250 rpm for 1 h. Spread 100 μL of each culture on a LB-Apr$^+$-DAP plate. Incubate the plates at 37 °C for 16 h until colonies appear.

3. Inoculate a single colony from each plate in 2 mL of LB supplemented with 50 μg/mL apramycin and 10 μL of 38 mg/mL DAP, and grow at 37 °C for ∼2 h until OD$_{600}$ reaches 0.6–0.8. Spin down 100 μL of cell culture in an Eppendorf tube and wash the cell pellet with 1 mL of fresh LB medium. Spin down the cells and wash once more. Resuspend the cell pellets each with 1 mL of LB. Mix 2 μL of the resuspended cells with 25 μL of *S. lividans* spores by pipetting and spot 2 μL aliquots onto R2 no-sucrose plates. Wait until all the spotted drops are absorbed into the plates. Incubate the plates at 30 °C for 16–18 h.

4. Flood the plates with 2 mL of a mixture of nalidixic acid and apramycin each at a concentration of 1 mg/mL. Nalidixic acid is used to kill *E. coli* after it donates the plasmid into *S. lividans*, and apramycin is used to select the successful *S. lividans* exconjugants. Incubate the plates at 30 °C for an additional 3–5 days until exconjugants appear, at which point exconjugants are picked and restreaked on ISP2 plates supplemented with 50 μg/mL apramycin and allowed to grow for 2 days.

5. Inoculate single colonies from the ISP2-Apr$^+$ plates into 10 mL of MYG supplemented with 50 µg/mL apramycin and grow the cultures at 30 °C for 2 days as seed cultures, of which 2.5 mL is subsequently inoculated to 250 mL of fresh MYG supplemented with 50 µg/mL apramycin and grown for another 84 h.

### 3.1.5 Detection of aureothin (and its derivatives)

1. Centrifuge the cultures at 4000 rpm for 10 min to remove the cells. Extract the supernatants with an equal volume of ethyl acetate and evaporate to dryness using a rotary evaporator.

2. Perform LC–MS on an Agilent 1100 series LC/MSD XCT plus ion-trap mass spectrometer with an Agilent SB-C18 reverse-phase column. HPLC parameters for detection of aureothin are as follows: solvent A, 1% acetic acid in water; solvent B, acetonitrile; gradient, 10% B for 5 min, to 100% B for 10 min, maintain at 100% B for 5 min, return to 10% B for 10 min, and finally maintain at 10% B for 7 min; flow rate 0.3 mL/min; detection by UV spectroscopy at 367 nm. Under such conditions, aureothin is eluted at 17.8 min. Mass spectra are acquired in ultra-scan mode using electrospray ionization with positive polarity. The MS system is operated using a drying temperature of 350 °C, a nebulizer pressure of 35 psi, a drying gas flow of 8.5 L/min, and a capillary voltage of 4500 V. The same parameters are applied to detect the aureothin derivative, which is eluted at 15.8 min.

## 3.2. Site-directed mutagenesis for generating new derivatives

The aureothin gene cluster contains three type I polyketide synthases (PKSs) that are composed of multiple domains including β-ketosynthase, acyl transferase, acyl carrier protein, ketoreductase, dehydratase (DH), and enoylreductase domains. Because of the importance of polyketides in medicine and the modular, programmable nature of the biosynthetic machinery, PKSs have been extensively investigated and engineered to produce new derivatives (Staunton & Weissman, 2001).

### 3.2.1 DNA preparation

Site-directed mutagenesis can be readily used to generate new derivatives of aureothin. Unlike the labor-intensive and time-consuming procedures used in the conventional native host-based methods and heterologous host-based methods (Blodgett et al., 2007; Ito et al., 2009; Karray, Darbon, Nguyen, Gagnat, & Pernodet, 2010), DNA assembler only requires adding site-specific mutation(s) into the PCR primers used to generate pathway fragments. As an example, the active sites of the DH domain of AurB will be targeted. The motifs HXXXGXXXXP and DXXX(Q/H) were found to be conserved among the DH domains of PKSs, and the histidine and the aspartic acid were identified as the catalytic residues (Keatinge-Clay, 2008; Moriguchi, Kezuka, Nonaka, Ebizuka, & Fujii, 2010; Pawlik, Kotowska, Chater, Kuczek, & Takano, 2007). Based on this information, AurB DH H964A mutant and AurB DH H964A/D1131A double mutant are designed. We expect them to produce compound **2** (Fig. 10.2B) if the corresponding domain is completely inactivated, or a mixture of both compound **2** and aureothin if the activity is only reduced.

The pathway fragment carrying the mutation(s) is split into two fragments (for AurB DH H964A mutant) or three fragments (for AurB DH H964A/D1131A double mutant) as shown in Fig. 10.2A. Mutations are added to the primers (Table 10.1). The assembly, verification, and heterologous expression of the pathway variants are performed using the same procedures as with the wild-type gene cluster.

### 3.2.2 Detection and analysis of the aureothin derivatives

The aureothin derivative is extracted and detected using the same procedures as with aureothin. The mutation will lead to significant changes in HPLC analysis in which several new peaks will appear and the aureothin peak will be dramatically decreased (Fig. 10.2B). The distinguishable peak at 15.8 min is compound **2** as confirmed by its MS and MS2 patterns. This result demonstrates that single/multiple site-directed mutagenesis can be easily performed without going through the complicated multistep procedures used by other conventional approaches (Blodgett et al., 2007; Ito et al., 2009; Karray et al., 2010).

## 4. CHARACTERIZATION OF NATURAL PRODUCT GENE CLUSTERS

The second example we chose to demonstrate the application of DNA assembler in characterization of natural product gene clusters is the spectinabilin biosynthetic gene cluster from *Streptomyces spectabilis* (Fig. 10.4A). Compared to aureothin, spectinabilin is a longer nitrophenyl-containing polyketide, which exhibits antimalarial and antiviral activities (Isaka, Jaturapat, Kramyu, Tanticharoen, & Thebtaranonth, 2002; Kakinuma, Hanson, & Rinehart, 1976). The cluster consists of four PKSs

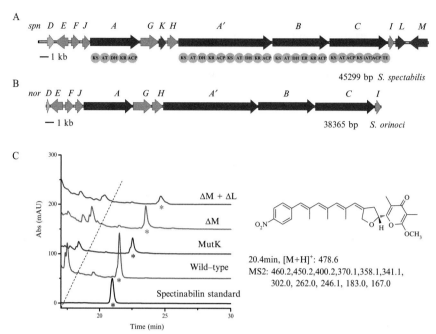

**Figure 10.4** Gene inactivation and scar-less deletion studies on the spectinabilin biosynthetic pathways. (A) The spectinabilin biosynthetic pathway from *S. spectabilis*. (B) The spectinabilin biosynthetic pathway from *S. orinoci*. (C) LC–MS analysis of *S. lividans* clones carrying the wild-type spectinabilin biosynthetic pathway and the mutant spectinabilin biosynthetic pathways. The asterisks indicate the spectinabilin peaks. MutK, the mutant with a stop codon introduced to SpnK; ΔM, the mutant with SpnM deleted; ΔM+ΔL, the mutant with SpnM and SpnL deleted. *Reproduced by permission of The Royal Society of Chemistry.* (For color version of this figure, the reader is referred to the online version of this chapter.)

(Choi et al., 2010) and shares a very high sequence homology with the spectinabilin biosynthetic pathway from *Streptomyces orinoci* (Traitcheva et al., 2007; Fig. 10.4B). One major difference between these two clusters is that the *S. spectabilis* cluster contains three extra genes, *spnK*, *spnL*, and *spnM*. To investigate their essentiality in spectinabilin biosynthesis, DNA assembler can be used to inactivate *spnK* by site-directed mutagenesis and completely delete *spnM* and *spnM*+*spnL*.

## 4.1. Heterologous expression of the spectinabilin biosynthetic gene cluster

Initially, we attempted to assemble the spectinabilin biosynthetic pathway using the same procedure as we did for the aureothin biosynthetic pathway. However, a low efficiency (<10%) was observed and most of the constructs underwent severe gene deletion(s). The failure was likely due to the high sequence similarity among PKS domains. Whether the high sequence similarity between genes could potentially be problematic varies case-by-case. It seemed not to be a problem when the smaller aureothin gene cluster was assembled. To address this issue, a modified two-step assembly strategy was devised (Fig. 10.5), through which we were able to assemble this biosynthetic pathway of 45 kb with an efficiency of 30%.

### 4.1.1 Construction of the spectinabilin biosynthetic gene cluster

1. Split the entire gene cluster into three intermediate plasmids, thus separating the four PKSs (Fig. 10.5A). The first contains *spnDEFJAGKH*; the second contains the last 400 bp of *spnH*, the full-length *spnA'*, and the first 400 bp of *spnB*; and the third contains *spnBCILM*. Two restriction sites, *Ssp*I and *Pac*I, are engineered at the appropriate positions of the intermediate plasmids by adding the corresponding recognition sites to the primers. The pathway fragments are amplified and cotransformed with the *S. cerevisiae* helper fragment and the *E. coli* helper fragment. In the first step of assembly, the *S. lividans* helper fragment can be omitted. The correctly assembled intermediate plasmids are identified using the same procedure as for the aureothin constructs.

2. Perform restriction digestion by *Ssp*I and *Pac*I to generate three intermediate fragments (Fig. 10.5B), which will subsequently be cotransformed with a fragment obtained from restriction digestion of the master helper plasmid by *Apa*LI and *Xho*I. Here, the lithium acetate/single-stranded carrier DNA/polyethylene glycol method (Gietz & Woods, 2006) is used instead of the electroporation protocol (Shao et al., 2009) because we did not obtain colonies through electroporation. This might be due to the large sizes of the intermediate fragments (the largest is ~20 kb). The master helper plasmid was constructed by cotransferring the three PCR-amplified helper fragments, which shared overlaps of 40–80 bp, by transformation into *S. cerevisiae*. E, Y, and S represent the corresponding *E. coli*, *S. cerevisiae*, and *Streptomyces* helper fragments. Identification of the correctly assembled constructs and heterologous expression of the cluster are performed using the same procedures as for the aureothin constructs.

### 4.1.2 Detection of spectinabilin

Spectinabilin is extracted and detected using the same procedure as with aureothin. The cell-produced spectinabilin has the same retention time (20.8 min) and MS/MS pattern as authentic spectinabilin (Fig. 10.4C).

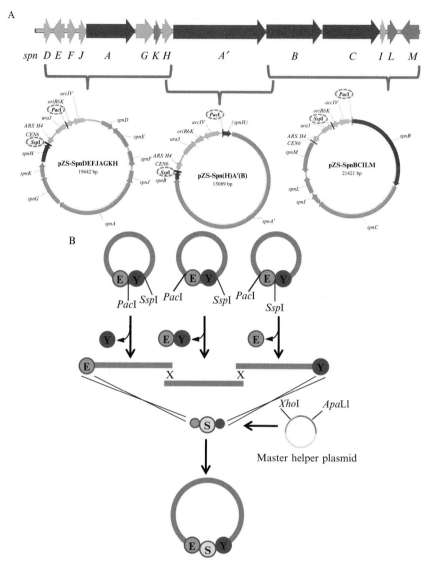

**Figure 10.5** The two-step strategy for assembling the spectinabilin biosynthetic pathway. (A) The first step involves construction of three intermediate plasmids, separating the four PKS genes into three plasmids. Two restriction sites, *Ssp*I and *Pac*I, are engineered at the appropriate positions of the intermediate plasmids. (B) In the second step, restriction digestion by *Ssp*I and *Pac*I generates three intermediate fragments, which are cotransformed with a fragment obtained from restriction digestion of the master helper plasmid by *Apa*LI and *Xho*I. *Reproduced by permission of The Royal Society of Chemistry.* (See Color Insert.)

## 4.2. Gene inactivation and scar-less gene deletion

To inactivate *spnK*, the codon TGC encoding Cys199 is changed to the stop codon TGA in the PCR primers. Introducing a stop codon into the target gene located in the middle of the gene cluster instead of completely removing it has the advantage of maintaining the pathway structure without affecting transcription of the neighboring genes. The genes *spnM* and *spnM+spnL* are individually removed from the biosynthetic pathway by redesigning the reverse primer for amplifying the last pathway fragment.

As a result, all three mutants will still produce spectinabilin (Fig. 10.4C), indicating that *spnK*, *spnL*, and *spnM* are not required for spectinabilin biosynthesis, supporting our previous hypothesis that they are only involved in upregulating the substrate concentration and product transportation (Choi et al., 2010).

# 5. CONSTRUCTION OF ARTIFICIAL GENE CLUSTERS

## 5.1. Creation of a hybrid spectinabilin–aureothin gene cluster

To further demonstrate the versatility of DNA assembler, a hybrid biosynthetic pathway between the aureothin and the spectinabilin pathways can be easily created. Because the aureothin pathway lacks the counterpart of the *spnA'* gene in the spectinabilin biosynthetic pathway, *spnA'* can be removed and a hybrid pathway comprising *spnDEFJAG* and *aurHBCI* can be generated using DNA assembler (Fig. 10.6A). The resulting construct will produce aureothin in the heterologous host (Fig. 10.6B), which is consistent with a previous study that was achieved using a much more complicated experimental protocol (Traitcheva et al., 2007).

## 5.2. Outlook for engineering of artificial gene clusters

The DNA assembler-based synthetic biology tool can overcome many of the limitations associated with conventional approaches for engineering of natural product gene clusters. With little modification, DNA assembler can be used for promoter/regulator replacement, artificial operon construction, and rapid library creation. Natural product biosynthetic pathways in organisms are often under a variety of complicated regulations. Replacing a sophisticated promoter or regulator from a native host with a well-studied one commonly used in a heterologous host will overcome the hurdle of promoter recognition across different species. Consequently, it can activate

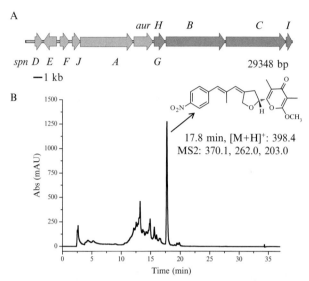

**Figure 10.6** Assembly of a hybrid biosynthetic pathway between the spectinabilin biosynthetic pathway and the aureothin biosynthetic pathway. (A) The hybrid biosynthetic pathway. (B) LC–MS analysis of the *S. lividans* clone carrying the hybrid biosynthetic pathway. *Reproduced by permission of The Royal Society of Chemistry.* (For color version of this figure, the reader is referred to the online version of this chapter.)

a silent cryptic pathway and possibly improve pathway expression, which in turn increases product yield. In addition to improving a pathway at the transcriptional level, a pathway can be optimized at translational level through codon optimization. All these sophisticated manipulations could be easily achieved through DNA assembler but are extremely difficult to accomplish by other conventional methods. On the other hand, a library of variant biochemical pathways containing promoters of different strengths or genes from a variety of sources can be readily generated through DNA assembler. Such a library will be very useful for metabolic engineering and synthetic biology applications, such as balancing pathway expression, optimizing metabolic carbon flux, and generating new compounds.

DNA assembler enables sophisticated genetic manipulations such as point mutagenesis and scar-less gene substitution and deletion, which can be used to confirm gene function, locate key amino acid residues, study biosynthetic mechanisms, express biosynthetic pathways heterologously, and generate new derivatives. Because the DNA fragments to be assembled are completely mobile and amenable to all sorts of sophisticated genetic manipulations accessible to PCR or can be chemically synthesized *de novo* with optimized codons, this genomics-driven, synthetic biology enabled method offers the ultimate versatility and flexibility in characterizing and engineering natural product biosynthetic pathways.

## ACKNOWLEDGMENTS

This work was supported by the National Academies Keck Futures Initiative on Synthetic Biology and the National Institutes of Health (GM077596).

## REFERENCES

Blodgett, J. A., Thomas, P. M., Li, G., Velasquez, J. E., van der Donk, W. A., Kelleher, N. L., et al. (2007). Unusual transformations in the biosynthesis of the antibiotic phosphinothricin tripeptide. *Nature Chemical Biology*, *3*, 480–485.

Bull, A. T., Goodfellow, M., & Slater, J. H. (1992). Biodiversity as a source of innovation in biotechnology. *Annual Review of Microbiology*, *46*, 219–252.

Challis, G. L. (2008). Mining microbial genomes for new natural products and biosynthetic pathways. *Microbiology (Reading, England)*, *154*, 1555–1569.

Choi, Y. S., Johannes, T. W., Simurdiak, M., Shao, Z., Lu, H., & Zhao, H. (2010). Cloning and heterologous expression of the spectinabilin biosynthetic gene cluster from *Streptomyces spectabilis*. *Molecular BioSystems*, *6*, 336–338.

Davies, J. (1999). Millennium bugs. *Trends in Cell Biology*, *9*, M2–M5.

Demain, A. L. (2006). From natural products discovery to commercialization: A success story. *Journal of Industrial Microbiology & Biotechnology*, *33*, 486–495.

Dewick, P. M. (2002). *Medical natural products. A biosynthetic approach* (2nd ed.). Chichester, UK: John Wiley and Sons.

Eliot, A. C., Griffin, B. M., Thomas, P. M., Johannes, T. W., Kelleher, N. L., Zhao, H., et al. (2008). Cloning, expression, and biochemical characterization of *Streptomyces rubellomurinus* genes required for biosynthesis of antimalarial compound FR900098. *Chemical Biology*, *15*, 765–770.

Gietz, R. D., & Woods, R. A. (2006). Yeast transformation by the LiAc/SS Carrier DNA/PEG method. *Methods in Molecular Biology*, *313*, 107–120.

Gross, H. (2007). Strategies to unravel the function of orphan biosynthesis pathways: Recent examples and future prospects. *Applied Microbiology and Biotechnology*, *75*, 267–277.

He, J., & Hertweck, C. (2003). Iteration as programmed event during polyketide assembly; molecular analysis of the aureothin biosynthesis gene cluster. *Chemical Biology*, *10*, 1225–1232.

Herbert, R. B. (1989). *The biosynthesis of secondary metabolites* (2nd ed.). London, UK: Chapman and Hall.

Isaka, M., Jaturapat, A., Kramyu, J., Tanticharoen, M., & Thebtaranonth, Y. (2002). Potent *in vitro* antimalarial activity of metacycloprodigiosin isolated from *Streptomyces spectabilis* BCC 4785. *Antimicrobial Agents and Chemotherapy*, *46*, 1112–1113.

Ito, T., Roongsawang, N., Shirasaka, N., Lu, W., Flatt, P. M., Kasanah, N., et al. (2009). Deciphering pactamycin biosynthesis and engineered production of new pactamycin analogues. *Chembiochem*, *10*, 2253–2265.

Kakinuma, K., Hanson, C. A., & Rinehart, K. L. (1976). Spectinabilin, a new nitro-containing metabolite isolated from *Streptomyces spectabilis*. *Tetrahedron*, *32*, 217–222.

Karray, F., Darbon, E., Nguyen, H. C., Gagnat, J., & Pernodet, J. L. (2010). Regulation of the biosynthesis of the macrolide antibiotic spiramycin in *Streptomyces ambofaciens*. *Journal of Bacteriology*, *192*, 5813–5821.

Keatinge-Clay, A. (2008). Crystal structure of the erythromycin polyketide synthase dehydratase. *Journal of Molecular Biology, 384*, 941–953.

Leadbetter, J. R. (2003). Cultivation of recalcitrant microbes: Cells are alive, well and revealing their secrets in the 21st century laboratory. *Current Opinion in Microbiology, 6*, 274–281.

Li, J. W. H., & Vederas, J. C. (2009). Drug discovery and natural products: End of an era or an endless frontier? *Science, 325*, 161–165.

Moriguchi, T., Kezuka, Y., Nonaka, T., Ebizuka, Y., & Fujii, I. (2010). Hidden function of catalytic domain in 6-methylsalicylic acid synthase for product release. *The Journal of Biological Chemistry, 285*, 15637–15643.

Newman, D. J., & Cragg, G. M. (2007). Natural products as sources of new drugs over the last 25 years. *Journal of Natural Products, 70*, 461–477.

Pawlik, K., Kotowska, M., Chater, K. F., Kuczek, K., & Takano, E. (2007). A cryptic type I polyketide synthase (cpk) gene cluster in *Streptomyces coelicolor* A3(2). *Archives of Microbiology, 187*, 87–99.

Shao, Z., Luo, Y., & Zhao, H. (2011). Rapid characterization and engineering of natural product biosynthetic pathways via DNA assembler. *Molecular BioSystems, 7*, 1056–1059.

Shao, Z., Zhao, H., & Zhao, H. (2009). DNA assembler, an *in vivo* genetic method for rapid construction of biochemical pathways. *Nucleic Acids Research, 37*, e16.

Staunton, J., & Weissman, K. J. (2001). Polyketide biosynthesis: A millennium review. *Natural Product Reports, 18*, 380–416.

Traitcheva, N., Jenke-Kodama, H., He, J., Dittmann, E., & Hertweck, C. (2007). Noncolinear polyketide biosynthesis in the aureothin and neoaureothin pathways: An evolutionary perspective. *Chembiochem, 8*, 1841–1849.

Woodyer, R. D., Shao, Z., Thomas, P. M., Kelleher, N. L., Blodgett, J. A., Metcalf, W. W., et al. (2006). Heterologous production of fosfomycin and identification of the minimal biosynthetic gene cluster. *Chemical Biology, 13*, 1171–1182.

Zerikly, M., & Challis, G. L. (2009). Strategies for the discovery of new natural products by genome mining. *Chembiochem, 10*, 625–633.

# CHAPTER ELEVEN

# Reassembly of Functionally Intact Environmental DNA-Derived Biosynthetic Gene Clusters

**Dimitris Kallifidas, Sean F. Brady[1]**
Laboratory of Genetically Encoded Small Molecules, Howard Hughes Medical Institute, The Rockefeller University, New York, USA
[1]Corresponding author: e-mail address: sbrady@rockefeller.edu

## Contents

## Abstract

Only a small fraction of the bacterial diversity present in natural microbial communities is regularly cultured in the laboratory. Those bacteria that remain recalcitrant to culturing cannot be examined for the production of bioactive secondary metabolites using standard pure-culture approaches. The screening of genomic DNA libraries containing DNA isolated directly from environmental samples (environmental DNA (eDNA)) provides an alternative approach for studying the biosynthetic capacities of these organisms. One drawback of this approach has been that most eDNA isolation procedures do not permit the cloning of DNA fragments of sufficient length to capture large natural product biosynthetic gene clusters in their entirety. Although the construction of eDNA libraries with inserts big enough to capture biosynthetic gene clusters larger than ~40 kb remains challenging, it is possible to access large gene clusters by reassembling them from sets of smaller overlapping fragments using transformation-associated

*Methods in Enzymology*, Volume 517
ISSN 0076-6879
http://dx.doi.org/10.1016/B978-0-12-404634-4.00011-5

recombination in *Saccharomyces cerevisiae*. Here, we outline a method for the reassembly of large biosynthetic gene clusters from captured sets of overlapping soil eDNA cosmid clones. Natural product biosynthetic gene clusters reassembled using this approach can then be used directly for functional heterologous expression studies.

# 1. INTRODUCTION

Culture-independent analyses of natural microbial populations, including 16S rRNA sequencing and shotgun sequencing of DNA extracted directly from environmental samples, have consistently shown that only a small fraction of bacterial species is readily cultivated under standard laboratory conditions (Rappe & Giovannoni, 2003; Torsvik, Ovreas, & Thingstad, 2002; Torsvik, Salte, Sorheim, & Goksoyr, 1990). By some estimates, as much as 99% of the bacterial diversity present in many environmental samples is not readily cultured, rendering these organisms ineffectual as sources of new bioactive small molecules. The cloning and subsequent analysis of DNA extracted directly from environmental samples, which has broadly been defined as the field of metagenomics, are now being used to unravel the hidden biosynthetic capacity of natural bacterial populations (Brady, Simmons, Kim, & Schmidt, 2009; Daniel, 2005; Handelsman, Rondon, Brady, Clardy, & Goodman, 1998; Simon & Daniel, 2009). In bacteria, genes responsible for the biosynthesis of secondary metabolites, including genes with biosynthetic, transport, resistance, and regulatory functions, are often found clustered on the bacterial chromosome (Bentley et al., 2002). This conserved architecture can be exploited to permit the cloning of complete secondary metabolite biosynthetic gene clusters directly from environmental samples. The analysis of eDNA libraries has now identified new biosynthetic enzymes, gene clusters, and bioactive small molecules.

While the construction of cosmid-based eDNA libraries containing 30- to 40-kb inserts is now routine, it remains technically challenging to construct larger insert libraries that would permit the cloning of biosynthetic gene clusters that span more than 40 kb of DNA. Bacterial artificial chromosome (BAC)-based metagenomic libraries offer the possibility of capturing DNA inserts longer than 100 kb. However, because of the difficulties associated with obtaining very large DNA fragments from environmental samples, they are often orders of magnitude smaller than comparable cosmid-based libraries (Daniel, 2005). An alternative strategy to capturing complete biosynthetic gene clusters on individual large-insert clones is to recover the clusters on multiple overlapping clones and then reassemble the pathway from these overlapping fragments. Restriction enzyme- and bacterial recombination-based approaches can be used to overcome simple reassembling problems (Binz, Wenzel, Schnell, Bechthold, & Muller, 2008), but these approaches are laborious and often not practical when faced with the need to reassemble very large gene clusters captured on multiple overlapping clones. Yeast has a more efficient DNA recombination system than most bacteria (Larionov et al., 1994; Nagano, Takao, Kudo, Iizasa, & Anai, 2007), and we have found it very useful for reassembling collections of eDNA cosmids into complete biosynthetic gene clusters.

Transformation-associated recombination (TAR) in *Saccharomyces cerevisiae* was initially developed as a way to permit the direct cloning of target DNA fragments from pools of crude genomic DNA (Gibson, Benders, Andrews-Pfannkoch, et al., 2008; Kouprina & Larionov, 2008; Kouprina, Noskov, & Larionov, 2006; Larionov et al., 1996; Mathee et al., 2008; Nagano et al., 2007). In TAR cloning protocols, genomic DNA and a capture vector, engineered to carry small homology arms corresponding to sequences flanking the target DNA of interest, are introduced into *S. cerevisiae* by cotransformation. The capture vector arms and the homologous genomic sequences undergo recombination to yield a stable plasmid containing the targeted genomic region. In more recent years, this technique has been exploited to assemble sets of overlapping synthetic DNA fragments into larger sequences (Gibson, Benders, Axelrod, et al., 2008; Shao, Zhao, & Zhao, 2009). The protocol outlined here describes the creation of large eDNA libraries from soil, the arraying of these libraries to facilitate the recovery of overlapping clones, and the use of TAR to reassemble overlapping clones into full-length gene clusters that can be used in downstream heterologous expression studies (Fig. 11.1).

## 2. PROTOCOL FOR CONSTRUCTING, ARRAYING, AND SCREENING ENVIRONMENTAL LIBRARIES

### 2.1. Isolation of high-molecular-weight eDNA and library construction

Methods for constructing large eDNA libraries from environmental samples have been described in detail elsewhere (Brady, 2007; Rondon et al., 2000; Simon & Daniel, 2010; Zhou, Bruns, & Tiedje, 1996). Here, we briefly

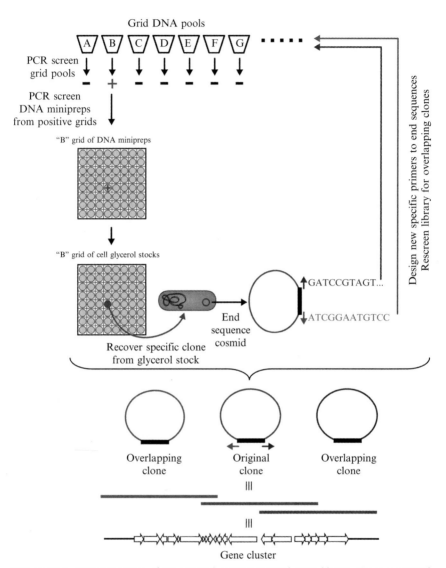

**Figure 11.1** PCR screening of an arrayed environmental DNA library. Once an initial clone of interest is recovered, new primers are designed based on the end sequencing of this clone. These new primers are then used to rescreen the library for overlapping clones. The screening cycle is repeated until the full biosynthetic pathway is recovered. (See Color Insert.)

outline a strategy for constructing very large eDNA cosmid libraries from freshly collected soil samples. In brief, 250 g of soil which has been passed through a 1/8″ (3–4 mm) screen is mixed 1:1 (w:v) with lysis buffer (100 m$M$ Tris–HCl, 100 m$M$ EDTA, 1.5 $M$ NaCl, 1% (w/v) CTAB (cetyltrimethylammonium bromide), 2% (w/v) SDS (sodium dodecyl sulfate), pH 8.0, and heated at 70 °C). After 2 h, soil particulates are removed by centrifugation (30 min, $4000 \times g$). Crude eDNA is then precipitated from the resulting supernatant by adding 0.7 volumes of isopropanol and subsequently collected by centrifugation (30 min, $4000 \times g$). The resulting pellet is washed with 70% ethanol, air-dried, and resuspended in a minimal volume of TE (10 m$M$ Tris, 1 m$M$ EDTA, pH 8.5). The remaining soil contaminants are removed by large-scale gel purification on a 1% agarose gel (16 h, 20 V). Purified high-molecular-weight eDNA is recovered by electroelution (2 h, 100 V), concentrated using centrifugal concentrators (MWCO 100 KDa) and end-repaired (End-It, Epicentre). Blunt-ended high-molecular-weight eDNA is ligated with a cosmid vector that has been digested with a blunt-end generating restriction enzyme and dephosphorylated. Ligation reactions are then packaged into lambda phage, and the packaging reaction is used to transfect *E. coli*. Based on the number of colonies obtained from an initial test packaging reaction, this procedure should be scaled to yield at least 10,000,000 individual eDNA cosmid clones for an individual soil sample. We have found that soil eDNA cosmid libraries must contain at least 10,000,000 unique members (eDNA megalibraries) before we are able to consistently recover overlapping clones for genetic loci of interest (Kim et al., 2010).

## 2.2. Arraying and screening eDNA megalibraries for overlapping clones

The procedure outlined below has been designed to facilitate the identification of clones in very large eDNA libraries. We have found that it is most effective to divide large libraries into multiple unique sublibraries grouped into 8 × 8 or 8 × 12 grids. While sublibraries of any size can be used, we have found that pools of approximately 5000 clones provide a good balance between the size of the arrayed library and the ability to efficiently recover individual clones of interest. A DNA miniprep and corresponding glycerol stock are created from each sublibrary and these are used for PCR screening and clone recovery, respectively.

1. Based on the titer obtained in the small-scale titering experiment described above, set up an appropriate number of ligation and packaging reactions to obtain at least 250,000 eDNA clones. This will be sufficient to generate one $8 \times 8$ grid of sublibraries containing 4000–5000 members each. We find it easiest to work with packaging reaction volumes that yield one grid at a time.

2. Confirm the titer of this large-scale reaction.

3. Carry out a large-scale phage infection reaction using the same conditions that were used for the small-scale titer reaction. Generally, this will consist of mixing the large-scale packaging reaction with a 10-fold excess of an *E. coli* culture with an $OD_{600}$ of 0.5–1.0.

4. After 1.5 h of shaking at 37 °C without any antibiotic selection, this mixture is used to inoculate 5 ml aliquots of LB (10 g/l tryptone, 5 g/l yeast extract, 10 g/l NaCl, pH 7.0) (+ antibiotic) broth with enough of the transfection reaction to obtain approximately 5000 unique transformants. Either 64 or 96 5-ml cultures should be set up depending on whether $8 \times 8$ or $8 \times 12$ grids are being used.

5. Incubate cultures overnight at 37 °C with shaking.

6. For each library pool, remove 500 µl to create a glycerol stock and use the remainder for a DNA miniprep. These matching DNA miniprep and glycerol stock pairs will be used for PCR screening and clone recovery, respectively.

7. Array the DNA minipreps in either $8 \times 8$ or $8 \times 12$ grids.

8. Pool small aliquots (10 µl) of DNA from each miniprep in a grid into a master aliquot for that grid. These "grid master pools" will be used for PCR screening. To facilitate higher-throughput screening, sublibraries containing pools of aliquots from each row and/or column may also be generated and used as templates in the PCR screening reactions.

This process is repeated until the arrayed library contains at least 10,000,000 unique cosmid clones. Clones containing sequences of interest can be identified in the arrayed library using the hierarchical pools miniprep DNA as templates in PCR reactions with primers designed to recognize sequences of interest. Once an amplicon-positive pool has been identified, it is possible to use either colony hybridization or dilution PCR to recover the specific clone of interest from the corresponding glycerol stock (Banik & Brady, 2008; Pham, Palden, & DeLong, 2007). The recovered cosmid clone is then end-sequenced using vector-specific primers and then primers designed to the end sequences are used to rescreen the arrayed library for overlapping clones. This process is continued until the full biosynthetic gene cluster is recovered from the library.

# 3. USE OF TAR TO REASSEMBLE OVERLAPPING CLONES INTO COMPLETE BIOSYNTHETIC PATHWAYS

## 3.1. Capture vector construction

We developed the TAR capture vector pTARa to facilitate the reassembly of gene clusters in yeast and the subsequent introduction of these clusters into *Streptomyces* spp. by conjugation for heterologous expression studies (Feng, Kim, & Brady, 2010). In addition to the elements required for selection and propagation in yeast and *E. coli*, pTARa has been equipped with a *Dra*I fragment from pOJ436 (Bierman et al., 1992) that contains an origin of transfer (*oriT*), an apramycin resistance marker, and the phage φC31 integration system needed for integration into diverse *Streptomyces* spp. The protocol outlined below is written with pTARa in mind. Other capture vectors could be used with minor modifications to the general protocol.

For each TAR experiment, a capture vector must be outfitted with homology arms that correspond to sequences flanking the pathway to be reassembled. The cloning of the homology arms into the pTARa capture vector is outlined below:

1. Design two sets of primers, an upstream set (UPS1 and UPS2) that will amplify ~1 kb of the proximal end of the left outermost clone and a downstream set (DWS1 and DWS2) that will amplify ~1 kb of the distal sequence of the right outermost clone (Table 11.1).

   a. The upstream forward primer (UPS1) should be designed to contain a *Bmt*I site (GCTAGC) for cloning into pTARa and therefore should read 5′-GCGC**GCTAGC** + 20 bp of the start of proximal targeting sequence-3′.

**Table 11.1** Primers used for capture vector construction

| Name | Primer description |
| --- | --- |
| UPS1 | 5′-GCGC**GCTAGC** + 20 bp of the start of proximal targeting sequence-3′[a] |
| UPS2 | Reverse complement of DWS1 |
| DWS1 | 5′-20 bp that is ~1 kb downstream of UPS1 + **GTTAAC** + 20 bp that is ~1 kb upstream of DWS2-3′[b] |
| DWS2 | 5′-GCGC**GCATGC** + 20 bp reverse complement of the end of the distal targeting sequence-3′[c] |

The added [a]*Bmt*I, [b]*Hpa*I, and [c]*Sph*I sites are shown in bold.

    **b.** The downstream reverse primer (DWS2) should be designed to contain an *Sph*I site for cloning into the pTARa vector and therefore should read 5′-GCGC**GCATGC**+20 bp reverse complement of the end of the distal targeting sequence-3′.

    **c.** The downstream forward primer, DWS1, is the reverse complement of the upstream reverse primer, UPS2. Conceptually, DWS1 is the easier of the two to design and therefore its construction is described here. DSW1 consists of a 20-bp stretch of sequence that is ∼1 kb downstream of UPS1 and a 20-bp stretch of sequence that is ∼1 kb upstream of DWS2 (Fig. 11.2). These two sequences are linked via an *Hpa*I site (**GTTAAC**) that will be used to linearize the capture vector prior to the TAR reaction.

**2.** Set up two first-round PCR reactions—one with primers UPS1 and UPS2 and the appropriate cosmid template and a second with primers DWS1 and DWS2 and the appropriate cosmid template.

**3.** Gel-purify the PCR products using a commercially available kit (e.g., Qiagen).

**4.** Set up a second round of PCR using primers UPS1 and DWS2 and 1 μl of each purified amplicon as a template. The first-round amplicons contain terminal complementary sequences that were introduced by the UPS2 and DWS1 primers and will therefore be linked together in the second-round PCR reaction.

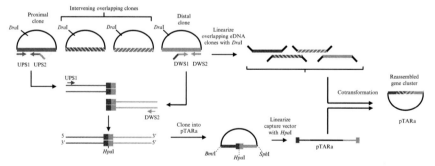

**Figure 11.2** TAR reassembly of gene clusters capture on overlapping cosmid clones. The proximal and the distal ends of the outermost clones are amplified using upstream and downstream sets of primers (UPS1/UPS2, DWS1/DWS2), respectively. Primers, UPS2 and DWS1, are reverse complement sequences and therefore these two amplicons can be linked with a second round of PCR. The resulting amplicons, consisting of clone-specific homology arms, are ligated into pTARa capture vector. Linearized capture vector (*Hpa*I) and linearized overlapping clones (*Dra*I) are cotransformed in yeast for the recombination reaction. (See Color Insert.)

5. Gel-purify the resulting PCR product.
6. Digest the gel-purified amplicon with *Bmt*I/*Sph*I and ligate this with similarly digested pTARa vector.
7. Introduce the ligation reaction by transformation into copy control *E. coli* (e.g., EPI300 Epicentre) and then check the newly created captured vector by *Hpa*I digestion to ensure you have the correct construct.
8. Use a copy control induction protocol (Epicentre) to obtain sufficient capture vector for the downstream TAR reaction. You will need 0.5 μg of digested capture vector for each TAR reaction.

## 3.2. Yeast spheroplast preparation

TAR requires the simultaneous cotransformation of yeast with the capture vector and the overlapping clones to be reassembled. Yeast transformation is accomplished by incubating DNA with spheroplasts, which are cells from which the cell wall has been almost completely removed. The following steps in the protocol outline the use of the yeast lytic enzyme, Zymolyase, to create the spheroplasts needed for carrying out the TAR reaction. This protocol has been modified from Kouprina and Larionov (2008).

1. Inoculate three uracil-deficient yeast CRY1-2 (MATa, ura3Δ, cyh2$^R$) colonies of different sizes in 50 ml YPD (yeast extract peptone dextrose: 10 g/l yeast extract, 20 g/l peptone, 20 g/l glucose) for overnight growth at 30 °C.
2. Next day pick the culture with an $OD_{660}$ of 3–5. This cell density is optimal for spheroplast preparation.
3. Harvest the cells by centrifugation at $1200 \times g$ for 10 min at 4 °C.
4. Wash the pellet once by resuspending with 40 ml of sterile ddH$_2$O and spin as before.
5. Wash the cell pellet twice with 40 ml 1 *M* sorbitol as before.
6. Resuspend the washed pellet in 20 ml SCE (1 *M* sorbitol; 100 m*M* Na citrate, pH 5.8; 10 m*M* EDTA, pH 8.0).
7. Add 40 μl of 2-mercaptoethanol along with 20 μl of a 10 mg/ml Zymolyase (Zymo Research) to the yeast suspension.
8. Incubate the mixture at 30 °C and monitor the spheroplast reaction every 20 min. Spheroplasting is monitored by diluting aliquots of the spheroplast mixture 1:10 in either 1 *M* sorbitol or 2% SDS and comparing the $OD_{660}$ values of these solutions. Spheroplasts are protected by the sorbitol solution but will lyse in SDS.

9. When the ratio between intact (sorbitol aliquot) and lysed (SDS aliquot) spheroplasts reaches approximately 4, stop the reaction by centrifugation ($500 \times g$, 5 min, 4 °C).

10. Remove the supernatant, gently resuspend the pellet in 20 ml of 1 $M$ sorbitol (do not vortex), and then centrifuge as in step 9. Repeat washing procedure once.

11. Resuspend the final spheroplast pellet in 2 ml STC (1 $M$ sorbitol; 10 m$M$ Tris, pH 7.5; 10 m$M$ CaCl$_2$). Spheroplasts can be stored at room temperature for up to 1 h before proceeding with the transformation protocol. Alternatively, spheroplasts can be aliquoted and stored at least up to 1 week at $-80$ °C.

## 3.3. Transformation of spheroplasts with pTARa capture vector along with overlapping clones capturing the entire biosynthetic pathway

Overlapping cosmid clones and the TAR vector must be linearized by restriction digestion prior to use in a TAR reaction. This section of the protocol outlines the restriction enzyme digestion of these vectors and the use of the linearized product to transform the yeast spheroplasts that were generated in the previous section of the protocol. Ultimately, the transformation reactions are plated on uracil dropout plates to select for yeast that potentially contain successful TAR-reassembled gene clusters.

1. In a 20-μl reaction, digest 1 μg of each overlapping cosmid clone to be used in the assembly reaction with *Dra*I. Stop the reaction by heating at 70 °C for 20 min. The digested cosmids do not need to be further purified.

2. Check an aliquot of the restriction digest on a 0.7% agarose gel to make sure *Dra*I alone does not cut in the eDNA insert. This gel should show a single fragment that comigrates with the 23-kb band of a lambda *Hin*dIII marker. We have found *Dra*I to be a good enzyme to start with because it cuts in many commonly used cosmid and fosmid vectors (Supercos, pWEB, pWEB:TNC, etc.) and is rarely found in eDNA inserts containing secondary metabolite biosynthetic gene clusters because they often arise from GC-rich genomes such as those of actinobacteria like *Streptomyces*. If *Dra*I cuts in the eDNA insert, other blunt-end cutting enzymes that cut in the vector should be explored for linearizing the clone.

3. Linearize 0.5 μg of pathway-specific capture vector with *Hpa*I.

4. Gel-purify the digested capture vector using a commercially available kit (Qiagen) according to the manufacturer's instructions.

5. Gently mix 200 ng of each linearized cosmid, an equimolar amount ($\sim$100 ng) of the linearized pathway-specific capture vector, and 200 $\mu$l of spheroplasts.

6. Incubate at room temperature for 10 min and then add 800 $\mu$l of PEG solution (20% polyethylene glycol (PEG) 8000; 10 m$M$ CaCl$_2$; 10 m$M$ Tris, pH 7.5).

7. Following a 10-min incubation at room temperature, spin the transformation mix at $300 \times g$ for 5 min at 4 °C.

8. Gently remove the supernatant and then resuspend the pellet in 800 $\mu$l SOS solution (1 $M$ sorbitol, 6.5 m$M$ CaCl$_2$, 0.25% yeast extract, 0.5% peptone).

9. Incubate the resuspended transformation mix at 30 °C for 40 min.

10. Add transformed spheroplasts to 7 ml of 50 °C synthetic complete (SC) top agar (1 $M$ sorbitol, 1.92 g/l SC uracil dropout supplement, 6.7 g/l yeast nitrogen base, 2% glucose, 2.5% agar).

11. Overlay the spheroplast top agar mixture onto SC uracil dropout agar plates (same composition as above but 2% agar).

12. Incubate plates at 30 °C. Colonies typically appear within 72 h.

## 3.4. Analysis of TAR-recombined clones

In order to identify pTARa clones containing reassembled gene clusters, individual yeast colonies are screened by PCR for the presence of fragments from each overlapping clone. Conveniently, the same primer pairs that were used to recover individual clones from the library can also be used in this PCR screen. The initial PCR analysis should be coupled with comparative restriction mapping and, if possible, full BAC sequencing to confirm the integrity of the reassembled DNA construct. In our experience, approximately 20% of yeast transformants carry the correctly reassembled biosynthetic pathway.

1. Pick yeast transformants with toothpicks and patch them onto SC uracil dropout plates for overnight growth at 30 °C.

2. Resuspend a small portion of the patch in 10 $\mu$l of 20 m$M$ NaOH and heat this mixture at 95 °C for 10 min.

3. Using 1.5 $\mu$l of the solution from step 2 as a template, PCR-screen each transformant with primer pairs that recognize DNA fragments found in each of the overlapping clones. It is often possible to use the same primers here that were used in the clone recovery step.

4. Grow yeast recombinants that are PCR positive for all parts of a pathway overnight in 2 ml of SC uracil dropout medium (30 °C, 225 rpm).

5. Isolate the pTARa-reassembled construct using ChargeSwitch™ Yeast Plasmid Isolation Kit (Invitrogen) per manufacturer's instructions.
6. Transform electrocompetent copy control *E. coli* (e.g., EPI300 from Epicentre) with 5 µl of the isolated pTARa construct and select on LB plates supplemented with 12.5 µg/ml chloramphenicol.
7. On the next day, inoculate a single transformant into 10 ml of LB with 12.5 µg/ml chloramphenicol supplemented with 60 µl of copy control BAC autoinduction solution (Epicentre).
8. Incubate at 37 °C for 20 h.
9. Isolate the reassembled pathway-containing BAC using alkaline lysis and then purify the DNA by phenol/chloroform extraction and isopropanol precipitation.
10. Digest the BAC clone using an enzyme that yields well-resolved DNA fragments by agarose gel electrophoresis. If the sequence data for the pathway of interest is available, compare the experimental restriction band pattern of the assembled BAC clone with the theoretically predicted band sizes of the clone. If the full sequence is not available, digest all overlapping clones with the same restriction enzyme and compare the restriction patterns produced by the reassembled and the individual clones. It may be necessary to use a 0.5% agarose gel run at 30 V overnight to resolve large restriction fragments.

## 4. DOWNSTREAM ANALYSIS

The ultimate aim of this protocol is to transfer faithfully reassembled gene clusters into model cultured bacterial hosts for heterologous expression studies in order to gain functional access to the metabolites encoded by these gene clusters. pTARa was specifically designed for use in *Streptomyces* and can be transferred into these bacteria by intergeneric conjugation using *E. coli* ET12567/pUZ8002 (Kan$^R$/Cm$^R$) as a donor strain following standard protocols (Kieser, Bibb, Buttner, Chater, & Hopwood, 2000). Organic extracts from cultures of *Streptomyces* spp. transformed with either the individual cosmid clones or the reassembled pathway can be compared by LCMS to look for the heterologous production of pathway-encoded metabolites. By exchanging the φC31 integrase system for any of a number of broad-host-range origins of replication, it could be used in a phylogenetically diverse collection of model cultured bacterial hosts.

## 5. EXAMPLE

Figure 11.3 outlines an example where this protocol was used to identify novel metabolites encoded by an eDNA gene cluster captured on two
overlapping clones (Feng et al., 2010). Degenerate primers based on conserved regions of minimal polyketide synthase (PKS) genes were used to amplify full-length ketosynthase$_\alpha$ genes (those encoding the alpha subunit of
the heterodimeric Type II ketosynthase) from DNA minipreps in an arrayed
eDNA library. Many of these genes exhibited low identity to known
ketosynthase$_\alpha$ sequences, suggesting that they were likely associated with
gene clusters distinct from previously sequenced clusters and might therefore
encode the biosynthesis of novel secondary metabolites. In the example outlined in Fig. 11.3, extracts from cultures of *Streptomyces albus* transformed
with the eDNA-derived Type II PKS-containing clone AB649 were found
to contain the tetracyclic polyketide intermediates rabelomycin and
dehydrorabelomycin. Complete sequencing of AB649 indicated that the
biosynthetic system responsible for the production of these metabolites
likely extended beyond the sequence captured on this clone. DNA

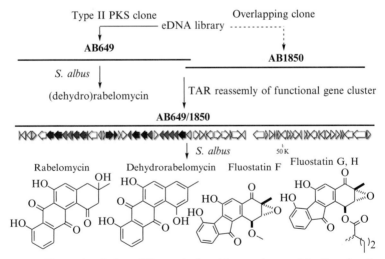

**Figure 11.3** Fluostatins F, G, and H were isolated from cultures of *S. albus* transformed
with a TAR-reassembled gene cluster captured on eDNA clones AB649 and AB1850. *S.
albus* transformed with clone AB649 alone produces only the intermediates
rabelomycin and dehydrorabelomycin. *S. albus* transformed with clone AB1950 produced no detectable clone-specific secondary metabolites. (For color version of this figure, the reader is referred to the online version of this chapter.)

minipreps from the arrayed eDNA library were rescreened with primers designed to recognize the terminal eDNA sequence captured in AB649, and the overlapping clone AB1850 was recovered in this screen. Using TAR, AB649 and AB1850 were reassembled into a large-insert BAC (AB649/1850) that was predicted to contain a complete biosynthetic gene cluster. By PCR analysis, restriction digestion, and full sequencing, this reassembled construct was found to be a faithful reassembly of the overlapping eDNA clones. Isolation and structure elucidation of metabolites obtained through heterologous expression of this intact gene cluster identified three novel fluostatins. The development of TAR as a tool for metagenomics makes it possible to routinely clone and functionally study even very large gene clusters directly from soil microbiomes.

## ACKNOWLEDGMENT

This work was supported by NIH GM077516. S. F. B. is a Howard Hughes Medical Institute early career scientist.

## REFERENCES

Banik, J. J., & Brady, S. F. (2008). Cloning and characterization of new glycopeptide gene clusters found in an environmental DNA megalibrary. *Proceedings of the National Academy of Sciences of the United States of America, 105,* 17273–17277.

Bentley, S. D., Chater, K. F., Cerdeno-Tarraga, A. M., Challis, G. L., Thomson, N. R., James, K. D., et al. (2002). Complete genome sequence of the model actinomycete Streptomyces coelicolor A3(2). *Nature, 417,* 141–147.

Bierman, M., Logan, R., O'Brien, K., Seno, E. T., Rao, R. N., & Schoner, B. E. (1992). Plasmid cloning vectors for the conjugal transfer of DNA from Escherichia coli to Streptomyces spp. *Gene, 116,* 43–49.

Binz, T. M., Wenzel, S. C., Schnell, H. J., Bechthold, A., & Muller, R. (2008). Heterologous expression and genetic engineering of the phenalinolactone biosynthetic gene cluster by using red/ET recombineering. *ChemBioChem, 9,* 447–454.

Brady, S. F. (2007). Construction of soil environmental DNA cosmid libraries and screening for clones that produce biologically active small molecules. *Nature Protocols, 2,* 1297–1305.

Brady, S. F., Simmons, L., Kim, J. H., & Schmidt, E. W. (2009). Metagenomic approaches to natural products from free-living and symbiotic organisms. *Natural Product Reports, 26,* 1488–1503.

Daniel, R. (2005). The metagenomics of soil. *Nature Reviews Microbiology, 3,* 470–478.

Feng, Z., Kim, J. H., & Brady, S. F. (2010). Fluostatins produced by the heterologous expression of a TAR reassembled environmental DNA derived type II PKS gene cluster. *Journal of the American Chemical Society, 132,* 11902–11903.

Gibson, D. G., Benders, G. A., Andrews-Pfannkoch, C., Denisova, E. A., Baden-Tillson, H., Zaveri, J., et al. (2008). Complete chemical synthesis, assembly, and cloning of a Mycoplasma genitalium genome. *Science, 319,* 1215–1220.

Gibson, D. G., Benders, G. A., Axelrod, K. C., Zaveri, J., Algire, M. A., Moodie, M., et al. (2008). One-step assembly in yeast of 25 overlapping DNA fragments to form a complete synthetic Mycoplasma genitalium genome. *Proceedings of the National Academy of Sciences of the United States of America, 105,* 20404–20409.

Handelsman, J., Rondon, M. R., Brady, S. F., Clardy, J., & Goodman, R. M. (1998). Molecular biological access to the chemistry of unknown soil microbes: A new frontier for natural products. *Chemical Biology, 5,* R245–R249.

Kieser, T., Bibb, M. J., Buttner, M. J., Chater, K. F., & Hopwood, D. A. (2000). *Practical streptomyces genetics.* Norwich, UK: The John Innes Foundation.

Kim, J. H., Feng, Z., Bauer, J. D., Kallifidas, D., Calle, P. Y., & Brady, S. F. (2010). Cloning large natural product gene clusters from the environment: Piecing environmental DNA gene clusters back together with TAR. *Biopolymers, 93,* 833–844.

Kouprina, N., & Larionov, V. (2008). Selective isolation of genomic loci from complex genomes by transformation-associated recombination cloning in the yeast Saccharomyces cerevisiae. *Nature Protocols, 3,* 371–377.

Kouprina, N., Noskov, V. N., & Larionov, V. (2006). Selective isolation of large chromosomal regions by transformation-associated recombination cloning for structural and functional analysis of mammalian genomes. *Methods in Molecular Biology, 349,* 85–101.

Larionov, V., Kouprina, N., Eldarov, M., Perkins, E., Porter, G., & Resnick, M. A. (1994). Transformation-associated recombination between diverged and homologous DNA repeats is induced by strand breaks. *Yeast, 10,* 93–104.

Larionov, V., Kouprina, N., Graves, J., Chen, X. N., Korenberg, J. R., & Resnick, M. A. (1996). Specific cloning of human DNA as yeast artificial chromosomes by transformation-associated recombination. *Proceedings of the National Academy of Sciences of the United States of America, 93,* 491–496.

Mathee, K., Narasimhan, G., Valdes, C., Qiu, X., Matewish, J. M., Koehrsen, M., et al. (2008). Dynamics of Pseudomonas aeruginosa genome evolution. *Proceedings of the National Academy of Sciences of the United States of America, 105,* 3100–3105.

Nagano, Y., Takao, S., Kudo, T., Iizasa, E., & Anai, T. (2007). Yeast-based recombineering of DNA fragments into plant transformation vectors by one-step transformation. *Plant Cell Reports, 26,* 2111–2117.

Pham, V. D., Palden, T., & DeLong, E. F. (2007). Large-scale screens of metagenomic libraries. *Journal of visualized experiments, 201.*

Rappe, M. S., & Giovannoni, S. J. (2003). The uncultured microbial majority. *Annual Review of Microbiology, 57,* 369–394.

Rondon, M. R., August, P. R., Bettermann, A. D., Brady, S. F., Grossman, T. H., Liles, M. R., et al. (2000). Cloning the soil metagenome: A strategy for accessing the genetic and functional diversity of uncultured microorganisms. *Applied and Environmental Microbiology, 66,* 2541–2547.

Shao, Z., Zhao, H., & Zhao, H. (2009). DNA assembler, an in vivo genetic method for rapid construction of biochemical pathways. *Nucleic Acids Research, 37,* e16.

Simon, C., & Daniel, R. (2009). Achievements and new knowledge unraveled by metagenomic approaches. *Applied Microbiology and Biotechnology, 85,* 265–276.

Simon, C., & Daniel, R. (2010). Construction of small-insert and large-insert metagenomic libraries. *Methods in Molecular Biology, 668,* 39–50.

Torsvik, V., Ovreas, L., & Thingstad, T. F. (2002). Prokaryotic diversity—magnitude, dynamics, and controlling factors. *Science, 296,* 1064–1066.

Torsvik, V., Salte, K., Sorheim, R., & Goksoyr, J. (1990). Comparison of phenotypic diversity and DNA heterogeneity in a population of soil bacteria. *Applied and Environmental Microbiology, 56,* 776–781.

Zhou, J., Bruns, M. A., & Tiedje, J. M. (1996). DNA recovery from soils of diverse composition. *Applied and Environmental Microbiology, 62,* 316–322.

CHAPTER TWELVE

# A Toolkit for Heterologous Expression of Metabolic Pathways in *Aspergillus oryzae*

**Khomaizon A. K. Pahirulzaman, Katherine Williams, Colin M. Lazarus[1]**
School of Biological Sciences, University of Bristol, Bristol, United Kingdom
[1]Corresponding author: e-mail address: c.m.lazarus@bristol.ac.uk

## Contents

## Abstract

Much has been learned about the activities of the key enzymes involved in eukaryotic natural product synthesis by isolating the relevant genes and expressing them in a suitable foreign host. *Aspergillus oryzae* has proved to be an amenable host for the functional analysis of megasynthases from other fungi, but secondary metabolites are often the products of suites of enzymes, and understanding their biosynthesis requires simultaneous expression of several genes. This chapter describes the development and use of a molecular toolkit that facilitates the rapid assembly of the genes constituting whole biosynthetic pathways in one or a few multiple gene expression plasmids designed to provide high-level expression in *A. oryzae*. Conventional DNA manipulation by restriction/ligation is replaced by homologous recombination in yeast and Gateway®-mediated site-specific recombination *in vitro*. The toolkit comprises an assembly vector used for the simple construction and modification of large genes from overlapping DNA fragments and three multigene expression vectors. Insertion of three tailoring enzyme genes by homologous recombination and one megasynthase gene

241

by Gateway® transfer into each of the expression vectors can be achieved in a little more than 1 week, and alternative selection markers in the expression plasmids permit cotransformation of *A. oryzae* with up to 12 genes.

# 1. INTRODUCTION

The recent explosion of data from microbial genome sequencing projects and their subsequent bioinformatic analysis have revealed a multitude of new natural product pathways. These have the potential to be exploited to produce new biologically active compounds, including agrochemicals and pharmaceuticals (Wilkinson & Micklefield, 2007). Heterologous expression systems for genes involved in natural product biosynthesis can be useful in elucidating their function as well as facilitating their further biochemical characterization and possible application. Heterologous expression of secondary metabolites has been achieved in both bacteria and fungi, with examples including expression of the actinorhodin biosynthetic gene cluster from *Streptomyces coelicolor* in *S. parvulus* (Malpartida & Hopwood, 1984) and expression of the 6-methylsalicylic acid synthase genes from *Penicillium patulum* in *S. coelicolor* (Bedford, Schweizer, Hopwood, & Khosla, 1995) and from *Aspergillus terreus in A. nidulans* (Fujii et al., 1996). However, finding a suitable, genetically tractable heterologous host can be challenging for a number of reasons. For example, differences in intron splicing, especially between less related organisms, may preclude the use of genomic fragments or require complicated removal of introns from the gene(s) prior to expression (Bedford et al., 1995), while differences in codon bias as well as incorrect protein folding can also be problematic. Some enzymes required for secondary metabolite synthesis, such as polyketide synthases (PKSs) and nonribosomal peptide synthetases (NRPSs), require posttranslational 4-phosphopantetheinylation catalyzed by 4-phosphopantetheinyl transferase (PPTase). If the heterologous host lacks endogenous PPTase activity, coexpression with a PPTase enzyme will be required (Lambalot et al., 1996). Additionally, specific primary metabolites may be required as starter or extender units for natural product synthesis. For these reasons, using a host that is similar to the donor organism may increase the likelihood of success; thus, heterologous expression of natural products from filamentous fungi may be more successful in a filamentous fungal host.

*Aspergillus* is a large and important genus within the filamentous fungi. It includes species which are human pathogens, such as *A. parasiticus* and

*A. fumigatus,* as well as mycotoxin-producing food contaminants such as *A. flavus.* It also includes *A. nidulans,* which is one of the most intensively studied species of filamentous fungi for basic genetic research. Species of commercial value include *A. terreus,* which produces the cholesterol-lowering secondary metabolite lovastatin, as well as *Aspergillus oryzae,* a fungus that has been widely used for over 2000 years in the Japanese fermentation industry to produce sake, miso, and soy sauce, and more recently in industrial enzyme production (Barbesgaard, Heldt-Hansen, & Diderichsen, 1992; Lubertozzi & Keasling, 2009). The genomes of several of the more important *Aspergillus* species, including *A. oryzae, A. nidulans,* and *A. fumigatus,* have been sequenced; the *A. oryzae* genome is 7–9 Mb larger than those of *A. nidulans* and *A. fumigatus* with the increase ascribed to genes for secretory hydrolytic enzymes as well as metabolic, degradation, and transport genes; this may be due to adaptations to domestication (Machida et al., 2005). Both *A. niger* and *A. oryzae* have GRAS (generally recognized as safe) status, which should facilitate approval of new food or drug products (Barbesgaard et al., 1992; Schuster, Dunn-Coleman, Frisvad, & van Dijck, 2002), but the ability of *A. oryzae* to secrete large amounts of protein makes it the more attractive heterologous host.

Transformation of filamentous fungi is inefficient compared to yeasts or *Escherichia coli,* mainly due to the presence of the fungal cell wall, as well as the lack of natural plasmids. Protoplast-mediated transformation, involving enzymatic removal of the cell wall, is the most commonly used method for *Aspergillus* species (Gomi, Iimura, & Hara, 1987). Nonhomologous or homologous integration of DNA may occur, and tandem insertions of plasmids are common (Weld, Plummer, Carpenter, & Ridgway, 2006). Other methods, including electroporation (Brown, Aufauvre-Brown, & Holden, 1998; Ward, Kodama, & Wilson, 1989), and *Agrobacterium*-mediated transformation (Meyer, Mueller, Strowig, & Stahl, 2003; Michielse, Hooykaas, van den Hondel, & Ram, 2008) have been utilized successfully in various *Aspergillus* species. Several auxotrophic strains of *A. oryzae* with corresponding selectable markers exist, including the arginine auxotroph M-2-3 that can be complemented by *arg*B from *A. nidulans* (Gomi et al., 1987), a uridine auxotroph complemented by the *pyr*G gene from *A. niger* (Mattern, Unkles, Kinghorn, Pouwels, & van den Hondel, 1987), and a nitrate assimilation mutant that can be complemented by the *nia*D gene from *A. oryzae* (Unkles et al., 1989). Dominant selectable markers such as the *bar* gene for use with glufosinate ammonium (Murakami et al., 1986) and the *ble* gene for use with the drug phleomycin (Mattern, Punt, & van den Hondel, 1988) can also be used.

## 2. FUNDAMENTAL ASPECTS OF VECTOR DEVELOPMENT

The fermentation process in *A. oryzae* is based on the abundant production of taka-amylase ($\alpha$-1,4-glucan-4-glucanhydrolase). The *amy*B gene encoding taka-amylase was cloned in 1989 (Tada et al., 1989); it is highly expressed in media containing substrate malto-oligosaccharides and is repressed by glycerol and glucose (Tada, Gomi, Kitamoto, Takahashi, et al., 1991). The inducible promoter, P*amy*B, was identified as consisting of the 613 bp upstream from the taka-amylase coding region (Tada, Gomi, Kitamoto, Kumagai, et al., 1991). The *amy*B expression cassette (promoter and terminator (T*amy*B)) was combined with the *A. nidulans arg*B selectable marker in the construction of plasmid pMAR5 (Shibuya, Tsuchiya, Tamura, Ishikawa, & Hara, 1992), from which pTAex3 was subsequently derived (Fujii, Yamaoka, Gomi, Kitamoto, & Kumagai, 1995). Conventional use of this vector involves ligation of the target coding region at unique *Eco*RI and/or *Sma*I recognition sequences in the expression site (i.e., between P*amy*B and T*amy*B). Its value in natural product research was first demonstrated by expression of the *A. nidulans w*A gene in *A. oryzae*, resulting in the identification of the novel naphthopyrone YWA1, a yellow precursor of the wild-type green spore pigment (Watanabe et al., 1999, 1998).

Use of the pTAex3/*A. oryzae* M-2-3 vector/host combination to functionally analyze a PKS gene isolated from *Phoma* sp. demonstrated the potential difficulties encountered in manipulating the large genes that encode megasynthases. Several steps of DNA fragment isolation and ligation were required to assemble the target coding region, whose eventual expression identified the gene product as squalestatin tetraketide synthase (Cox et al., 2004). The isolation of genes encoding even larger hybrid PKS–NRPSs, with coding regions of approximately 12 kb (Eley et al., 2007; Song, Cox, Lazarus, & Simpson, 2004), prompted the introduction of the Gateway® gene transfer mechanism into the expression system, greatly facilitating movement of coding regions from a Gateway® entry vector into the expression site (Section 4.1.4). This alone, however, did not address the difficulties of assembling megasynthase coding regions from fragments of less-than-full-length sequences and placing the product in a Gateway® entry vector. These problems were overcome by converting a basic Gateway® entry vector into a yeast-*E. coli* shuttle vector, in which genes could be assembled by homologous recombination in *Saccharomyces cerevisiae* (Section 4.1.1), and converting pTAex3 to a Gateway® destination vector (Fig. 12.1).

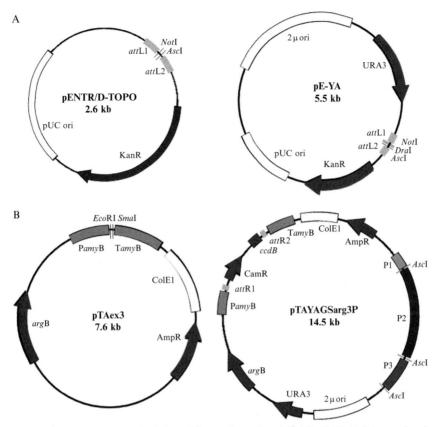

**Figure 12.1** Structures and origins of the toolbox plasmids. (A) pE-YA (right) was developed from pENTR/D-TOPO (left) as a basic gene assembly/modification vector. Coding regions constructed by homologous recombination in yeast can be transferred to expression cassettes by Gateway®-mediated site-specific recombination *in vitro*. See text for construction details. (B) pTAYAGSarg3P (right) was developed from pTAex3 (left) as a multigene expression vector. Coding regions can be placed under the control of all four promoters (P*adh* (P1), P*gpd*A (P2), P*eno* (P3), P*amy*B) by homologous recombination in yeast, but the *amy*B expression cassette was designed to be filled by Gateway®-mediated site-specific recombination *in vitro*. Note that in alternative versions of this plasmid, pTAYAGSbar3P and pTAYAGSble3P, the *arg*B selectable marker is replaced by P*trp*C::*bar* and P*trp*C::*ble*, respectively. See text for construction details. (See Color Insert.)

*S. cerevisiae* is one of the most intensively studied eukaryotic organisms, and therefore much molecular technology is available (Barr, 2003; Forsburg, 2001). One very useful property of *S. cerevisiae* is its propensity to perform homologous recombination, a mechanism of double-stranded-break repair that uses sequence homology to recombine DNA fragments (Orr-Weaver & Szostak,

1985). This process can be used to efficiently join a linearized vector and DNA fragment(s), if there is overlap of homologous sequence between them. Recombination events can be arranged to convert the linear vector to a circular recombinant plasmid that will transform yeast cells at high efficiency (Raymond, Pownder, & Sexson, 1999). In *S. cerevisiae*, less than 30 bp of overlap is required for efficient homologous recombination. This allows for incredible versatility (Hua, Qiu, Chan, Zhu, & Luo, 1997; Oldenburg, Vo, Michaelis, & Paddon, 1997) as 30 bases can easily be added to a standard PCR primer, allowing precise control over where recombination events should occur and bypassing the common problems associated with assembling multiple DNA fragments into large expressible genes. Homologous recombination in *S. cerevisiae* can have added advantages; for example, this method can be used to produce gene fusions with markers such as enhanced green fluorescent protein (eGFP) (Chalfie, Tu, Euskirchen, Ward, & Prasher, 1994; Cubitt et al., 1995) and to splice out introns (Section 4.1.2).

The largest *Dra*I fragment of shuttle vector pYES-DEST52 (Invitrogen) contains the 2 μ origin of replication and the *URA3* gene for plasmid maintenance and selection in *ura3* mutant strains of yeast (essentially all the standard cloning hosts). Ligation of this fragment into the unique *Hpa*I site of pENTR/D-TOPO (Invitrogen) and slight modification of the cloning site created pE-YA (Fig. 12.1A). Assembly of a coding region between the *Not*I and *Asc*I sites of this vector places it between the *att*L1 and *att*L2 sites necessary for LR-recombination-mediated transfer into any Gateway® destination vector (Section 4.1.4). Conversion of the expression vector involved insertion of a Gateway® destination cassette, comprising a chloramphenicol resistance gene and the toxic *ccd*B gene flanked by *att*R1 and *att*R2 sites (Invitrogen), into the expression site of pTAex3 (Fig. 12.1B). Chloramphenicol-resistant transformants of *E. coli* strain DB3.1 (Invitrogen), which is refractory to the toxic effects of the *ccd*B gene, were screened for insert orientation, and the plasmid with the *att*R1 and *att*R2 sites adjacent to P*amy*B and T*amy*B, respectively, was designated pTAex3GS. Variants of this plasmid were made by replacing the *arg*B selectable marker with glufosinate ammonium (pTAbarGS) and phleomycin (pTAbleGS) resistance markers.

## 3. HETEROLOGOUS EXPRESSION OF A FUNGAL NATURAL PRODUCT PATHWAY IN *A. oryzae*

### 3.1. Pathway reconstruction by sequential transformation

Although it can be informative to investigate the heterologous expression of a single gene, many natural products arise from biosynthetic pathways. In the case of polyketides, the basic skeleton elaborated by a megasynthase is modified by one or more tailoring enzymes to produce the final compound. Elucidation of the biosynthesis of tenellin, an amino-acylated polyketide produced by *Beauveria bassiana*, was achieved by heterologous expression in *A. oryzae*, culminating in the successful reconstruction of the whole biosynthetic pathway (Heneghan et al., 2010). This work will be used to exemplify the use of pE-YA and pTAex3GS (and derivatives thereof) in step-by-step pathway reconstruction.

The tenellin biosynthetic gene cluster, encoding a hybrid PKS–NRPS, a *trans*-acting enoyl reductase (ER) and two cytochrome P450s (P450-1 and P450-2), was isolated from *B. bassiana* (Eley et al., 2007). The *tenS* gene encoding the megasynthase (TENS) was assembled in pE-YA by homologous recombination in yeast (Section 4.1.1), transferred to pTAex3GS by Gateway® LR site-specific recombination *in vitro* (Fig. 12.4 inset), and the resultant plasmid (pTAtenS) was used to transform *A. oryzae* (Halo, Marshall, et al., 2008). Expressed alone, the TENS produced aberrant compounds related to tenellin. In addition to confirming that correct programming may be lost when a megasynthase is isolated from other members of its biosynthetic pathway (Kennedy et al., 1999), the results identified the C-terminal domain of TENS as a Dieckmann cyclase responsible for releasing the elaborated compound from the enzyme. Correct programming was restored to TENS by coexpression with the ER encoded by *tenC*. The *tenC* coding region was PCR-amplified, cloned in pENTR/D-TOPO, and transferred to the *amyB* expression cassette of pTAbarGS by Gateway® LR recombination; the resultant plasmid, pBar*ten*C, was used to transform a (prototrophic) TENS transformant to glufosinate ammonium resistance. Coexpression of ER with TENS resulted in identification of the first authentic precursor of tenellin, pretenellin A (Fig. 12.2A) (Halo, Marshall, et al., 2008). The same scheme was followed to place each of the P450 genes (*tenA* and *tenB*) individually into pTAbleGS and to transform a TENS + ER transformant to phleomycin resistance. Pretenellin A production was unchanged in the P450-2 transformant, but P450-1 catalyzed an unprecedented ring expansion reaction to convert pretenellin A to pretenellin B (Fig. 12.2A) (Halo, Heneghan, et al., 2008). To complete the tenellin pathway in *A. oryzae*, in the absence of further selection markers, the entire P450-2 expression cassette (P*amy*B-*ten*B-T*amy*B) was PCR-amplified from pBle*ten*B and ligated into pBle*ten*A. The resultant plasmid, pBle*ten*A*ten*B, was used to transform the TENS + ER transformant to phleomycin resistance. The production of tenellin, at five times the level observed for the native organism (*B. bassiana*) in culture, confirmed that P450-2 catalyzed the *N*-hydroxylation of pretenellin B (Fig. 12.2A) (Heneghan et al., 2010).

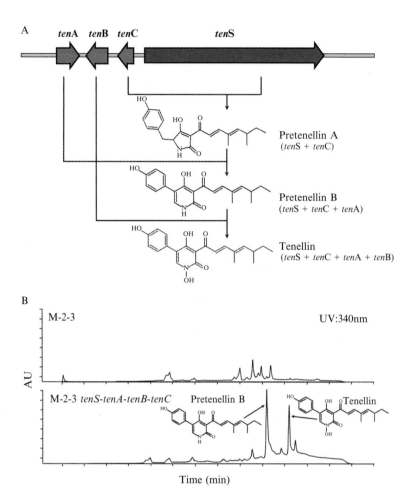

**Figure 12.2** Tenellin biosynthesis. (A) The tenellin biosynthetic pathway. Combined activities of the tenellin synthase (PKS–NRPS encoded by *ten*S) and a *trans*-acting enoyl reductase (encoded by *ten*C) produce the first pathway intermediate, pretenellin A. Ring expansion by a first cytochrome P450 (encoded by *ten*A) converts pretenellin A to pretenellin B, which is *N*-hydroxylated by a second cytochrome P450 (encoded by *ten*B) to tenellin. (B) Tenellin production in *A. oryzae* mediated by one-step transformation. The four chimaeric genes in the multigene expression plasmid pTAYAargTenellin are P*amy*B::*ten*S, P*adh*::*ten*A, P*gpd*A::*ten*C, and P*eno*::*ten*B. (See Color Insert.)

## 3.2. Pathway reconstruction in a single plasmid

Selection marker imposed limitation on *A. oryzae* transformation prompted the development of a multiple gene expression vector, capable of transferring four genes in a single transformation event. Homologous recombination in yeast was employed in vector construction and is a central feature of its application. In a first step, pTAex3GS was converted to a yeast–*E. coli* shuttle vector by transforming yeast with the *Stu*I-cut plasmid and a PCR product containing the 2 μ origin of replication and *URA3* gene amplified from pE-YA. The PCR primers were extended by 30 b sequences corresponding to the cut ends of pTAex3GS, allowing circularization to occur by homologous recombination and selection of colonies containing the shuttle vector pTAYAGS. The second step of development was to introduce three additional promoters to produce the four-site vector. Since both construction of this vector and its subsequent deployment (i.e., introduction of genes to be expressed) would be by recombination in yeast, the new promoters had to be nonhomologous to each other and to the *amy*B promoter already present. An initial construct used three *A. nidulans* promoters that are frequently used to drive resistance marker expression in a range of fungi: P*gpd*A (Punt et al., 1990), P*trp*C (Hamer & Timberlake, 1987), and P*oli*C (Turner et al., 1989). On functional testing, only P*gpd*A was found to drive expression of eGFP at an acceptable level in *A. oryzae* and, while quantitative RT-PCR showed it to be a stronger promoter than P*amy*B, the other two promoters were much weaker. Alternatives to P*trp*C and P*oli*C were selected by reference to an *A. oryzae* EST analysis (Akao et al., 2007), which showed that alcohol dehydrogenase (*adh*) and enolase (*eno*) genes are expressed at very high levels under the culture conditions used for heterologous gene expression from P*amy*B. To construct the four-site vector, pTAYAGS was linearized with *Ngo*MIV and combined with the PCR-amplified promoters P*adh*, P*gpd*A, and P*eno* to transform yeast cells. The *A. oryzae* promoters were assumed to be contained within approximately 500 bp upstream of the start of translation of their respective genes. All promoters were amplified with primer extensions to allow overlap with the cut ends of the plasmid (P*adh* and P*eno*) and/or each other, and to introduce an *Asc*I site downstream of each promoter. Figure 12.1B shows a map of the resultant plasmid, designated pTAYAGSarg3P, and a description of its use is given below (Sections 4.1.3 and 4.1.4). Its effectiveness was demonstrated by introducing the four genes that comprise the tenellin biosynthetic pathway and generating tenellin-producing *A. oryzae* in a single transformation (Fig. 12.2B). To further extend capability to pathways comprising more than four genes, such as the fumonisin biosynthetic pathway (Alexander, Proctor, & McCormick, 2009), the *arg*B selection marker was replaced by *bar* and *ble* genes to create pTAYAGSbar3P and pTAYAGSble3P, respectively. Co- or sequential transformation with constructs based on all three vectors should allow the heterologous expression of up to 12 genes in *A. oryzae*.

# 4. METHODS

The essential processes involved in biosynthetic pathway reconstruction are outlined in the following sections. Descriptions are generic, rather than detailed to the level of experimental protocols, with practical tips included where appropriate. The assumption is also made that most future applications will arise directly from genome sequence determinations, and all genes will be obtained by PCR; thus it should be noted that it is equally possible to manipulate restriction fragments of cloned genes available in libraries. The following procedures mirror those used to assemble the four-gene tenellin biosynthetic pathway from *B. bassiana*.

## 4.1. Biosynthetic pathway construction in the multigene expression vector

### 4.1.1 Reassembly of the megasynthase gene (Fig. 12.3)

1. Linearize pE-YA ($\sim$1 µg per transformation) by digestion with *Dra*I.
2. Amplify megasynthase gene from genomic DNA (or cDNA) as a number of overlapping fragments. A 12-kb coding region is conveniently amplified as three $\sim$4-kb fragments with $>$30-bp overlaps. The coding region start and end primers should be 5′-extended by 30 b of vector sequence upstream (includes *Not*I site) and downstream (includes *Asc*I site), respectively, of the *Dra*I site. A reliable high-fidelity DNA polymerase, such as Phusion (Thermo Scientific), must be used.
3. Analyze samples of the restriction digest and the PCRs by agarose gel electrophoresis. Ensure that the vector is fully digested; residual circular plasmid will generate a high background of nonrecombinant yeast transformants. Check that the gene fragments are free of contaminating off-target products and gel-purify target products if necessary; reactions yielding only target PCR products do not require any cleanup. Roughly estimate relative vector and PCR product concentrations.
4. Prepare and transform competent yeast cells by the high-efficiency TRAFO methods (http://home.cc.umanitoba.ca/~gietz/) (Gietz & Woods, 2002). The transforming DNA should include $\sim$1 µg of linearized vector and roughly equimolar amounts of PCR products in a total volume of 34 µl. Perform a parallel control transformation lacking one of the PCR products (which should prevent circle formation) to assess the transformation background. Any standard yeast gene expression host

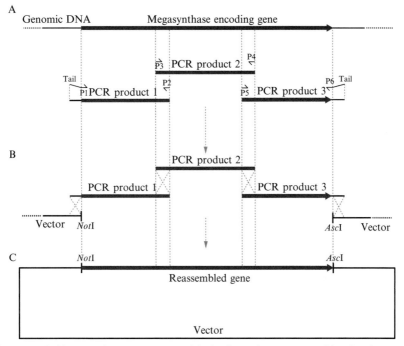

**Figure 12.3** Megasynthase gene assembly by homologous recombination in yeast. (A) Amplification of the coding region as a convenient number of overlapping PCR products. (B) Transformation of yeast with linearized shuttle vector (*Dra*I-cut pE-YA) and overlapping gene fragments. (C) Recombinant plasmid generated by homologous recombination. In pE-YA, the assembled gene is flanked by *att*L sites and is thus available for Gateway®-mediated transfer to any destination vector. It is also available for modification by further rounds of homologous recombination in yeast. (For color version of this figure, the reader is referred to the online version of this chapter.)

strain can be used. After 2–3 days' incubation on a minimal medium lacking uracil, visual comparison of test and control plates should indicate the likelihood of successful gene assembly. While the presence of a fair number of colonies on the control plate is usually observed, colonies on the test plate can be close to confluent.

5. Shuttle recombinant plasmids to *E. coli* for analysis. Selection of individual yeast transformant colonies is unnecessary. Plasmids can be prepared from multiple colonies *en masse* by scraping a toothpick across the test plate surface to collect a heavy bead of cells. Plasmid yields from yeast cells tend to be low, but the Zymoprep™ yeast miniprep II kit (Zymo Research) is reliable; when using this kit, elute the DNA with 10 μl

water and use 1 μl (or more if necessary) to transform *E. coli* to kanamycin resistance (most efficiently by electroporation).

6. Screen a small number of *E. coli* colonies for the presence of the expected gene fragments by PCR (optional) or proceed directly to plasmid preparation for screening by restriction analysis. A correctly assembled plasmid, comprising a complete megasynthase coding region between *Not*I and *Asc*I sites in pE-YA, is designated pE-YA-mega. It may be modified as required by further rounds of homologous recombination in yeast as in the following examples.

### 4.1.2 Modification of the megasynthase gene

1. *C-terminal tagging with eGFP*: Transformation of *A. oryzae* to produce a megasynthase may yield a mixture of expressing and nonexpressing transformants. The latter may arise because nonhomologous plasmid integration into the host genome can occur by crossing over within the megasynthase coding region, thereby disrupting it. C-terminal tagging and selection of green fluorescent transformants for expression analysis ensure that only clones producing the full-length megasynthase are analyzed. The observation of only nonfluorescing transformants can indicate that the megasynthase coding region contains a catastrophic mutation—a potential hazard of gene isolation by PCR—resulting in truncated protein production.

    i. Digest pE-YA-mega with *Asc*I.

    ii. PCR amplify the eGFP-coding region (including stop codon) from a source plasmid; the forward primer should have a 30-b 5′ extension corresponding to the last 10 codons (excluding the stop codon) of "mega," and the 3′ primer should have a 30-b 5′ extension corresponding to sequence downstream of the vector *Asc*I site.

    iii. Transform yeast with the linearized plasmid and PCR product as above.

2. *Intron removal/domain swapping*: In general, use of *A. oryzae* as a host for the heterologous expression of fungal genes allows constructs to be made using genomic fragments. However, examples of aberrant splicing have been encountered and overcome by intron removal in yeast. Similarly, functional analysis of catalytic domains within a megasynthase has been achieved by excision of a native domain and replacement by the equivalent domain from another megasynthase (Fisch et al., 2011). In both cases, one or two restriction sites must be identified that flank or lie within the region to be replaced. Unfortunately, digestion of

pE–YA-mega with the selected enzymes can result in cuts at additional points in the insert or vector sequence. These can be repaired by short (e.g., 250-bp) patches generated by PCR performed on the intact plasmid with primers that flank the nontarget restriction sites. In the case of intron removal:

    **i.** Digest pE–YA-mega with the selected enzyme(s).

    **ii.** Amplify the deleted intron-containing region from cDNA (e.g., library template or by RT-PCT) with primers that flank the cut sites of pE–YA-mega by at least 30 b. Alternatively, use overlap-extension PCR to create the intron-free fragment.

    **iii.** Amplify any patch sequences required.

    **iv.** Transform yeast with the cut vector and PCR product(s).

### 4.1.3 Assembling tailoring enzyme genes in the multigene expression vector (Fig. 12.4)

**1.** Digest pTAYAGSarg3P ($\sim$1 μg per transformation) with *Asc*I. This generates three fragments: the pTAYAGS vector plus P*adh* (11.6 kb), P*gpd*A (2.2 kb), and P*eno* (0.5 kb).

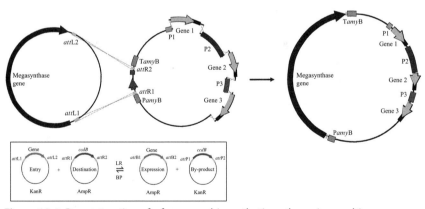

**Figure 12.4** Reconstruction of a four-gene biosynthetic pathway in a multigene expression vector. The vector (middle) is shown digested with *Asc*I to produce three fragments and ready to accept three coding regions (+UTRs) by homologous recombination in yeast. Site-specific recombination between the vector *att*R sites and the *att*L sites of an entry vector (left) is used to introduce a fourth coding region (typically of a megasynthase) into the *amy*B expression cassette. The two recombination reactions can be performed in either order to produce the fully constructed plasmid (right). The inset sketches the Gateway® reactions. (See Color Insert.)

2. Amplify the tailoring enzyme coding regions ($+3'$ UTRs) from genomic DNA. These coding regions are generally in the range 1–2 kb and so are amplified as single fragments. The coding-region-start primer should be $5'$-extended by the 30 b corresponding to the end of the relevant promoter. The reverse primer should prime 250–300 bp downstream of the end of the coding region to provide a "terminator" directing polyadenylation of transcripts when expressed in *A. oryzae*; it should be $5'$-extended by 30 b to overlap with the start of the next promoter or with the vector end sequence. If only one or two tailoring enzymes are to be used, the downstream overlap sequences can be adjusted accordingly. For example, to insert a single coding region ($+$UTR), it should be amplified with the forward primer extension that overlaps the $3'$ end of P*adh* (at one end of the vector sequence) and the reverse primer extension that overlaps with the other end of the vector.

3. Analyze samples of the restriction digest and PCRs by agarose gel electrophoresis. Estimation of the relative concentrations of plasmid and PCR products is made easier by the fact that the PCR products migrate with roughly the same mobility as P*gpd*A.

4. Transform yeast with a total of six DNA fragments (three from the plasmid digest plus three PCR products). Use the same background control and posttransformation procedures as described above. Note, however, that the presence of the Gateway® destination module within the *amy*B expression cassette dictates the use of *ccd*B-resistant cells, such as the DB3.1 or ccdB Survival (Invitrogen) strains of *E. coli*, when shuttling plasmids from yeast. The plasmid containing the three tailoring enzyme genes correctly assembled in pTAYAGSarg3P is designated pTAYAGSarg3Genes.

### 4.1.4 Insertion of the megasynthase gene in the multigene expression vector (Fig. 12.4)

1. The system is designed to utilize site-specific recombination *in vitro* (the Gateway® LR reaction—see inset in Fig. 12.4) to place the reassembled megasynthase coding region into the *amy*B expression cassette. This can be achieved in half the reaction volume recommended by the enzyme manufacturer (10 µl, Invitrogen). The reaction mixture comprises 1 µl each of the entry and destination vectors (standard minipreps of pE-YA-mega and pTAYAGSarg3Genes), 1 µl Clonase II enzyme mixture (Invitrogen), and 2 µl TE buffer. After incubation at 25 °C for at least 1 h (4–5 h is common), and without the need for any reaction

termination procedure (Invitrogen recommends incubation with pro-teinase K, but this is unnecessary), 1 μl of the reaction mix is used to transform any standard *E. coli* host strain (e.g., TOP10) to ampicillin resistance (most efficiently by electroporation). After incubation, the reaction mix will contain four plasmids: the two unreacted parental plasmids and two recombinants in which sequences originally between *att*L and *att*R sites have exchanged places. Of these, only the desired recombinant plasmid (designated pTAYAarg4Genes, with the megasynthase gene between *att*B sites in the *amy*B cassette) should yield colonies, as it is the only one that both lacks the toxic *ccd*B gene and has an ampicillin resistance marker. A small number of colonies should be analyzed for correct plasmid structure.

2. Gateway® technology provides an extremely efficient means to move any sequence located between *att*L sites of an entry vector to any destination vector with complementary *att*R sites. Thus, the reassembly of the megasynthase in an entry vector affords great flexibility in its subsequent deployment; despite its considerable length, it can be moved easily with-out risk of disruption, mutation, or loss of orientation to any Gateway®-adapted expression vector. If this convenience is not required, however, then homologous recombination in yeast can be used to reassemble the megasynthase directly in the *amy*B expression cassette of pTAYAGSarg3P before or after inserting the tailoring enzyme genes. *Eco*RI, *Kpn*I, and *Not*I can all be used to linearize pTAYAGSarg3P between P*amy*B and T*amy*B (but pTAYAGSarg3Genes may contain additional sites), and the start and end fragments of the megasynthase are amplified with primer extensions corresponding to the end of P*amy*B and the start of T*amy*B, respectively. Prior insertion of the tailoring enzyme genes may be more convenient in the unlikely event of the 8-base recognition sequence of *Asc*I being pre-sent in the megasynthase gene, but any such problem can be overcome by introducing a suitable patch fragment together with the tailoring enzyme genes.

## 4.2. Transformation of *A. oryzae* and analysis of transformants

The aim of this chapter has been to describe the molecular toolkit developed for the rapid assembly of biosynthetic pathways, in one or more multiple gene expression vectors, for heterologous expression in *A. oryzae*. Detailed fungal transformation protocols are outside the scope of the chapter, but this section is intended to provide some notes relating to the deployment of the completed plasmid constructs.

1. Sporulation of *A. oryzae* M-2-3 can be intermittent, and the appearance of cultures on solid spore-production medium may be deceptive. However, spore-less aerial mycelium scraped from the surface of plates can be used in place of spores for protoplast production.

2. Use approximately 10 µg of plasmid per transformation. Standard plasmid minipreps may be used, but midi-prepped plasmids tend to be purer and are preferable.

3. Selection of transformants by nutritional complementation commences directly on plating, whereas antibiotics are added by top layering after a recovery period of about 24 h postplating. Putative transformants are taken through two rounds of hyphal subculturing followed by streaking to single colonies on selective medium to ensure genetic purity. Note that *A. oryzae* is sensitive to zeocin (growth fully inhibited at 25 µg/ml in a minimal medium), which is much cheaper than bleomycin or phleomycin and can be used to select *ble* transformants.

4. Selected transformants are cultured in rich medium for metabolite production. Organic extracts are made from mycelia homogenized in their growth media, and initial analysis is by LC–MS. Figure 12.2B shows the HPLC trace of metabolites extracted from *A. oryzae* transformed with plasmid pTAYAargTenellin (=pTAYAarg4Genes, where the four genes encode the PKS–NRPS, ER, and two cytochrome P450s that comprise the tenellin biosynthetic pathway).

## 5. SUMMARY

The GRAS status of *A. oryzae,* as well as its long history of domestication, makes it a particularly attractive host for heterologous expression of natural product pathways from other filamentous fungi. The fungus is susceptible to a variety of transformation protocols using a range of nutritional and dominant selectable markers. Natural product research at the molecular level has tended to focus on the megasynthases responsible for polyketide or non-ribosomal peptide assembly, and much has been learned by expressing megasynthase genes in *A. oryzae.* Some investigations have highlighted the importance of other genes in the biosynthetic pathway, leading to coexpression experiments culminating in the heterologous expression of whole biosynthetic pathways. This approach has prompted the development of a molecular toolkit, based on homologous recombination in *S. cerevisiae* and Gateway®-mediated site-specific recombination *in vitro*, that addresses the problems associated with the efficient manipulation of multiple genes, some of which are very long. By replacing conventional restriction and ligation with homologous recombination, the toolkit makes it possible to transfer all the large and small genes encoding a complex eukaryotic biosynthetic pathway from genomic DNA to expression cassettes in one or a few plasmids in a little over a week. The whole assembly system is a simple, fast, and adaptable technique for cloning and expressing genes. The versatility of this system means that techniques such as gene fusions, domain swaps, and multigene assembly can all be easily incorporated into simple procedures. Furthermore, the system is not restricted to fungal natural products but can be adapted to express primary or secondary metabolic pathways from any source.

## ACKNOWLEDGMENTS

This work was supported by BBSRC grant BB/E007791/1. K. W. was supported by a BBSRC Doctoral Training Grant and K. A. K. P. by Majlis Amanah Rakyat and Institute for Medical Research (Malaysia). We thank past and present members of the Bristol Polyketide Group whose work has provided the background to this chapter.

## REFERENCES

Akao, T., Sano, M., Yamada, O., Akeno, T., Fujii, K., Goto, K., et al. (2007). Analysis of expressed sequence tags from the fungus *Aspergillus oryzae* cultured under different conditions. *DNA Research, 14*, 47–57.

Alexander, N. J., Proctor, R. H., & McCormick, S. P. (2009). Genes, gene clusters, and biosynthesis of trichothecenes and fumonisins in *Fusarium. Toxin Reviews, 28*, 198–215.

Barbesgaard, P., Heldt-Hansen, H. P., & Diderichsen, B. (1992). On the safety of *Aspergillus oryzae*—A review. *Applied Microbiology and Biotechnology, 36*, 569–572.

Barr, M. M. (2003). Super models. *Physiological Genomics, 13*, 15–24.

Bedford, D. J., Schweizer, E., Hopwood, D. A., & Khosla, C. (1995). Expression of a functional polyketide synthase in the bacterium *Streptomyces coelicolor* A3(2). *Journal of Bacteriology, 177*, 4544–4588.

Brown, J. S., Aufauvre-Brown, A., & Holden, D. W. (1998). Insertional mutagenesis of *Aspergillus fumigatus. Molecular & General Genetics, 259*, 327–335.

Chalfie, M., Tu, Y., Euskirchen, G., Ward, W. W., & Prasher, D. C. (1994). Green fluorescent protein as a marker for gene expression. *Science, 263*, 802–805.

Cox, R. J., Glod, F., Hurley, D., Lazarus, C. M., Nicholson, T. P., Rudd, B. A. M., et al. (2004). Rapid cloning and expression of a fungal polyketide synthase gene involved in squalestatin biosynthesis. *Chemical Communications*, 2260–2261.

Cubitt, A. B., Heim, R., Adams, S. R., Boyd, A. E., Gross, L. A., & Tsien, R. Y. (1995). Understanding, improving and using green fluorescent proteins. *Trends in Biochemical Sciences, 20*, 448–455.

Eley, K. L., Halo, L. M., Song, Z. S., Powles, H., Cox, R. J., Bailey, A. M., et al. (2007). Biosynthesis of the 2-pyridone tenellin in the insect pathogenic fungus *Beauveria bassiana. ChemBioChem, 8*, 289–297.

Fisch, K. M., Bakeer, W., Yakasai, A. A., Song, Z., Pedrick, J., Wasil, Z., et al. (2011). Rational domain swaps decipher programming in fungal highly reducing polyketide synthases and resurrect an extinct metabolite. *Journal of the American Chemical Society, 133*, 16635–16641.

Forsburg, S. L. (2001). The art and design of genetic screens: Yeast. *Nature Reviews. Genetics*, *2*, 659–668.

Fujii, I., Ono, Y., Tada, H., Gomi, K., Ebizuka, Y., & Sankawa, U. (1996). Cloning of the polyketide synthase gene *at*X from *Aspergillus terreus* and its identification as the 6-methylsalicylic acid synthase gene by heterologous expression. *Molecular & General Genetics*, *253*, 1–10.

Fujii, T., Yamaoka, H., Gomi, K., Kitamoto, K., & Kumagai, C. (1995). Cloning and nucleotide sequence of the ribonuclease T1 gene (*rnt*A) from *Aspergillus oryzae* and its expression in *Saccharomyces cerevisiae* and *Aspergillus oryzae*. *Bioscience, Biotechnology, and Biochemistry*, *59*, 1869–1874.

Gietz, R. D., & Woods, R. A. (2002). Transformation of yeast by the LiAc/SS carrier DNA/PEG method. *Methods in Enzymology*, *350*, 87–96.

Gomi, K., Iimura, Y., & Hara, S. (1987). Integrative transformation of *Aspergillus oryzae* with a plasmid containing the *Aspergillus nidulans arg*B gene. *Agricultural and Biological Chemistry*, *51*, 2549–2555.

Halo, L. M., Heneghan, M. N., Yakasai, A. A., Song, Z., Williams, K., Bailey, A. M., et al. (2008). Late stage oxidations during the biosynthesis of the 2-pyridone tenellin in the entomopathogenic fungus *Beauveria bassiana*. *Journal of the American Chemical Society*, *130*, 17988–17996.

Halo, L. M., Marshall, J. W., Yakasai, A. A., Song, Z., Butts, C. P., Crump, M. P., et al. (2008). Authentic heterologous expression of the tenellin iterative polyketide synthase nonribosomal peptide synthetase requires coexpression with an enoyl reductase. *ChemBioChem*, *9*, 585–594.

Hamer, J. E., & Timberlake, W. E. (1987). Functional organization of the *Aspergillus nidulans trp*C promoter. *Molecular and Cellular Biology*, *7*, 2352–2359.

Heneghan, M. N., Yakasai, A. A., Halo, L. M., Song, Z., Bailey, A. M., Simpson, T. J., et al. (2010). First heterologous reconstruction of a complete functional fungal biosynthetic multigene cluster. *ChemBioChem*, *11*, 1508–1512.

Hua, S.-B., Qiu, M., Chan, E., Zhu, L., & Luo, Y. (1997). Minimum length of sequence homology required for *in vivo* cloning by homologous recombination in yeast. *Plasmid*, *38*, 91–96.

Kennedy, J., Auclair, K., Kendrew, S. G., Park, C., Vederas, J. C., & Hutchinson, C. R. (1999). Modulation of polyketide synthase activity by accessory proteins during lovastatin biosynthesis. *Science*, *284*, 1368–1372.

Lambalot, R., Gehring, A. M., Fluge, R. S., Zuber, P., LaCelle, M., Marahiel, M. A., et al. (1996). A new enzyme superfamily—The phosphopantetheinyl transferases. *Chemistry & Biology*, *3*, 923–936.

Lubertozzi, D., & Keasling, J. D. (2009). Developing *Aspergillus* as a host for heterologous expression. *Biotechnology Advances*, *27*, 53–75.

Machida, M., Asai, K., Sano, M., Tanaka, T., Kumagai, T., Terai, G., et al. (2005). Genome sequencing and analysis of *Aspergillus oryzae*. *Nature*, *438*, 1157–1161.

Malpartida, F., & Hopwood, D. A. (1984). Molecular cloning of the whole biosynthetic pathway of a *Streptomyces* antibiotic and its expression in a heterologous host. *Nature*, *309*, 462–464.

Mattern, I. E., Punt, P. J., & van den Hondel, C. A. M. J. J. (1988). A vector of *Aspergillus* transformation conferring phleomycin resistance. *Fungal Genetics Newsletter*, *35*, 25.

Mattern, I. E., Unkles, S., Kinghorn, J. R., Pouwels, P. H., & van den Hondel, C. A. M. J. J. (1987). Transformation of *Aspergillus oryzae* using the *Aspergillus niger pyr*G gene. *Molecular & General Genetics*, *210*, 460–461.

Meyer, V., Mueller, D., Strowig, T., & Stahl, U. (2003). Comparison of different transformation methods for *Aspergillus giganteus*. *Current Genetics*, *43*, 371–377.

Michielse, C. B., Hooykaas, P. J. J., van den Hondel, C. A. M. J. J., & Ram, A. F. J. (2008). *Agrobacterium*-mediated transformation of the filamentous fungus *Aspergillus awamori*. *Nature Protocols*, *3*, 1671–1678.

Murakami, T., Anzai, H., Imai, S., Satoh, A., Nagaoka, K., & Thompson, C. (1986). The bialaphos biosynthetic genes of *Streptomyces hygroscopicus*: Molecular cloning and characterization of the gene cluster. *Molecular & General Genetics*, *205*, 42–53.

Oldenburg, K. R., Vo, K. T., Michaelis, S., & Paddon, C. (1997). Recombination-mediated PCR-directed plasmid construction *in vivo* in yeast. *Nucleic Acids Research*, *25*, 451–452.

Orr-Weaver, T. L., & Szostak, J. W. (1985). Fungal recombination. *Microbiological Reviews*, *49*, 33–58.

Punt, P. J., Dingemanse, M. A., Kuyvenhoven, A., Soede, R. D. M., Pouwels, P. H., & van den Hondel, C. A. M. J. J. (1990). Functional elements in the promoter region of the *Aspergillus nidulans gpd*A gene encoding glyceraldehyde-3-phosphate dehydrogenase. *Gene*, *93*, 101–109.

Raymond, C. K., Pownder, T. A., & Sexson, S. L. (1999). General method for plasmid construction using homologous recombination. *BioTechniques*, *26*, 134–141.

Schuster, E., Dunn-Coleman, N., Frisvad, J. C., & van Dijck, P. W. (2002). On the safety of *Aspergillus niger*—A review. *Applied Microbiology and Biotechnology*, *59*, 426–435.

Shibuya, I., Tsuchiya, K., Tamura, G., Ishikawa, T., & Hara, S. (1992). Overproduction of an alpha-amylase/glucoamylase fusion protein in *Aspergillus oryzae* using a high expression vector. *Bioscience, Biotechnology, and Biochemistry*, *56*, 1674–1675.

Song, Z. S., Cox, R. J., Lazarus, C. M., & Simpson, T. J. (2004). Fusarin C biosynthesis in *Fusarium moniliforme* and *Fusarium venenatum*. *ChemBioChem*, *5*, 1196–1203.

Tada, S., Gomi, K., Kitamoto, K., Kumagai, C., Tamura, G., & Hara, S. (1991). Identification of the promoter region of the taka-amylase A gene required for starch induction. *Agricultural and Biological Chemistry*, *55*, 1939–1941.

Tada, S., Gomi, K., Kitamoto, K., Takahashi, K., Tamura, G., & Hara, S. (1991). Construction of a fusion gene comprising the taka-amylase A promoter and the *Escherichia coli* beta-glucuronidase gene and analysis of its expression in *Aspergillus oryzae*. *Molecular & General Genetics*, *229*, 301–306.

Tada, S., Iimura, Y., Gomi, K., Takahashi, K., Hara, S., & Yoshizawa, K. (1989). Cloning and nucleotide-sequence of the genomic taka-amylase A gene of *Aspergillus oryzae*. *Agricultural and Biological Chemistry*, *53*, 593–599.

Turner, G., Brown, J., Kerry—Williams, S., Bailey, A. M., Ward, M., Punt, P. J., & van den Hondel, C. A. M. J. J. (1989). J., Analysis of the *oli*C promoter of *Aspergillus nidulans*. Nevalainen, H. and M. Penttila (Ed.). Foundation for Biotechnical and Industrial Fermentation Research, *6. Molecular Biology of Filamentous Fungi. Proceedings of Embo (European Molecular Biology Organization)—Alko Workshop, Espoo, Finland, July 2–7, 268p*. Foundation for Biotechnical and Industrial Fermentation Research: Helsinki, Finland. Illus. Paper, 101–110.

Unkles, S. E., Campbell, E. I., Ruiter-Jacobs, Y. M. J. T., Broekhuijsen, M., Macro, J. A., Carrez, D., et al. (1989). The development of a homologous transformation system for *Aspergillus oryzae*; based on the nitrate assimilation pathway: A convenient and general selection system for filamentous fungal transformation. *Molecular & General Genetics*, *218*, 99–104.

Ward, M., Kodama, K. H., & Wilson, L. J. (1989). Transformation of *Aspergillus awamori* and *A. niger* by electroporation. *Experimental Mycology*, *13*, 289–293.

Watanabe, A., Fujii, I., Sankawa, U., Mayorga, M. E., Timberlake, W. E., & Ebizuka, Y. (1999). Re-identification of *Aspergillus nidulans wA* gene to code for a polyketide synthase of naphthopyrone. *Tetrahedron Letters*, *40*, 91–94.

Watanabe, A., Ono, Y., Fujii, I., Sankawa, U., Mayorga, M. E., Timberlake, W. E., et al. (1998). Product identification of polyketide synthase coded by *Aspergillus nidulans* wA gene. *Tetrahedron Letters*, *39*, 7733–7736.

Weld, R. J., Plummer, K. M., Carpenter, M. A., & Ridgway, H. J. (2006). Approaches to functional genomics in filamentous fungi. *Cell Research*, *16*, 31–44.

Wilkinson, B., & Micklefield, J. (2007). Mining and engineering natural-product biosynthetic pathways. *Nature Chemical Biology*, *3*, 379–386.

CHAPTER THIRTEEN

# *De Novo* Synthesis of High-Value Plant Sesquiterpenoids in Yeast

**Trinh-Don Nguyen, Gillian MacNevin, Dae-Kyun Ro[1]**
Department of Biological Sciences, University of Calgary, Calgary, Alberta, Canada
[1]Corresponding author: e-mail address: daekyun.ro@ucalgary.ca

## Contents

## Abstract

Terpenoids comprise a structurally diverse group of natural products. Despite various and important uses of terpenoids (e.g., flavors, drugs, and nutraceuticals), most of them are, however, still extracted from plant sources, which suffer from high cost and low yield. Alternatively, terpenoids can be produced in microbes using their biosynthetic genes. With the explosion of sequence data, many genes for terpenoid metabolism can be characterized by biochemical approaches and used for the microbial production of terpenoids. However, substrates for *in vitro* studies of terpene synthases are costly, and the enzymatic synthesis of terpenoids *in vitro* using recombinant enzymes is insufficient to meet the chemical characterization need. Here, we describe the use of engineered yeast (EPY300) to evaluate *in vivo* production of sesquiterpenoids. Two

*Methods in Enzymology*, Volume 517
ISSN 0076-6879
http://dx.doi.org/10.1016/B978-0-12-404634-4.00013-9

261

sesquiterpene synthase genes (for valencene and 5-*epi*-aristolochene synthases) were expressed in EPY300 in native and N-terminal thioredoxin fusion forms. By using the thioredoxin fusion, valencene biosynthesis was slightly decreased; however, the production of 5-*epi*-aristolochene was increased by 10-fold, producing 420 µg mL$^{-1}$ of 5-*epi*-aristolochene. Accordingly, the thioredoxin-fused 5-*epi*-aristolochene was coexpressed with 5-*epi*-aristolochene dihydroxylase (cytochrome P450 monooxygenase) and its reductase in EPY300. This combinatorial expression yielded hydroxylated sesquiterpene, capsidiol, at $\sim$250 µg mL$^{-1}$. Detailed experimental procedures and other considerations for this work are given.

# 1. INTRODUCTION

Sesquiterpenoids are a class of enormously diverse natural products derived from the 15-carbon precursor, farnesyl pyrophosphate (FPP). The chemical diversity of sesquiterpenoids starts from the diverse sesquiterpene hydrocarbon backbones, which are created by carbocation cascade reactions programmed in sesquiterpene synthases (STSs). The C15 sesquiterpene skeletons are then often oxygenated by regio- and stereo-selective cytochrome P450 monooxygenases (P450s). Both STS and P450 are supergene families widely present in bacteria, fungi, and plants, and they play critical roles in generating terpenoid diversity in nature. Specifically, the coordinated reactions of STSs and P450s synthesize activated terpenoid backbones on which other chemical decorations (e.g., acetylation, methylation, lipidation, and further oxidations) can occur to increase the chemical diversity of sesquiterpenoids (Fraga, 2006). Commercial applications of sesquiterpenoids are as diverse as their structures. A few examples among many include nootkatone (grapefruit-flavor; Dolan et al., 2009), lactucopicrin (antifeedant and analgesic; Rees & Harborne, 1985; Wesołowska, Nikiforuk, Michalska, Kisiel, & Chojnacka-Wójcik, 2006), drimanes (deterrent, antimicrobial, and insecticidal; Gershenzon & Dudareva, 2007), thapsigargin (smooth endoplasmic reticulum Ca$^{2+}$-ATPase inhibitor; Furuya, Lundmo, Short, Gill, & Isaacs, 1994), hernandulcin (low-calorie sweetener; Compadre, Pezzuto, Kinghorn, & Kamath, 1985), artemisinin (antimalarial drug; Eckstein-Ludwig et al., 2003), and valerenic acid (mild sedative; Yuan et al., 2004). Furthermore, in recent years, the volatile and combustible properties of the sesquiterpenes inspired the development of terpene-based biofuel from bisabolene and farnesene (Peralta-Yahya et al., 2011; Renniger & McPhee, 2008). Despite the diverse and critical uses of sesquiterpenoids in our lives, as well as their roles in interactions of plants with other organisms (e.g., allelochemicals and insecticides), the majority of their biosynthetic pathways remain unknown, and most terpenoids are still purified from plant sources. However, recent progress in implementing the synthetic metabolic pathways in heterologous hosts (e.g., *Escherichia coli* and yeast) has provided the potential to manufacture sesquiterpenoids in alternative bioplatforms. In addition, various next-generation sequencing technologies have generated the sequences of thousands of transcripts, which display high homology to known enzymes involved in terpenoid metabolism, such as terpene synthases (including STSs) and P450s. They are awaiting proper biochemical characterizations for accurate annotation. All these new genes can be useful components in synthetic metabolism and will certainly guide us to create novel systems for terpenoid production in the future.

The obvious first step to build a reliable gene/enzyme catalogue is the biochemical characterization of the enzymes encoded in cDNAs, many of which are currently listed in the transcript and genome database without accurate annotation. Although molecular and biochemical techniques for the characterization of cDNA and its recombinant enzymes have matured, such studies are limited by the lack or high cost of terpenoid standards and substrates. Most of the terpene synthase products are intermediates in metabolic pathways. Therefore, they do not accumulate in plants, making it practically impossible to purify these terpenoids from plants. Chemical synthesis of terpenoids is extremely difficult owing to multiple chiral centers and regio-specificities. Although a large-scale *in vitro* reaction can be performed to obtain terpenes at a milligram scale, starting substrates (e.g., FPP for sesquiterpenoids) are very costly, and hence, many laboratories cannot afford the expense associated with the enzymatic production of terpenes *in vitro* to the amount required for structural elucidation.

In order to overcome these problems, the yeast strain EPY300 has been developed to produce an elevated level of FPP (Ro et al., 2006; Ro et al., 2008). Expressing an STS separately or in combination with sesquiterpene-modifying P450s (sesquiterpene oxidase, STO, or sesquiterpene hydroxylase, STH) in a metabolically engineered yeast thus allows synthesis of the desired sesquiterpenoids to an amount necessary for standard chemical characterization. A simple sugar (galactose) is used as a carbon source, and so the cost of terpenoid production by microbial *de novo* biosynthesis is negligible. Here, we describe a simple procedure to express an STS in the EPY300 strain to produce a high abundance of valencene, an important aromatic component of citrus (Furusawa, Toshihiro, Yoshiaki, & Yoshinori, 2005), and 5-*epi*-aristolochene, a precursor of several phytoalexins against fungal pathogens. Furthermore, the same yeast platform was used to produce the phytoalexin capsidiol by additionally coexpressing a P450 and cytochrome P450 reductase (CPR). Capsidiol is a dihydroxy-5-*epi*-aristolochene and a critical defensive compound in tobacco and chili pepper (Maldonando-Bonilla, Betancourt-Jiménez, & Lozoya-Gloria, 2008).

## 2. YEAST CULTURE AND METABOLITE ANALYSIS

### 2.1. Yeast and plasmid used for sesquiterpenoid production

Microbial *de novo* synthesis of sesquiterpenoids is achieved by using the yeast strain EPY300 (S288C, MATα his3Δ1 leu2Δ0 $P_{GAL1}$–tHMG1::δ1 $P_{GAL1}$–upc2-1::δ2 erg9::$P_{MET3}$–ERG9::HIS3 $P_{GAL1}$–ERG20::δ3 $P_{GAL1}$–tHMG1::δ4). EPY300 was metabolically engineered to synthesize an elevated level of the central sesquiterpenoid precursor, FPP. The overproduction of FPP in this strain was achieved by altering the mevalonate biosynthetic pathway (Ro et al., 2006; Fig. 13.1). The genetic modifications introduced in EPY300 include:

i. genomic integrations of two copies of 3-hydroxy-3-methyl glutaryl-CoA reductase (HMG-CoA reductase) as an N-terminal truncated form under the *GAL1* promoter (Donald, Hampton, & Fritz, 1997);

ii. a genomic integration of a mutant transcription factor upc2-1 under the *GAL1* promoter: upc2-1 upregulates several genes in the ergosterol biosynthetic pathway in yeast; the exact mechanism of how the point mutation in upc2-1 activates the mevalonate pathway is not fully understood, but it is likely that the mutation increases the stability of this transcription factor (Davies, Wang, & Rine, 2005);

iii. replacement of the *ERG9* (*squalene synthase*) promoter with the methionine-repressible promoter ($P_{MET3}$); and

iv. genomic integration of *ERG20* (*FPP synthase*) under the *GAL1* promoter.

Common δ-sequences (long repeats of yeast retro-transposons) were used to integrate the truncated *HMG-CoA reductase, upc2-1,* and *ERG20,* and the native *ERG9* promoter was replaced with the methionine-repressible promoter. Therefore, upon addition of galactose and methionine, overall mevalonate flux significantly increases while the metabolic flux to ergosterol decreases. Plant genes for STSs and STOs/STHs are then expressed under galactose-inducible promoters in EPY300 to redirect the abundant FPP pool

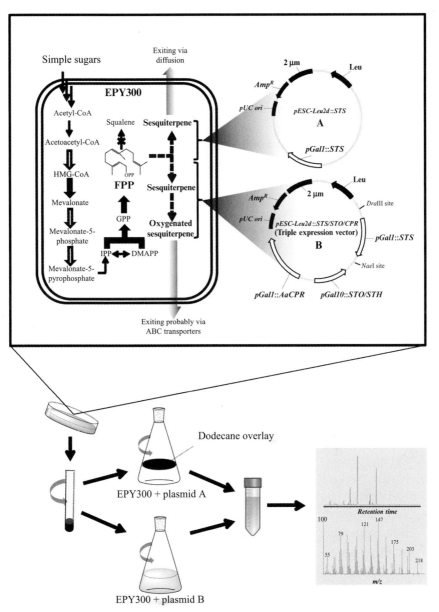

**Figure 13.1** A simple system for *de novo* production of sesquiterpenoids in engineered yeast. The EPY300 strain was engineered to overproduce farnesyl pyrophosphate (FPP), a common substrate for sesquiterpene synthase (STS). Thick, solid arrows indicate the genes (*HMG-CoA reductase* and *FPP synthase*) directly upregulated by *GAL1*-driven overexpression. Empty arrows are the indirectly upregulated genes by the transcription factor upc2-1, including *ERG13* (*HMG-CoA synthase*), *ERG12* (*mevalonate kinase*), and *ERG8* (*phosphomevalonate kinase*). The cross indicates the *squalene synthase* gene controlled

to the desired sesquiterpenoids. In order to increase the plasmid copy numbers, the *Leu2d* selection marker was used for gene expression in EPY300 (Erhat & Hollenberg, 1983). The majority of the *Leu2d* promoter was deleted, leaving only 29 bp of promoter, resulting in a decrease of *Leu2* transcript level. To compensate for the weak promoter activity, the yeast increases the copy number of the *Leu2d* plasmid, and the cDNA coded in the plasmid thus has a higher level of expression than the standard *Leu2* selectable marker. We have routinely used EPY300 and the *Leu2d* selectable marker to produce complex sesquiterpenoids with a yield of 10–500 mg $L^{-1}$ yeast culture in shaken flasks without much optimization (Ikezawa et al., 2011; Nguyen et al., 2010; Ro et al., 2008).

When sesquiterpenoids which are prone to acid-induced rearrangement are synthesized using EPY300 in synthetic medium, caution needs to be taken because yeast medium is acidified (pH $\sim$ 3) after 3 days of cultivation, and the sesquiterpenoid structures can thus be altered. In this case, we often reduce medium acidification (to pH 6 after the same period of culture) by adding HEPES buffer to synthetic complete (SC) medium to a final concentration of 100 m$M$ (pH 7.0–7.5) (Ikezawa et al., 2011; Nguyen et al., 2010). Tris–HCl and phosphate buffers are found to interfere with yeast growth.

## 2.2. Medium preparation

Yeast can be cultured in either selection (SC dropout) or rich (YPA) media. Components of 1 L SC dropout liquid medium are:

1. 6.7 g yeast nitrogen base without amino acids (Difco Microbiology).
2. 21 mg adenine.
3. 85.6 mg of each of 16 standard amino acids (without histidine, leucine, methionine, and tryptophan) (Sigma-Aldrich).
4. 173.4 mg leucine, 85.6 mg of each of the other three amino acids (histidine, methionine, and tryptophan), and 85.6 mg of uracil (Sigma-

---

under the methionine-repressible promoter. Dashed arrows indicate the enzymatic steps catalyzed by recombinant enzymes. STS-harboring vector (plasmid A) and triple expression vector (plasmid B) are introduced by transformation into EPY300 to produce sesquiterpene and oxygenated sesquiterpenes, respectively. The transgenic EPY300 is then cultivated in flasks. For sesquiterpene-producing cultures, dodecane (10% of culture volume) is overlaid to trap volatile products, whereas no dodecane is overlaid to produce oxygenated sesquiterpenes (see Section 5.1). Metabolite profiles are analyzed with GC–MS and GC–FID. STS, sesquiterpene synthase; STO/STH, sesquiterpene oxidase/hydroxylase; CPR, cytochrome P450 reductase.

Aldrich) separately or in combination, depending on the selection markers of the transformed plasmid.

 *Note*: EPY300 is selected in SC–His–Met medium. If the transformed yeast has the *Leu2* or *Leu2d* selection marker (as in this specific procedure), SC–His–Met–Leu would be used for the cultivation of transgenic EPY300.

**5.** Double distilled water to 900 mL.

Components of 1 L rich (YPA) liquid medium are:

**1.** 10 g Bacto<sup>TM</sup> yeast extract (BD Bioscience)
**2.** 20 g Bacto<sup>TM</sup> peptone (BD Bioscience)
**3.** 80 mg adenine hemisulfate (Sigma-Aldrich)
**4.** Double distilled water to 900 mL

For solid medium, 15 g Difco Bacto agar (Difco Microbiology) is added.

 Solutions of 20% (w/v) glucose and galactose are prepared separately as carbohydrate sources for yeast medium. Both carbohydrate sources and the medium are sterilized by filtration or autoclaving. Before use, glucose and galactose solutions are added separately or in combination (see Section 2.3.) to the above medium to a final total sugar concentration of 2% (w/v).

## 2.3. Culture conditions

Owing to extensive genetic modifications and cytotoxicity of FPP, EPY300 may not be genetically stable. It is recommended that EPY300 be freshly streaked out from glycerol stocks. All yeast inoculations and cultures are performed in a controlled-environment incubator shaker at 30 °C, 200 rpm.

## 2.4. Metabolite analysis

A 6890 N gas chromatography coupled with either a 5975B mass spectrometer (GC–MS) or a 6850 flame ionization detector (GC–FID) (Agilent) is used to analyze the production of sesquiterpenes and hydroxylated sesquiterpenes from yeast cultures.

**1.** Injection port temperature: 250 °C
**2.** Separation column: DB-5MS (30-m length × 250-μm inner diameter × 0.25-μm film thickness)
**3.** GC oven temperature program: 80 °C for 1 min, followed by linear increases of 10 °C min$^{-1}$ to 250 °C and then of 30 °C min$^{-1}$ to 300 °C
**4.** Injection volume: 1–2 μL of metabolite extract

*De novo* synthesized valencene and capsidiol in yeast cultures were analyzed with authentic standards of commercial valencene (Sigma-Aldrich) and capsidiol extracted from green pepper (*Capsicum annuum*) using the method reported by Zhao, Schenk, Takahashi, Chappell, and Coates (2004). For *de novo* synthesized 5-*epi*-aristolochene, its chemical identity was confirmed with published retention index and mass spectrum (Adams, 2007), and its abundance was estimated as the equivalent of authentic valencene by GC–FID. Similarly, the abundance of hydroxylated products other than capsidiol was estimated using capsidiol standard as an equivalent hydroxylated sesquiterpene.

## 3. EXPRESSION PLASMID CONSTRUCT AND TRANSFORMATION

### 3.1. Plasmid construct for sesquiterpene production

We used the high-copy *pESC-Leu2d* plasmid to express valencene and 5-*epi*-aristolochene synthases under the GAL1 promoter. The vector *pESC-Leu2d* only differs from the *pESC-Leu* plasmid (Agilent) at the promoter of the *Leu* selection marker (see Section 2.1.).

1. Isolate STS cDNA by PCR from plant tissues: the *5-epi-aristolochene synthase* gene (*NtEAS*; GenBank: L04680) (Facchini and Chappell, 1992) was isolated from tobacco (*Nicotiana tabaccum*) leaf cDNA by the forward primer 5′-ATCGA<u>GGGCCC</u>GCCATGGCCTCAGCAGCA GTTGCAAAC-3′ and the reverse primer 5′-ACCAT<u>CTCGAG</u>TCAA ATTTTGATGGAGTCCACAAGT-3′; and the *valencene synthase* gene (*CpVS*) was isolated from grapefruit (*Citrus* × *paradisi*) flavedo cDNA by the forward primer 5′-ATCGA<u>GGGCCC</u>GCCATGTCGTCTGG AGAAACATTTCG-3′ and the reverse primer 5′-A<u>CTCGAGTCA</u> AAATGGAACGTGGTCTCCTAG-3′ designed based on the *C. sinensis valencene synthase* gene (GenBank: AF441124) (Sharon–Asa et al., 2003).

2. Digest the amplicons and *pESC-Leu2d* with *Apa*I and *Xho*I and ligate the digested products to generate *pESC-Leu2d::NtEAS* and *pESC-Leu2d:: CpVS* (under the GAL1 promoter) (plasmid A, Fig. 13.1).

To evaluate if the STS activity can be improved by the N-terminal thioredoxin fusion, the multiple cloning site of *pESC-Leu2d* was modified to include the thioredoxin gene.

1. Amplify the thioredoxin gene (*Trx*) from pET-48b(+) (EMD Biosciences) by the forward primer 5′-AATGT<u>AGATCT</u>GCCATGAG CGATAAAATTATTCACC-3′ and the reverse primer 5′-TCAG T<u>AGATCT</u>GGCCAGGTTAGCGTCGAGGAACTCTTTC-3′.

2. Digest the amplicons and *pESC-Leu2d* with *Bgl*II and *Bam*HI, respectively, and ligate the digested product to the *pESC-Leu2d* under the *GAL1* promoter.

3. Digest *pESC-Leu2d::Trx* and the aforementioned STSs with *Apa*I and *Xho*I, and ligate the digested products to generate *pESC-Leu2d::Trx-NtEAS* and *pESC-Leu2d::Trx-CpVS*.

   *Note*: After *Trx* is cloned under the *GAL1* promoter of *pESC-Leu2d*, the reading frame will be changed as follows: ... *GAC GCT AAC CTG GCC AGA TCC GTA ATA CGA CTC ACT ATA* <u>GGG CCC</u> ... with the *italicized* and underlined letters indicating the 3′-end of *Trx* and *Apa*I recognition site, respectively. If the start codon of the cDNA is placed immediately after the *Apa*I site, the cDNA will be inframe with *Trx*, and the expressed protein will be fused with Trx by a 10-residue linker of Arg-Ser-Val-Ile-Arg-Leu-Thr-Ile-Gly-Pro.

## 3.2. Plasmid construct for oxygenated sesquiterpene production

Capsidiol, or 1,3-dihydroxy-5-*epi*-aristolochene, is a dihydroxy-sesquiterpenoid. To produce capsidiol in yeast, the *pESC-Leu2d* plasmid was modified to include an additional expression cassette under the *GAL1* promoter between its *Nae*I and *Dra*III sites. The triple expression construct harboring three genes: tobacco *5-epi-aristolochene dihydroxylase* (*NtEAH* or *CYP71D20*) under the *GAL10* promoter, *CPR* from *Artemisia annua* (encoding for the redox partner of NtEAH) under the *GAL1* promoter, and *NtEAS* in the additional expression cassette.

1. Amplify the *NtEAH* gene (GenBank: AF368376) (Takahashi et al., 2005) from tobacco leaf cDNA with the forward primer 5′-ACTTAT<u>TCTAGA</u>GCCATGCAATTCTTCAGCTTGGTTTCC-3′ and the reverse primer 5′-TAGTTT<u>TCTAGA</u>GCCTCTCGAG AAGGTTGATAAGGAGTGG-3′.

2. Digest the amplicons and *pESC-Leu2d* vector with *Xba*I and *Spe*I, respectively, and ligate the digested products to generate *pESC-Leu2d::NtEAH* under the *GAL10* promoter.

3. Digest *pESC-Ura::CPR* (under the *GAL1* promoter) (Ro et al., 2006) and *pESC-Leu2d::NtEAH* with *Bam*HI and *Nhe*I, and ligate the digested products to generate *pESC-Leu2d::NtEAH/CPR*.

4. Amplify the expression cassette of *Trx-NtEAS* (under the *GAL1* promoter) in *pESC-Leu2d::Trx-NtEAS* (see Section 3.1.) with the forward primer 5′-GTCAAT<u>CACTACGTG</u>AGTACGGATTAGAAG CCGC

CGA-3′ and the reverse primer 5′-GTCAATG<u>CCGGC</u>CTTCGAG CGTCCCAAAACCT- 3′.

5. Digest the amplicons and *pESC-Leu2d::NtEAH/CPR* with *Dra*III and *Nae*I, and ligate the digested products to generate *pESC-Leu2d:: NtEAH/CPR/Trx-NtEAS* (plasmid B, Fig. 13.1).

## 3.3. Yeast transformation

The method for introducing the generated plasmids by transformation into EPY300 is based on the LiAc/ssDNA/PEG method (Gietz & Schiestl, 2007).

1. Inoculate EPY300 in 1 mL SC–His–Met with 2% glucose overnight.
2. Transfer the inoculation to a 125- to 150-mL Erlenmeyer flask containing 30 mL SC–His–Met with 2% glucose and allow the yeast to grow for 2–4 h.
3. Prepare carrier DNA by boiling 2 mg mL$^{-1}$ single-stranded DNA from salmon testes (Sigma–Aldrich) for 5 min and quickly chill on ice.
4. After 2–4 h of growing EPY300, transfer the culture to a Falcon tube, centrifuge at $3500 \times g$ for 5 min, and discard the supernatant.
5. Wash the cells by adding 10–20 mL of sterile water, vortex, centrifuge at $3500 \times g$ for 5 min, and discard the supernatant.
6. Add 1 mL of sterile water and mix well by pipetting up and down with a 1-mL micropipette tip.
7. Transfer the cell solution to a 1.5-mL microcentrifuge tube, centrifuge at $20{,}000 \times g$ for 30 s, and discard the supernatant.
8. Add 120 μL 50% polyethylene glycol, 18 μL 1 *M* lithium acetate, and 25 μL 2 mg mL$^{-1}$ carrier DNA to the microcentrifuge tube, and mix with the cells by vortexing.
9. Add about 500 μg plasmid DNA to the cell solution, and mix by stirring with the micropipette tip.
10. Incubate the microcentrifuge tube in a water bath at 42 °C for 30–40 min.
11. Centrifuge at $20{,}000 \times g$ for 30 s, and discard the supernatant.
12. Add 200 μL of sterile water, and mix gently by stirring and/or pipetting up and down a few times with a 1-mL micropipette tip.
13. Spread the cells gently on a SC–His–Met–Leu plate with three or four strokes, and dry the plate.
14. Incubate at 30 °C: yeast transformed with the standard *Leu2* marker will develop colonies in 2 days, but yeast transformed with *Leu2d* requires 3–5 days to develop colonies, and the number of colonies will be fewer than standard *Leu2* transformants.

## 4. *DE NOVO* PRODUCTION OF SESQUITERPENES

### 4.1. Culture and metabolite analysis

Four plasmids, *pESC-Leu2d::NtEAS*, *pESC-Leu2d::Trx-NtEAS*, *pESC-Leu2d::CpVS*, and *pESC-Leu2d::Trx-CpVS*, were used to individually transform EPY300 to produce either 5-*epi*-aristolochene or valencene.

1. Inoculate yeast in a test tube with 1 mL of SC–His–Met–Leu with 2% glucose for about 15–18 h.
2. Transfer 600 μL of the inoculation to a 125- to 150-mL Erlenmeyer flask containing 30 mL of SC–His–Met–Leu with 1.8% galactose, 0.2% glucose, and 1 m*M* methionine.
3. Overlay 3 mL of dodecane (Sigma-Aldrich) (about 10% of total culture volume) to trap the volatile sesquiterpenes produced during culture.
4. Allow yeast to grow for 48 h.
5. Pour the yeast culture into a polypropylene Falcon tube, centrifuge at $3500 \times g$ for 5 min, and collect the dodecane overlay.
6. Make a 100-fold dilution of the dodecane with hexane or ethyl acetate and inject 1 μL for GC–MS analysis.

### 4.2. Production of valencene and 5-*epi*-aristolochene

GC–MS analyses of the dodecane fraction showed that EPY300 expressing *CpVS* under the *GAL1* promoter synthesized 5.3 μg mL$^{-1}$ of valencene (**1**), while EPY300 expressing *NtEAS* produced 39.2 μg mL$^{-1}$ of 5-*epi*-aristolochene (**2**). In the previous experiment, sunflower δ-cadinene synthase could only be expressed properly in *E. coli* and yeast after translational fusion of thioredoxin to the N-terminus of STS (Göpfert, MacNevin, Ro & Spring, 2009). Thioredoxin fusion reduces the formation of inclusion bodies in *E. coli* and may also help proper folding or stability of STSs in yeast. In order to assess if the same modification can improve STS expression and thus *in vivo* sesquiterpene production in yeast, the thioredoxin gene was fused to the N-terminus of CpVS and NtEAS with a 10-amino acid linker (ASVIRLTIGP). When the sesquiterpenes were measured, a remarkable increase of 5-*epi*-aristolochene synthesis (about 10-fold) was observed from EPY300 expressing the thioredoxin-fused NtEAS. Approximately, 400 μg of 5-*epi*-aristolochene was synthesized from a 1-mL culture. In contrast, the EPY300 expressing the thioredoxin-fused CpVS decreased valencene production to 50% of the yeast expressing CpVS (Fig. 13.2). Although thioredoxin may benefit the folding or solubility of the NtEAS recombinant enzyme, it did not help the terpene productivity of recombinant CpVS and appeared to interfere with CpVS activities.

## 5. *DE NOVO* PRODUCTION OF HYDROXYLATED SESQUITERPENE

### 5.1. Culture and metabolite analysis

EPY300 transformed with the *pESC-Leu2d* plasmid coding *NtEAH*, *CPR*, and *Trx-NtEAS* was cultured for capsidiol production. Although the overlay of dodecane significantly helped capture volatile sesquiterpene from the yeast culture, we observed that the dodecane overlay interfered with the production of oxygenated sesquiterpenoids. By using the dodecane overlay, the amount of sesquiterpene olefins increases, whereas the amount of oxygenated sesquiterpenoids decreases, relative to the culture without dodecane overlay. Despite the fact that the sesquiterpene hydrocarbons show very low solubility in water, sesquiterpene hydrocarbons synthesized by STSs appear to stay in the medium for some time and diffuse in and out of yeast cells. Dodecane in the medium seems to sequester sesquiterpene hydrocarbons and hence prevents further oxidation by cytochrome P450 inside the yeast cells. Based on this observation, dodecane was not overlaid in the culture when oxygenated sesquiterpenoids are synthesized in the EPY300 platform.

1. Inoculate yeast in a test tube with 1 mL of SC–His–Met–Leu with 2% glucose for about 15–18 h.
2. Transfer 600 µL of the inoculation to a 125- to 150-mL Erlenmeyer flask containing 30 mL of either SC–His–Met–Leu or YPA media with 1.8% galactose, 0.2% glucose, and 1 m$M$ methionine.
3. Allow yeasts to grow for 120 h. To determine the yield of capsidiol at different time-points, 2 mL from the culture was collected and extracted with 2 mL ethyl acetate in a polypropylene Falcon tube every 24 h.
4. Inject 2 µL of the ethyl acetate fractions for GC–MS and GC–FID analysis.

### 5.2. Production of capsidiol

As the thioredoxin-fused NtEAS produced a significantly higher level of 5-*epi*-aristolochene, this fusion enzyme was further used to coexpress *NtEAH* and *CPR* from the triple expression plasmid. The coordinated action of

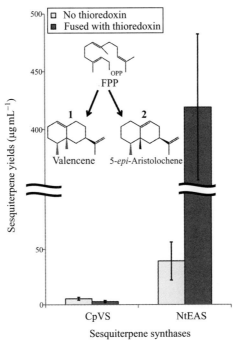

**Figure 13.2** Effect of thioredoxin fusion on *in vivo* sesquiterpene synthase (STS) productivity in yeast. Values are mean ± SD (*n* = 3). CpVS, valencene synthase from grapefruit (*Citrus* × *paradisi*); NtEAS, 5-*epi*-aristolochene synthase from tobacco (*Nicotiana tabaccum*).

these three enzymes is expected to synthesize 1,3-dihydroxy-5-*epi*-aristolochene (Fig. 13.3A). In the GC–MS analysis of culture extract, capsidiol (**3**) was clearly detected, together with the nonhydroxylated 5-*epi*-aristolochene. In addition to capsidiol, an unknown compound (**4**), possibly a hydroxy-5-*epi*-aristolochene (Fig. 13.3B), and farnesol (**5**), an FPP derivative, were clearly detected. The accumulation pattern of capsidiol was different, depending on the culture medium used (Fig. 13.3C). In the SC medium, capsidiol yield increased relatively fast in the first 48 h but reached a plateau at 72 h. The final yield after 120 h of cultivation was 35.7 $\mu g \, mL^{-1}$, which was comparable to 33.4 $\mu g \, mL^{-1}$ at 72 h. Therefore, there was only a marginal increase of oxygenated terpene production after 48 h of cultivation. However, when the transgenic yeast was cultivated in YPA medium, capsidiol yield was about eightfold higher than in SC medium, reaching up to 250 $\mu g \, mL^{-1}$. An exponential increase of capsidiol was observed between 48 and 72 h of cultivation in YPA medium. The

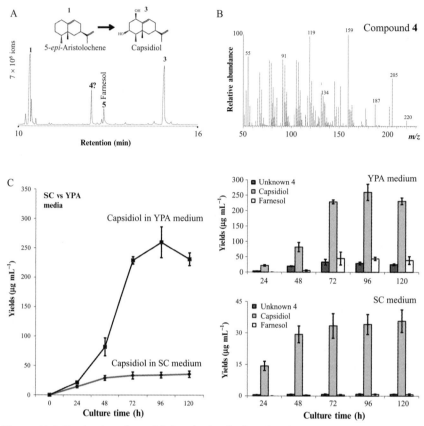

**Figure 13.3** Production of capsidiol and other hydroxylated compounds in engineered yeast. (A) GC–MS metabolite profile of EPY300 expressing *NtEAH/CPR/Trx-NtEAS* after 48 h of culture in YPA medium reveals the *de novo* synthesis of 5-*epi*-aristolochene (**1**), capsidiol (**3**), unknown compound (**4**), and farnesol (**5**). (B) Mass fragmentation spectrum of compound (**4**), putatively a hydroxy-5-*epi*-aristolochene. (C) Time-course yields and relative abundances of capsidiol and other hydroxylated compounds in two different media (mean ± SD, $n = 5$). Quantification is based on total ion counts in GC–FID analysis.

accumulation of compound **4** and farnesol was also much higher when yeast was cultivated in YPA medium.

## 6. GENERAL DISCUSSION

In this chapter, we demonstrate that significant amounts of sesquiterpene hydrocarbons and oxygenated sesquiterpenes can be *de novo* synthesized in yeast, using a high-copy plasmid, *pESC-Leu2d*, and metabolically engineered EPY300 yeast strain. As observed here, it should be noted that the level of *in vivo* sesquiterpene production dramatically varied between 2 µg mL$^{-1}$ (CpVs) and 420 µg mL$^{-1}$ (Trx-NtEAS) using two STSs and their respective thioredoxin-fused enzymes. Although further work is necessary to understand what caused such remarkable variations, we speculate that the innate STS kinetic capacity and *in vivo* enzyme folding, stability, and/or solubility contribute to the variations. It was previously reported and discussed that the kinetic efficiency ($k_{cat}/K_m$) of NtEAS is 10-fold higher than that of valencene synthase from citrus (Takahashi et al., 2007). The valencene synthase from grapefruit (CpVS) used in this work was 99% identical in sequence to the published valencene synthase, and CpVS is also likely to have a much lower kinetic efficiency than NtEAS. The difference in kinetic efficiencies may explain the low yield of valencene. However, enzyme folding, stability, and/or solubility are more difficult to assess than enzyme kinetic efficiency. Nonetheless, the observed 10-fold increase of 5-*epi*-aristolochene *in vivo* activity in EPY300 expressing thioredoxin-fused NtEAS clearly showed the usefulness of using thioredoxin fusion to increase the sesquiterpene yield in yeast. Unfortunately, this thioredoxin-fusion strategy does not seem to be generally applicable to all STSs since the same thioredoxin fusion decreased the *in vivo* activity of CpVS. We reasoned that the low yield of valencene in EPY300 might not be caused by the problems associated with enzyme folding, stability, and/or solubility but is fundamentally limited by the poor kinetic efficiency of CpVS. The thioredoxin located at the N-terminus of CpVS could distort the overall CpVS structure, and therefore, the fusion may interfere with the optimal CpVS reaction in yeast. In the case of NtEAS, non-optimal folding, stability, and solubility might occur without thioredoxin fusion, and thioredoxin could solve those problems of NtEAS in yeast. As a result, improvement of the folding, stability, and/or solubility of NtEAS could increase the yield of 5-*epi*-aristolochene in yeast. Certainly, more experiments are required to address these questions, but the thioredoxin fusion is one strategy worthy of testing to increase sesquiterpene production in yeast. The thioredoxin-harboring *pESC-Leu2d* plasmid made in this work can be a useful tool to create the fusion enzyme.

Another simple change that could improve the yield of sesquiterpenoids in yeast is the medium. With its high stability in yeast, *pESC-Leu2d* can be used to reconstitute one-gene (STS alone) and three-gene (STS, P450, and CPR) biosynthetic pathways as described here and elsewhere (Nguyen et al., 2010; Ro et al., 2008), or even a four-gene (STS, two P450s, and CPR) pathway (Ikezawa et al., 2011) in both selection SC and rich YPA media. However, there is no selection pressure when YPA medium is used, and whether the transgenic yeast maintains the *Leu2d* plasmid or not will primarily depend on the cytotoxicity of the sesquiterpenoids endogenously produced in yeast. In our previous study, when the anionic sesquiterpenoid molecule artemisinic acid was produced in EPY300, a massive transcriptional induction of ATP-Binding Cassette (ABC)-transporter genes was observed, and accordingly, the plasmid stability of *Leu2d* became very low (Ro et al., 2008). The use of YPA medium can be one of the simple parameters to evaluate the sesquiterpenoid yield and also possible cytotoxicity of the sesquiterpenoids in yeast.

The methods described here use a cheap sugar source and easily synthesize more than 10 mg of sesquiterpene and oxygenated sesquiterpene. Many uncharacterized STS and P450 genes can be plugged into this system so that their biochemical identity can be deciphered, and the structures of sesquiterpenoid products can be elucidated by various analytical techniques such as nuclear magnetic resolution spectroscopy.

## ACKNOWLEDGMENTS

The authors would like to thank Dr. David Hart, at the Institute of Food Research (Biotechnology and Biological Sciences Research Council, UK), for providing the authentic capsidiol standard. This work is supported by the Discovery Grant from the National Sciences and Engineering Research Council of Canada (NSERC), Canada Research Chair program, and the Next-Generation BioGreen 21 Program (SSAC grant PJ008108), Rural Development Administration, Republic of Korea (to D.-K. R.), and the Bettani Bahlsen Memorial Graduate Scholarship (to T.-D. N.).

## REFERENCES

Adams, R. P. (2007). Identification of essential oil components by gas chromatography/mass spectrometry (4th ed.). Calrol Stream, IL: Allured Publishing Corporation.

Compadre, C. M., Pezzuto, J. M., Kinghorn, A. D., & Kamath, S. K. (1985). Hernandulcin: An intensely sweet compound discovered by review of ancient literature. *Science, 227,* 417–419.

Davies, B. S. J., Wang, H. S., & Rine, J. (2005). Dual activators of the sterol biosynthetic pathway of *Saccharomyces cerevisiae*: Similar activation/regulatory domains but different response mechanisms. *Molecular and Cellular Biology, 25,* 7375–7385.

Dolan, M. C., Jordan, D. A., Schulze, T. L., Schulze, C. J., Manning, M. C., Ruffolo, D., et al. (2009). Ability of two natural products, nootkatone and carvacrol, to suppress *Ixodes scapularis* and *Amblyomma americanum* (Acari: Ixodidae) in a lyme disease endemic area of New Jersey. *Journal of Economic Entomology, 102,* 2316–2324.

Donald, K. A. G., Hampton, R. Y., & Fritz, I. B. (1997). Effects of overproduction of the catalytic domain of 3-hydroxy-3-methylglutaryl coenzyme A reductase on squalene synthesis in *Saccharomyces cerevisiae*. *Applied and Environmental Microbiology, 63,* 3341–3344.

Eckstein-Ludwig, U., Webb, R. J., van Goethem, I. D. A., East, J. M., Lee, A. G., Kimura, M., et al. (2003). Artemisinins target the SERCA of *Plasmodium falciparum*. *Nature, 424,* 957–961.

Erhat, E., & Hollenberg, C. P. (1983). The presence of a defective *LEU2* gene on 2μ DNA recombinant plasmids of *Saccharomyces cerevisiae* is responsible for curing and high copy number. *Journal of Bacteriology, 156*, 625–635.

Facchini, P. J., & Chappell, J. (1992). Gene family for an elicitor-induced sesquiterpene cyclase in tobacco. *Proceedings of the National Academy of Sciences of the United States of America, 89*, 11088–11092.

Fraga, B. M. (2006). Natural sesquiterpenoids. *Natural Product Reports, 24*, 1350–1381.

Furusawa, M., Toshihiro, H., Yoshiaki, N., & Yoshinori, A. (2005). Highly efficient production of nootkatone, the grapefruit aroma from valencene, by biotransformation. *Chemical & pharmaceutical bulletin, 53*, 1513–1514.

Furuya, Y., Lundmo, P., Short, A. D., Gill, D. L., & Isaacs, J. T. (1994). The role of calcium, pH, and cell proliferation in the programmed (apoptotic) death of androgen-independent prostatic cancer cells induced by thapsigargin. *Cancer Research, 54*, 6167–6175.

Gershenzon, J., & Dudareva, N. (2007). The function of terpene natural products in the natural world. *Nature Chemical Biology, 3*, 408–414.

Gietz, R. D., & Schiestl, R. H. (2007). High-efficiency yeast transformation using the LiAC/SS carrier DNA/PEG method. *Nature Protocols, 2*, 31–34.

Göpfert, J. C., MacNevin, G., Ro, D.-K., & Spring, O. (2009). Identification, functional characterization and developmental regulation of sesquiterpene synthases from sunflower capitate glandular trichomes. *BMC Plant Biology, 9*, 86–103.

Ikezawa, N., Göpfert, J. C., Nguyen, T.-D., Kim, S.-U., O'Maille, P. E., Spring, O., et al. (2011). Lettuce costunolide synthase (*CYP71BL2*) and its homolog (*CYP71BL1*) from sunflower catalyze distinct regio- and stereo-selective hydroxylations in sesquiterpene lactone metabolism. *Journal of Biological Chemistry, 286*, 21601–21611.

Maldonando-Bonilla, L. D., Betancourt-Jiménez, M., & Lozoya-Gloria, E. (2008). Local and systematic gene expression of sesquiterpene phytoalexin biosynthetic enzymes in plant leaves. *European Journal of Plant Pathology, 121*, 439–449.

Nguyen, T.-D., Göpfert, J. C., Ikezawa, N., MacNevin, G., Kathiresan, M., Conrad, J., et al. (2010). Biochemical conservation and evolution of germacrene A oxidase in Asteraceae. *Journal of Biological Chemistry, 285*, 16588–16598.

Peralta-Yahya, P. P., Ouellet, M., Chan, R., Mukhopadhyay, A., Keasling, J. D., & Lee, T. S. (2011). Identification and microbial production of a terpene-based advanced biofuel. *Nature Communications, 2*, 483–490.

Rees, S. B., & Harborne, J. B. (1985). The role of sesquiterpene lactones and phenolics in the chemical defence of the chicory plant. *Phytochemistry, 24*, 2225–2231.

Renniger, N., & McPhee, D. (2008). Fuel compositions comprising farnesane and farnesane derivatives and method of making and using same. U.S. Patent No. 7399323.

Ro, D.-K., Ouellet, M., Paradise, E. M., Burd, H., Eng, D., Paddon, C. J., et al. (2008). Induction of multiple pleiotropic drug resistance genes in yeast engineered to produce an increased level of anti-malarial drug precursor, artemisinic acid. *BMC Biotechnology, 8*, 83–96.

Ro, D.-K., Paradise, E. M., Ouellet, M., Fisher, K. J., Newman, K. L., Ndungu, J. M., et al. (2006). Production of the anti-malarial drug precursor artemisinic acid in engineered yeast. *Nature, 440*, 940–943.

Sharon-Asa, L., Shalit, M., Frydman, A., Bar, E., Holland, D., Or, E., et al. (2003). Citrus fruit flavor and aroma biosynthesis: Isolation, functional characterization, and developmental regulation of *CsTPS1*, a key gene in the production of the sesquiterpene aroma compound valencene. *The Plant Journal, 36*, 664–674.

Takahashi, S., Yeo, Y., Greenhagen, B. T., McMullin, T., Song, L., Maurina-Brunker, J., et al. (2007). Metabolic engineering of sesquiterpene metabolism in yeast. *Biotechnology and Bioengineering, 97*, 170–181.

Takahashi, S., Zhao, Y., O'Maille, P. E., Greenhagen, B. T., Noel, J. P., Coates, R. M., et al. (2005). Kinetic and molecular analysis of 5-*epi*-aristolochene 1,3-dihydroxylase, a cytochrome P450 enzyme catalyzing successive hydroxylations of sesquiterpenes. *Journal of Biological Chemistry*, *280*, 3686–3696.

Wesołowska, A., Nikiforuk, A., Michalska, K., Kisiel, W., & Chojnacka-Wójcik, E. (2006). Analgesic and sedative activities of lactucin and some lactucin-like guaianolides in mice. *Journal of Ethnopharmacology*, *107*, 254–258.

Yuan, C.-S., Mehendale, S., Xiao, Y., Aung, H. H., Xie, J.-T., & Ang-Lee, M. K. (2004). The gamma-aminobutyric acidergic effects of valerian and valerenic acid on rat brainstem neuronal activity. *Anesthesia & Analgesia*, *98*, 353–358.

Zhao, Y., Schenk, D. J., Takahashi, S., Chappell, J., & Coates, R. M. (2004). Eremophilane sesquiterpenes from capsidiol. *Journal of Organic Chemistry*, *69*, 7428–7435.

# CHAPTER FOURTEEN

# *Streptomyces coelicolor* as an Expression Host for Heterologous Gene Clusters

## Juan Pablo Gomez-Escribano[1], Mervyn J. Bibb

Department of Molecular Microbiology, John Innes Centre, Norwich, United Kingdom
[1]Corresponding author: e-mail address: juan-pablo.gomez-escribano@jic.ac.uk

## Contents

## Abstract

The expression of a gene or a set of genes from one organism in a different species is known as "heterologous expression." In actinomycetes, prolific producers of natural products, heterologous gene expression has been used to confirm the clustering of secondary metabolite biosynthetic genes, to analyze natural product biosynthesis, to produce variants of natural products by genetic engineering, and to discover new compounds by screening genomic libraries. Recent advances in DNA sequencing have enabled the rapid and affordable sequencing of actinomycete genomes and revealed a large number of secondary metabolite gene clusters with no known products. Heterologous expression of these cryptic gene clusters combined with comparative metabolic profiling provides an important means to identify potentially novel compounds. In this chapter, the methods and strategies used to heterologously express actinomycete gene clusters, including the techniques used for cloning secondary metabolite gene clusters, the *Streptomyces* hosts used for their expression, and the techniques employed to analyze their products by metabolic profiling, are described.

# 1. INTRODUCTION

Actinomycetes, Gram-positive mycelial bacteria of terrestrial and marine origin, are prolific producers of natural products (often referred to as secondary metabolites) with a wide range of biological activities. The genes required for natural product regulation, biosynthesis, and transport (and for antibiotics, those for immunity) are almost invariably clustered together in the genome of the producing organism. While the genetic manipulation of individual producer strains has played a crucial role in deciphering natural product biosynthesis, heterologous gene expression has also proved extremely useful. For example, it has been used to:

demonstrate that an entire biosynthetic gene cluster has been cloned (e.g., Malpartida & Hopwood, 1984);

study the function of individual genes present in a cluster of interest (Gust, 2009);

analyze the biosynthesis of a natural product in a host that is more genetically tractable than the natural producer (Gust, 2009);

make the metabolite at higher levels than in the natural producer (e.g., chloramphenicol in Gomez-Escribano & Bibb, 2011);

create new compounds by combinatorial biosynthesis (i.e., by combining genes from different biosynthetic pathways generally from different microorganisms; e.g., Alt, Burkard, Kulik, Grond, & Heide, 2011) or by mutasynthesis (i.e., by feeding unnatural precursors to a mutant whose biosynthetic pathway has been genetically manipulated to prevent synthesis of the natural precursor, thus resulting in the incorporation of new structural motifs; Heide, 2009);

clone the gene cluster for a particular metabolite by screening a genomic library of the natural producer (e.g., cloning of the actinorhodin and prodiginine gene clusters of *Streptomyces coelicolor* in *Streptomyces venezuelae*; Mervyn J. Bibb, unpublished data); and

discover novel biologically active compounds by expressing genomic libraries from consortia of microorganisms (essentially "metagenomic libraries"; Banik & Brady, 2010).

Recently, genome sequencing has revealed that actinomycetes (and other groups of differentiating microorganisms) contain a large number of cryptic secondary metabolite gene clusters with no known products, and thus, these organisms have the potential to produce many more natural products than previously thought (Baltz, 2008). Many of these cryptic gene clusters are not expressed at significant levels in their natural hosts under typical laboratory screening conditions, but their expression can be activated by genetic manipulation of pathway-specific regulatory genes (e.g., Gottelt, Kol, Gomez-Escribano, Bibb, & Takano, 2010; Laureti et al., 2011). However, when a strain is difficult to culture or genetically intractable, then heterologous expression may be the only efficient approach for the identification of the products of these newly discovered gene clusters.

Actinomycetes possess DNA with a high GC content (on average about 73 mol% G + C) and have a correspondingly biased codon usage. The regulatory sequences and mechanisms that these organisms employ to activate secondary metabolism are quite distinct from those observed for gene regulation in other well-studied and manipulable bacterial systems, such as *Escherichia coli* and *Bacillus subtilis*. Consequently, in the absence of extensive genetic engineering, heterologous expression of these cryptic gene clusters will most likely be readily achieved by using an actinomycete expression host.

Here, we describe strategies used to express a gene cluster heterologously and to analyze its product(s). This involves (1) cloning the gene cluster in a suitable vector, (2) selecting an appropriate host strain for expression, (3) introducing and propagating the gene cluster in the chosen host, (4) culturing the recombinant strain under appropriate growth conditions, and (5) analyzing the host with and without the cloned gene cluster by comparative metabolic profiling.

## 2. *S. coelicolor* STRAINS FOR HETEROLOGOUS EXPRESSION

This chapter focuses on derivatives of *S. coelicolor* as hosts for the heterologous expression of secondary metabolite gene clusters. However, other *Streptomyces* species have also been used. *Streptomyces lividans*, a close relative of *S. coelicolor*, is frequently used as host, mostly because of its ease of genetic manipulation and the lack of secondary metabolite production when using particular growth media; strain TK24 has proved particularly popular since it contains a mutation *rpsL[K88E]* (Shima, Hesketh, Okamoto, Kawamoto, & Ochi, 1996) that enhances the level of secondary metabolite production. In our experience, actinorhodin production is often induced when *S. lividans* (particularly TK24) is cultured in a range of different liquid media or when introducing an exogenous gene cluster; the use of *S. lividans* TK24 derivatives with deletions of the actinorhodin and prodiginine gene clusters circumvents this problem (Martinez et al., 2004; Ziermann & Betlach, 1999). Other strains used include *Streptomyces avermitilis* and derivatives improved for heterologous expression (Komatsu, Uchiyama, Omura, Cane, & Ikeda, 2010) and *Streptomyces albus* (e.g., Chen, Wendt-Pienkowski, & Shen, 2008). The strategies and tools used with these strains are essentially the same as those described below for *S. coelicolor*.

*S. coelicolor* is the most genetically characterized actinomycete, its genome has been sequenced (Bentley et al., 2002) and annotated to a high standard (http://strepdb.streptomyces.org.uk), and a large array of tools are available for genetic manipulation (Gust, Challis, Fowler, Kieser, & Chater, 2003; 2004; Gust et al., 2004; Kieser, Bibb, Buttner, Chater, & Hopwood, 2000). *S. coelicolor* produces a variety of secondary metabolites (Table 14.1), including polyketides and nonribosomally synthesized peptides (Challis & Hopwood, 2003), indicating a plentiful supply of the precursors required for these important classes of secondary metabolites.

The wild-type strain of *S. coelicolor* A3(2), John Innes Centre strain 1147, produces four secondary metabolites with antibiotic activity: the chromosomally encoded actinorhodin, prodiginine complex, calcium-dependent antibiotic (CDA), and the plasmid (SCP1)-determined methylenomycin. M145, a plasmid-free derivative of 1147 (and thus unable to produce methylenomycin; Kieser et al., 2000) was used to construct *S. coelicolor* M512 (Floriano & Bibb, 1996) in which actinorhodin and prodiginine production were abolished by deletion of pathway-specific activator genes (*actII*-orf4 and *redD*, respectively). *S. coelicolor* CH999 (McDaniel, Ebert-Khosla, Hopwood, & Khosla, 1993) is another plasmid-free strain unable to make actinorhodin (after deletion of the *act* gene cluster) or prodiginines (after introduction of mutations into the *red* gene cluster). Both M512 and CH999 have been used as hosts for heterologous expression (Heide, 2009; Rodriguez, Menzella, & Gramajo, 2009).

*S. coelicolor* M145 was also used for the construction of M1152 and M1154, developed specifically for the heterologous expression of secondary metabolite gene clusters (Gomez-Escribano & Bibb, 2011). The

**Table 14.1** Summary of the metabolites produced by *Streptomyces coelicolor* M145 for which the chemical structure is known

| Gene cluster (Sco nos.) | Metabolite | Mass (Da) [calculated] | References |
|---|---|---|---|
| *act* (5071–5092) | Actinorhodin | 634.1323 | Bystrykh et al. (1996) |
| | Alpha-actinorhodin | 588.1268 | Taguchi et al. (2000) |
| | Beta-actinorhodin | 672.1479 | |
| | Gamma-actinorhodin | 630.1010 | |
| | Epsilon-actinorhodin | 632.1166 | |
| | Actinorhodinic acid | 666.1221 | |
| | Kalafungin | 300.0634 | |
| | Aloesaponarin II | 254.0579 | |
| *red* (5877–5898) | Undecylprodigiosin | 393.2780 | Tsao, Rudd, He, Chang, and Floss (1985) |
| | Streptorubin B | 391.2624 | Mo et al. (2008) |
| *cda* (3210–3249) | CDA1b | 1562 | Kempter et al. (1997) |
| | CDA2a | 1574 | Hojati et al. (2002) |
| | CDA2b | 1576 | |
| | CDA3a | 1480 | |
| | CDA3b | 1482 | |
| | CDA4a | 1494 | |
| | CDA4b | 1496 | |
| *cpk* (6269–6288) | Coelimycin P1 | 348 | Gomez-Escribano et al., 2012 |

| Gene cluster (Sco nos.) | Metabolite | Mass (Da) [calculated] | References |
|---|---|---|---|
| *gcs* (7221) | Germicidin A | 196.1099 | Song et al. (2006) |
| | Isogermicidin A | | |
| | Germicidin B | 182.0943 | |
| | Isogermicidin B | | |
| | Germicidin C | | |
| *cch* (0489–0499) | (desferri) Coelichelin | 565.2708 | Lautru, Deeth, Bailey, and Challis (2005) |
| | (ferri)Coelichelin | 618.1822 | |
| *des* (2780–2785) | Desferrioxamine B | 560.3534 | Barona-Gómez et al. (2006) |
| | Ferrioxamine B | 613.2648 | |
| | Desferrioxamine E | 600.3483 | Barona-Gómez, Wong, Giannakopulos, Derrick, and Challis (2004) |
| | Ferrioxamine E | 653.2597 | |
| | Desferrioxamine G$_1$ | 618.3588 | |
| | Ferrioxamine G$_1$ | 671.2703 | |
| (5222–5223) | Albaflavenone | 218.1671 | Lin, Hopson, and Cane (2006) |
| | Albaflavenol | 220.1827 | Zhao et al. (2008) |
| *cyc2* (6073) | Geosmin | 182.1671 | Gust et al. (2003); Jiang, He, and Cane (2007) |

Metabolites not produced by *S. coelicolor* M1152 and M1154 are in shaded boxes. For Sco numbers, see http://streptomyces.org.uk/.

actinorhodin, prodiginine, CDA, and the cryptic Type I polyketide (*cpk*, Sco6269–6288) gene clusters were deleted from both strains, thus removing all detectable antibiotic activity and reducing potential competition for precursors between endogenous and cloned biosynthetic pathways. It also mitigates the opportunity for cross talk between biosynthetic pathways that might modify and potentially inactivate the products of cloned gene clusters. In addition, these strains carry mutations in *rpoB* (encoding the β-subunit of RNA polymerase) and *rpsL* (encoding ribosomal protein S12) reported previously to increase levels of secondary metabolite production (Hu, Zhang, & Ochi, 2002; Shima et al., 1996). Both M1152 and M1154 contain the *rpoB* [C1298T, S433L] mutation, while M1154 contains, in addition, *rpsL* [A262G, K88E]. Both strains express heterologous gene clusters at significantly higher levels than M145, and sometimes in media not permissive for secondary metabolite production by M145. Neither M1152 nor M1154 contain an introduced antibiotic resistance marker and thus all of the genes commonly used to select for recombinant clones are available. Importantly, the metabolite profiles of these strains (see Section 7) are much simpler than those of M145 and even M512, markedly facilitating the analysis of culture supernatants and identification of the products of cloned gene clusters (Figs. 14.1 and 14.2;

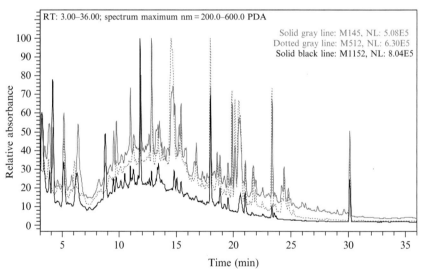

**Figure 14.1** Simplified HPLC profile of a culture supernatant of *S. coelicolor* M1152. Overlay of HPLC chromatograms (spectrum maximum, i.e., the maximum absorbance detected at any wavelength at any time point) illustrating the simplified chromatogram of M1152 compared with that of M512 (actinorhodin and prodiginine nonproducer) and the parent of both strains, M145. For details see legend of Fig. 14.3.

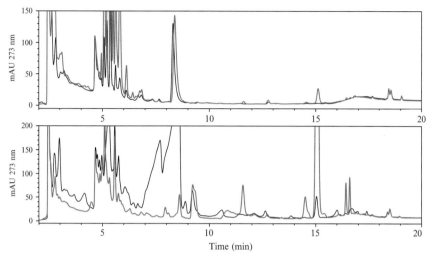

**Figure 14.2** HPLC analyses of *S. coelicolor* strains expressing the chloramphenicol gene cluster from *Streptomyces venezuelae*. Top panel, comparative metabolic profiling of a strain expressing the chloramphenicol gene cluster (gray) and a nonproducing control (black); the differentially produced peak at 15 min corresponds to chloramphenicol. Bottom panel, chromatograms of M145 (black) and M1152 (gray) both expressing the gene cluster and showing elevated levels of antibiotic production in, and simpler chromatograms for, M1152. The method was modified from that in Section 7.2 to allow separation of chloramphenicol from other compounds (see Gomez-Escribano and Bibb, 2011 for methods).

Table 14.1). Thus, *S. coelicolor* M1152 and M1154 are currently our preferred choices as heterologous expression hosts. Although M1152 grows and sporulates slightly better than M1154, the latter strain sometimes yields higher levels of production, and we recommend using both strains, at least initially, as expression hosts.

Gene clusters from several actinomycete genera have been expressed in streptomycetes (e.g., those for actagardine (from *Actinoplanes garbadinensis*; Boakes, Cortés, Appleyard, Rudd, & Dawson, 2009) and capreomycin (from *Saccharothrix mutabilis*; Felnagle, Rondon, Berti, Crosby, & Thomas, 2007) in *S. lividans*, for thiocoraline (from a *Micromonospora* species; Lombó et al., 2006) in *S. albus*, and for erythromycin (from *Saccharopolyspora erythraea*) and pradimicin (from *Actinomadura hibisca*) in *S. venezuelae* (Mervyn J. Bibb, unpublished data). When intergeneric transfer into a streptomycete failed to result in product formation, heterologous expression was achieved by using a more closely related actinomycete host (e.g., Foulston & Bibb, 2010, where a *Nonomuraea* strain was used to express the microbisporicin gene cluster from *Microbispora corallina*).

## 3. CLONING OF SECONDARY METABOLITE GENE CLUSTERS

While clustering greatly facilitates cloning of genes for an entire bio-synthetic pathway, some of these gene clusters are very large, particularly those encoding Type I modular polyketide synthases or nonribosomal peptide synthetases (NRPS), frequently extending beyond 50 kb in size. For example, the gene cluster for *S. coelicolor*'s CDA, a lipopeptide, contains three NRPSs and is greater than 80 kb (Hojati et al., 2002). This has consequences for the vector chosen for cloning. Generally, we recommend first cloning the targeted gene cluster into *E. coli* because of its higher growth rate and ease of manipulation.

### 3.1. Cosmids

Cosmid vectors can accommodate up to 42 kb of DNA. They are multicopy plasmids in *E. coli*, facilitating DNA isolation and *in vitro* manipulation. However, deletion of inserted DNA sometimes occurs with cosmid clones, perhaps indicative of sequences that are deleterious to *E. coli*, particularly at high copy number. SuperCos1 (Agilent), which was used to make an ordered *S. coelicolor* cosmid library (Redenbach et al., 1996), has proved to be a very effective vector, but its derivatives require modification before they can be transferred to and stably maintained in a *Streptomyces* host (see Section 4.4).

### 3.2. Bacterial Artificial Chromosomes and P1-derived Artificial Chromosomes

Larger fragments of DNA (up to 200 kb) can be cloned in bacterial artificial chromosome (BAC) vectors that are based on the origin of replication of the F plasmid or in P1-derived artificial chromosome (PAC) vectors based on the P1 phage origin of replication. Both vectors are maintained at a single copy per chromosome in *E. coli* and appear to be structurally more stable than cosmid clones, perhaps a reflection of their lower copy number. A modified version of pBACe3.6 (Osoegawa, de Jong, Frengen, & Ioannou, 2001), in which the chloramphenicol resistance gene was replaced with *neo* (encoding kanamycin resistance), was used to construct a partial genome library of *S. coelicolor* (Bentley et al., 2002; Helen Kieser and Nick Bird, personal communication). However, in our experience, PAC libraries exhibit higher levels of structural stability.

A series of PAC vectors (the ESAC (*E. coli-Streptomyces* Artificial Chromosome) vectors) based on pCYPAC2 (Osoegawa et al., 2001) were developed specifically for use with *Streptomyces*. They integrate at the phage φC31 attachment site present in the genomes of most *Streptomyces* species (Sosio et al., 2000). While clones derived from the early vectors had to be introduced into *Streptomyces* by transformation, pESAC13 clones can be transferred by conjugation from *E. coli* (Margherita Sosio, personal communication).

The isolation of high-molecular-weight DNA from actinomycetes and the construction of large insert libraries are not trivial, and the development of the required expertise and technology can be difficult and time consuming. We have obtained PAC libraries of a wide range of actinomycetes with average insert sizes ranging from 125 to 150 kb from Dr. Changhe Yu (Bio S&T Inc., Montreal, Canada; http://www.biost.com).

## 3.3. Assembling large gene clusters

Even when using BAC and PAC vectors, the desired gene cluster may sometimes be split between two or more clones. It will often be necessary or desirable to join the two segments together in a single construct that can be used readily for heterologous expression. Achieving this by restriction enzyme digestion and ligation is usually not possible due to the lack of appropriate restriction sites. Alternative strategies involve homologous (Sosio, Bossi, & Donadio, 2001) and Red/ET-mediated recombination (Gust, 2009).

Further details on the construction of genomic libraries can be found in the instructions provided by the suppliers of large insert vectors (e.g., for SuperCos 1, from Agilent) and in Osoegawa et al. (2001) and Sosio et al. (2000, 2001).

Genomic libraries made in *E. coli* can be screened by Southern or PCR analysis to identify clones carrying the gene cluster of interest. When sequence information is available, the probes or primers used can be specific (Bibb & Hesketh, 2009; Claesen & Bibb, 2010; Foulston & Bibb, 2010); alternatively, they can be based on conserved amino acid sequences found in some biosynthetic enzymes.

## 4. INTRODUCTION AND PROPAGATION OF CLONED GENE CLUSTERS IN *S. coelicolor* HOSTS

There are two ways of introducing plasmid DNA into *S. coelicolor*: protoplast transformation (Bibb, Ward, & Hopwood, 1978) and conjugation from *E. coli* (Flett, Mersinias, & Smith, 1997). Both techniques have been used widely, but for convenience conjugation from *E. coli* is now generally preferred (protocols in Kieser et al., 2000, p. 249).

## 4.1. Conjugation from *E. coli*

Mobilization of constructs by conjugation from *E. coli* requires the presence of an *origin of transfer* (*oriT*) in the construct to be transferred as well as the conjugation machinery, which can be provided *in trans* by genes carried on a different plasmid or integrated into the *E. coli* chromosome. The most commonly used *oriT* and transfer machinery are derived from RP4 (RK2); the transfer machinery is usually provided by either the self-transferable (i.e., with *oriT*) pUB307 or the nontransferable (i.e., without *oriT*) pUZ8002 (Gust et al., 2004).

## 4.2. DNA methylation

*S. coelicolor* possesses a methylation-dependent restriction system that markedly reduces the efficiency of introduction of methylated DNA (González-Cerón, Miranda-Olivares, & Servín-González, 2009). Consequently, DNA used for protoplast transformation is usually isolated from another *Streptomyces* species (usually *S. lividans*, which lacks the methylation-dependent restriction system and which can be transformed efficiently with methylated DNA) or from a methylation-deficient *E. coli* strain, usually *E. coli* ET12567 (MacNeil et al., 1992). ET12567 is also the most commonly used donor strain for conjugation in conjunction with pUZ8002 (Gust et al., 2004). Genomic libraries are usually constructed in DNA-methylating *E.coli* strains (e.g., DH10B), which are generally more robust than methylation-deficient strains. Triparental conjugation can also be performed between *E. coli* DH10B containing a cosmid, BAC or PAC clone, *E. coli* ET12567 [pUB307], and a *S. coelicolor* recipient strain. The protocol is similar to that for a biparental conjugation (Kieser et al., 2000, p. 249), but the, efficiency of plasmid transfer into the streptomycete recipient is lower.

## 4.3. Propagation of constructs in *Streptomyces*

Once the construct is in *Streptomyces*, it must be stably maintained. Self-replicative vectors have been developed, both low and high copy number, and used for heterologous expression (Malpartida & Hopwood, 1984). pIJ86, a high copy-number vector with the strong constitutive *ermE*★ promoter (Helen Kieser, personal communication), has been used to express single-operon gene clusters (Anyarat Thanapipatsiri, personal communication). Antibiotic selection should generally be applied continuously to ensure maintenance of the expression construct, but this can adversely affect subsequent analysis of the culture supernatant, particularly if the cloned gene cluster encodes a compound with antibiotic activity.

The preferred option is to integrate the construct into the chromosome by site-specific recombination using the integration machinery of either of two temperate phages: φC31 and φBT1. This requires the presence of a specific nucleotide sequence (*attB*) in the chromosome of the *Streptomyces* host and a corresponding sequence (*attP*) in the vector; an *integrase* enzyme encoded by *int*, usually included in the vector, performs site-specific recombination between the two attachment sequences, resulting in integration of the construct into the chromosome. φC31 *attB* is in *S. coelicolor Sco3798* (Combes, Till, Bee, & Smith, 2002) and φBT1 *attB* is located in *S. coelicolor Sco4848* (Gregory, Till, & Smith, 2003). PCR or Southern analysis (e.g., Eustáquio et al., 2005) should be used to confirm insertion of the transferred plasmid at the expected site. For PCR analysis, one of the two primers should correspond to chromosomal sequences adjacent to *attB* and the other to vector sequences within the integrated construct (see fig. S2 in Supporting Information of Foulston & Bibb, 2010).

Integration at either the φC31 or φBT1 *attB* sites appears to have no deleterious effect on growth or sporulation, and constructs are maintained without antibiotic selection. Integration of a gene cluster at the φC31 *attB* site may result in precocious production of the targeted metabolite, perhaps reflecting an effect of chromosomal location on the expression of pathway-specific regulatory genes. The two *attB* sites function independently, and it is possible to use both simultaneously to integrate two different constructs into the *S. coelicolor* chromosome.

## 4.4. Modification of constructs for conjugation and integration

Most vectors used for genomic library construction do not contain functions for conjugation or for integration into the *Streptomyces* genome. However, these can be incorporated into any construct by Red/ET-mediated recombination (Gust et al., 2004). A cassette that encodes the integration machinery (*attB* and *int*), a RP4 (RK2) *oriT*, and a selectable marker, and flanked by sequences homologous to the backbone of the vector used for genomic library construction, is used for PCR-targeting. Sources of frequently used cassettes include pIJ10702 (previously known as pMJCOS1; Yanai, Murakami, & Bibb, 2006), which contains the apramycin resistance gene *aac(3)IV* that is frequently used to replace the kanamycin resistance gene *neo* of SuperCos1 derivatives, and pIJ787 and pIJ788 (Gust et al., 2004) that contain a tetracycline resistance gene (*tet*) that is frequently used to replace the ampicillin resistance gene *bla* of SuperCos1 (and many comparable plasmids).

Replacement results in conjugative constructs that integrate at the φC31 *attB* site. pRT801 (Gregory et al., 2003) can be used similarly as a source of a φBT1 integrative cassette. All of these cassettes can be readily adapted for use with other vectors and their derivatives by simply designing PCR primers with sequences corresponding to the different plasmids (Gust et al., 2004).

## 5. PRESERVATION OF SELECTED CLONES

We have sometimes observed variability in production levels from different exconjugants or transformants (e.g., Flinspach et al., 2010) containing ostensibly the same expression construct; therefore, we recommend working with at least three independent clones when initially examining metabolite production.

Three exconjugants or transformants are streaked onto SFM agar (Kieser et al., 2000) containing the antibiotic used for construct selection and, if a conjugation with *E. coli* has been performed, nalidixic acid (25 μg/ml), which is used to kill the *E. coli* donor strain. After harvesting (Kieser et al., 2000), the spores are diluted and spread on SFM plates supplemented with the same antibiotics to obtain single colonies. One colony from each of the three exconjugants or transformants is streaked onto separate SFM plates (supplemented with the appropriate antibiotics) to obtain independent spore stocks which are resuspended in 20% glycerol and stored at − 80 °C (a working stock can be kept at − 20 °C).

*Note 1: Streptomyces* mycelium consists of mycelial compartments containing several chromosomes, not all of which will necessarily carry the gene cluster immediately following conjugation with *E. coli*. Consequently, it is important to isolate a single colony after a round of sporulation (spores are unigenomic) on selective medium to ensure chromosomal homogeneity.

*Note 2*: Some *S. coelicolor* clones may not sporulate well, and it may be necessary to store them as frozen mycelial stocks in 20% glycerol at −80 °C. For preparing mycelial glycerol stocks, 50 ml of TSB:YEME 50:50 in a 250-ml baffled flask are inoculated with a 1 cm$^2$ mycelial patch and incubated at 250 rpm at 30 °C until late exponential growth phase (24–36 h). 0.9 ml of sterile 40% glycerol is mixed with 0.9 ml of the resulting mycelial culture in a 2-ml tube and stored at −80 °C. If the mycelium grows in clumps, it can be dispersed with a glass homogeniser before mixing with glycerol.

## 6. GROWTH IN LIQUID CULTURE FOR METABOLITE ANALYSIS

Growth in liquid culture can be carried out in a range of volumes, influenced in part by the number of clones to be analyzed. With large metagenomic libraries, cultivation in small volumes coupled with the use of robotics permits screening of thousands of clones (Baltz, 2008). In a medium-sized project, vials with 5–10 ml of medium (and glass beads or small springs to facilitate aeration) can be used to analyze hundreds of strains; plates with 24-deep square wells with 3 ml of growth medium per well have been used for heterologous expression in *S. coelicolor* (Flinspach et al., 2010; Siebenberg, Bapat, Lantz, Gust, & Heide, 2010). For studying tens of clones, we prefer to use flasks with 50–100 ml of growth medium; this allows sampling at different times during growth and provides enough mycelium to accurately quantify growth by determining dry-cell weight and sufficient material for metabolite analysis.

Streptomycetes require good aeration for rapid growth, and this can be achieved by using baffles, springs, or glass beads in the bottom of the flask, together with orbital shaking at 220–250 rpm. *S. coelicolor* grows optimally at 30 °C and in our experience springs promote more dispersed growth than baffles or glass beads. Comprehensive methods for cultivation of *Streptomyces* can be found in Kieser et al. (2000).

### 6.1. Preparation of flasks

To prevent the mycelium sticking to the flask wall, flasks are rinsed with a 5% solution of dimethyldichlorosilane (5% DMDCS in toluene (Sylon CT), Supelco 33065-U) and allowed to dry overnight in a fume hood. This is carried out before sterilization and addition of springs or glass beads. Typically, culture medium represents no more than 1/5 of the total volume of the flask (we usually use 250-ml flasks containing stainless steel springs and 50 ml of culture medium).

## 6.2. Culture medium

Many secondary metabolites are only produced in particular media, and hence the choice of culture medium may be crucial; even when a metabolite is produced in a screening medium, optimization will usually be required. The use of M1152 and M1154 somewhat alleviates this problem (Section 2), and for these strains, GYM medium (which yields rapid growth rates and cultures that reach stationary phase within 24–36 h of inoculation) and R3 medium (Shima et al., 1996) have proved particularly effective.

## 6.3. Inoculation from spores

$10^7$–$10^8$ spores are germinated in 10 ml of $2 \times$ YT medium in glass Universals for 7–8 h; germlings are harvested by centrifugation, resuspended in fresh medium by vortexing, and dispersed by sonication in a water bath for 1 min. The germlings are used to inoculate up to 50 ml of growth medium in a 250-ml flask, followed by incubation at 250 rpm and 30 °C. A more detailed description can be found in Kieser et al., 2000; page 52.

## 6.4. Inoculation from frozen mycelium

One glycerol stock (see Section 5; Note 2) is used to inoculate 50 ml of medium in a 250-ml flask, followed by incubation at 250 rpm at 30 °C for about 24 h or until mid-late exponential growth phase; the optical density of this *seed-culture* or *pre culture* is measured at 600 nm ($OD_{600}$) and the volume required for inoculation of 50 ml of medium to an $OD_{600}$ of 0.25 calculated (e.g., if the seed-culture has an $OD_{600} = 4$, then $(50 \times 0.25)/4 = 3.125$ ml of seed-culture will be needed); the calculated volume is harvested by centrifugation, and the mycelium is resuspended in culture medium and used to inoculate up to 50 ml of medium in a 250-ml flask, followed by incubation at 250 rpm and 30 °C.

## 6.5. Sample harvest and process

Below, we describe the method we use when expressing gene clusters with no known products (i.e., cryptic gene clusters) and where the metabolite is produced at a readily detectable level; lower levels of production would require larger culture volumes and organic extraction and concentration.

Typically, duplicate 1 ml samples are harvested and transferred to 1.5-ml Eppendorf tubes; after centrifugation at 13,000 rpm for 5 min at 4 °C, the supernatant is pipetted to a new tube and stored at − 80 °C. To extract compounds within or attached to the mycelium, the latter is resuspended in 500 μl of methanol by vortexing and pelleted by centrifugation, and the methanol is transferred to a new tube and stored at − 80 °C; methanol extracts can be concentrated in a centrifuge-vacuum system (e.g., Speed-Vac). If an estimation of growth is required, 0.5 ml of culture can be used to determine optical density (OD; if the culture is not pigmented, then the $OD_{600}$ provides a reasonable estimate of growth). For a more accurate determination, 5–10 ml of culture is removed, the mycelium is harvested by centrifugation, washed with 0.9% NaCl, and filtered onto a preweighed cellulose acetate filter (0.45 μm pore). After drying to completeness in an oven at 30–50 °C, the filter is weighed and the amount of dried mycelium, and hence growth, determined.

## 7. COMPARATIVE METABOLIC PROFILING

To identify the product of a cloned gene cluster, comparative metabolic profiling (Challis, 2008) is performed: samples are analyzed using high performance liquid chromatography (HPLC) usually coupled to mass spectrometry (LC–MS). Compounds present in the heterologous host expressing the gene cluster, but not in the same strain lacking the foreign DNA, arc likely to be derived from the cluster (Fig. 14.2); this can then be confirmed by mutational analysis. Note that the same procedure can be used with the natural producer and a nonproducing mutant to identify the product of a gene cluster of interest.

### 7.1. Controls

The nonproducing control strain in a heterologous expression study is usually the host containing the vector but not the cloned gene cluster. While the host strain can be used, this does not allow for the possible induction of host metabolite production by the cloning vector itself. Confirmation that any identified product results from the cloned gene cluster can be achieved by making specific mutations in candidate biosynthetic genes.

Another useful negative control to analyze is the initial culture medium: this will serve to identify components of the culture medium that have not been completely metabolized during growth.

## 7.2. Liquid chromatography

For most purposes, we use a method adapted from that developed by Prof. Gregory Challis and Dr. Lijiang Song at the University of Warwick, UK. Some metabolites may require different chromatographic methods, especially when detection is only by UV/visible absorbance and the compounds are not well resolved (e.g., Fig. 14.2). Examples of chromatograms are given in Figs. 14.1–14.3. We generally use the following conditions.

*Stationary phase*: C18 reverse phase analytical column, for example, Spherisorb ODS2 C18, $4.6 \times 250$ mm and 5 μm particle size (Waters); Luna C18, $100 \times 2$ mm and 3 μm particle size (Phenomenex).

*Mobile phases*: 18 MΩ water with 0.1% formic acid (A) and 100% methanol with 0.1% formic acid (B).

*Sample preparation*: Prior to injection, the sample is allowed to thaw on ice and cleared by centrifugation at 13 rpm for 5 min at 4 °C, and 2/3 of the supernatant (to avoid carrying over insoluble material) is carefully pipetted into a HPLC vial.

*Injection, flow, and gradient*: For a small column ($100 \times 2$ mm and 3 μm particle size), up to 20 μl of sample are injected and analyzed at a flow rate of 230 μl/min with the following gradient: 0 min—1% B; 4 min—2% B; 32 min—100% B; 36.1 min—1% B; 8.2 min—1% B for equilibration. For a larger column ($4.6 \times 250$ mm and 5 μm particle size), up to 100 μl of sample are injected at a flow rate of 1 ml/min and with the following gradient: 0 min—0% B; 2 min—5% B; 35 min—100% B; 40 min—100% B; 41 min—0% B; 10 min—0% B for equilibration.

*Detection*: UV/visible absorbance is measured, preferably with a photodiode array detector (PDA) set to record spectra between 190 and 700 nm. When possible, chromatograms are recorded at wavelengths of 210 nm (at which most compounds absorb strongly, but so do methanol and other solvents), 240 nm (for detection of compounds with double carbon–carbon bonds, e.g., polyketides), and 280 nm (for aromatic motifs). Siderophores produced by *S. coelicolor* absorb maximally at around 435 nm (Lautru et al., 2005); thus, monitoring at this wavelength can be useful, particularly when mass spectrometric detection is not available. For LC–MS, the flow after the PDA is introduced into an electrospray ionization mass spectrometry detector. Typically, we record data in positive and negative mode to determine ions with a mass-to-charge ($m/z$) ratio between 100 and 2000 units, which should cover most natural products.

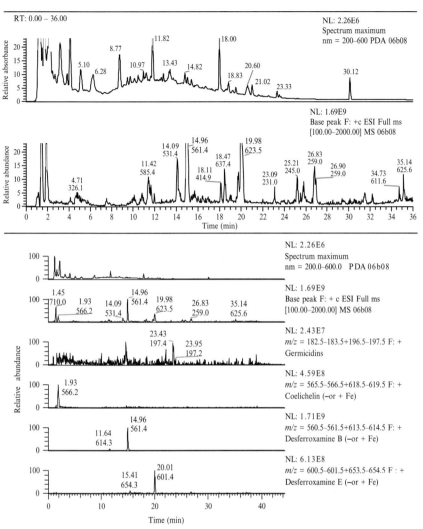

**Figure 14.3** LC–MS analysis of *S. coelicolor* M1152. Typical result expected from a control LC–MS run of a culture supernatant of M1152 grown in R3 medium for 5 days. Ten microliters of sample were analyzed in a Luna C18(2) 100 × 2 mm 3 μm particle size column (Phenomenex) fitted with a guard column in a Surveyor HPLC attached to a PDA UV/visible detector and a DecaXPplus Ion Trap MS (Thermo Scientific). The top two panels show an expanded view of the spectrum maximum (maximum absorbance detected at any wavelength at any time point) and base peak chromatogram, shown as relative percentage of the absorbance or intensity of the largest peak (indicated as normalization level, NL). The elution times and mass-to-charge ratios (*m/z*) of the most abundant ion in relevant peaks are included for reference. The bottom panel shows the spectrum maximum and base peak chromatogram, followed by the extracted ion chromatograms for the ions corresponding to proton adducts $[M+H]^+$ of the most important metabolites produced by M1152 (see Table 14.1). In the case of the germicidins, only germicidin A and isogermicidin A are clearly detected in this sample. Coelichelin, but not ferricoelichelin, is clearly detected. Both desferroxamine B and E are detected in their free form or bound to iron (+53 *m/z* units).

## 7.3. Data analysis

Chromatograms of control samples and of the host expressing the cloned gene cluster are compared to identify differential peaks (Fig. 14.2); peaks present only in the latter are likely to represent the metabolites of interest. An array of informatic tools may be available with the software of the HPLC manufacturer; for example, it may be possible to plot chromatograms for a specific wavelength or range of wavelengths, or a chromatogram where the maximum absorbance detected at any wavelength at any time point is plotted (called *Spectrum Maximum* by Thermo Scientific software and *Max Plot* by Waters software).

If MS detection has been performed, the total ion chromatogram (TIC, which plots the cumulative abundances of all of the ions detected at each time point) and the base peak chromatogram (BPC, which plots the intensity of the most abundant ion at each time point) are most useful for comparing the metabolic profiles of two samples. To confirm differentially observed ions as candidate products of gene clusters, extracted ion chromatograms (EIC) are plotted, that is the intensity of the selected ions or $m/z$ range at each time point for each sample; the presence of the candidate ions in the sample from the host expressing the gene cluster, and their absence from the control samples confirm the compounds as candidates for further study. This will generally require the purification of larger amounts of the compound for determination of structure and biological activity. The accuracy of modern mass spectrometers often allows the prediction of a single molecular formula that can be used, in conjunction with MS fragmentation data, UV spectra and an appropriate database, to determine whether an identified compound is likely to be novel or to correspond to a previously identified molecule.

It is during this comparative process that the markedly simplified metabolic profiles of *S. coelicolor* M1152 and M1154 are particularly useful (Figs. 14.1 and 14.2); the base line of the chromatogram is also lower and more stable than that observed for M145. This greatly facilitates the identification of small peaks that may inevitably result from low levels of expression of some heterologous gene clusters, particularly since initial monitoring is unlikely to be performed at the optimal wavelength for an unknown compound. A compilation of known *S. coelicolor* metabolites is provided in Table 14.1 for reference, highlighting the metabolites still produced by strains M1152 and M1154.

## ACKNOWLEDGMENTS

We thank Lionel Hill and Andrew Hesketh (John Innes Centre, Norwich, UK), and Lijiang Song and Gregory Challis (University of Warwick, Coventry, UK) for many useful discussions and for the provision of protocols that have contributed to the methods described in this chapter. The research reported was funded by the UK Biotechnology and Biological Sciences Research Council (BBSRC) and by the European Commission.

## REFERENCES

Alt, S., Burkard, N., Kulik, A., Grond, S., & Heide, L. (2011). An artificial pathway to 3,4-dihydroxybenzoic acid allows generation of new aminocoumarin antibiotic recognized by catechol transporters of *E. coli*. *Chemistry & Biology*, *18*, 304–313.

Baltz, R. H. (2008). Renaissance in antibacterial discovery from actinomycetes. *Current Opinion in Pharmacology*, *8*, 557–563.

Banik, J. J., & Brady, S. F. (2010). Recent application of metagenomic approaches toward the discovery of antimicrobials and other bioactive small molecules. *Current Opinion in Microbiology*, *13*, 603–609.

Barona-Gómez, F., Lautru, S., Francou, F.-X., Leblond, P., Pernodet, J.-L., & Challis, G. L. (2006). Multiple biosynthetic and uptake systems mediate siderophore-dependent iron acquisition in *Streptomyces coelicolor* A3(2) and *Streptomyces ambofaciens* ATCC 23877. *Microbiology*, *152*, 3355–3366.

Barona-Gómez, F., Wong, U., Giannakopulos, A. E., Derrick, P. J., & Challis, G. L. (2004). Identification of a cluster of genes that directs desferrioxamine biosynthesis in *Streptomyces coelicolor* M145. *Journal of the American Chemical Society*, *126*, 16282–16283.

Bentley, S. D., Chater, K. F., Cerdeño-Tárraga, A.-M., Challis, G. L., Thomson, N. R., James, K. D., et al. (2002). Complete genome sequence of the model actinomycete *Streptomyces coelicolor* A3(2). *Nature*, *417*, 141–147.

Bibb, M. J., & Hesketh, A. (2009). Analyzing the regulation of antibiotic production in streptomycetes. *Methods in Enzymology*, *458*, 93–116.

Bibb, M. J., Ward, J. M., & Hopwood, D. A. (1978). Transformation of plasmid DNA into *Streptomyces* at high frequency. *Nature*, *274*, 398–400.

Boakes, S., Cortés, J., Appleyard, A. N., Rudd, B. A. M., & Dawson, M. J. (2009). Organization of the genes encoding the biosynthesis of actagardine and engineering of a variant generation system. *Molecular Microbiology*, *72*, 1126–1136.

Bystrykh, L. V., Fernández-Moreno, M. A., Herrema, J. K., Malpartida, F., Hopwood, D. A., & Dijkhuizen, L. (1996). Production of actinorhodin-related "blue pigments" by *Streptomyces coelicolor* A3(2). *Journal of Bacteriology*, *178*, 2238–2244.

Challis, G. L. (2008). Mining microbial genomes for new natural products and biosynthetic pathways. *Microbiology (Reading, England)*, *154*, 1555–1569.

Challis, G. L., & Hopwood, D. A. (2003). Synergy and contingency as driving forces for the evolution of multiple secondary metabolite production by *Streptomyces* species. *Proceedings of the National Academy of Sciences of the United States of America*, *100*, 14555–14561.

Chen, Y., Wendt-Pienkowski, E., & Shen, B. (2008). Identification and utility of FdmR1 as a *Streptomyces* antibiotic regulatory protein activator for fredericamycin production in *Streptomyces griseus* ATCC 49344 and heterologous hosts. *Journal of Bacteriology*, *190*, 5587–5596.

Claesen, J., & Bibb, M. (2010). Genome mining and genetic analysis of cypemycin biosynthesis reveal an unusual class of posttranslationally modified peptides. *Proceedings of the National Academy of Sciences of the United States of America*, *107*, 16297–16302.

Combes, P., Till, R., Bee, S., & Smith, M. C. M. (2002). The *Streptomyces* genome contains multiple pseudo-*attB* sites for the phiC31-encoded site-specific recombination system. *Journal of Bacteriology*, *184*, 5746–5752.

Eustáquio, A. S., Gust, B., Galm, U., Li, S. M., Chater, K. F., & Heide, L. (2005). Heterologous expression of novobiocin and clorobiocin biosynthetic gene clusters. *Applied and Environmental Microbiology*, *71*, 2452–2459.

Felnagle, E. A., Rondon, M. R., Berti, A. D., Crosby, H. A., & Thomas, M. G. (2007). Identification of the biosynthetic gene cluster and an additional gene for resistance to the antituberculosis drug capreomycin. *Applied and Environmental Microbiology*, *73*, 4162–4170.

Flett, F., Mersinias, V., & Smith, C. P. (1997). High efficiency intergeneric conjugal transfer of plasmid DNA from *Escherichia coli* to methyl DNA-restricting streptomycetes. *FEMS Microbiology Letters*, *155*, 223–229.

Flinspach, K., Westrich, L., Kaysser, L., Siebenberg, S., Gomez-Escribano, J. P., Bibb, M., et al. (2010). Heterologous expression of the biosynthetic gene clusters of coumermycin A(1), clorobiocin and caprazamycins in genetically modified *Streptomyces coelicolor* strains. *Biopolymers*, *93*, 823–832.

Floriano, B., & Bibb, M. (1996). *afsR* is a pleiotropic but conditionally required regulatory gene for antibiotic production in *Streptomyces coelicolor* A3(2). *Molecular Microbiology*, *21*, 385–396.

Foulston, L. C., & Bibb, M. J. (2010). Microbisporicin gene cluster reveals unusual features of lantibiotic biosynthesis in actinomycetes. *Proceedings of the National Academy of Sciences of the United States of America*, *107*, 13461–13466.

Gomez-Escribano, J. P., & Bibb, M. J. (2011). Engineering *Streptomyces coelicolor* for heterologous expression of secondary metabolite gene clusters. *Microbial Biotechnology*, *4*, 207–215.

Gomez-Escribano, J. P., Song, L., Fox, D. J., Yeo, V., Bibb, M. J., & Challis, G. L. (2012). Structure and biosynthesis of the unusual polyketide alkaloid coelimycin P1, a metabolic product of the cpk gene cluster of *Streptomyces coelicolor* M145. *Chemical Science*, *3*, 2716–2720.

González-Cerón, G., Miranda-Olivares, O. J., & Servín-González, L. (2009). Characterization of the methyl-specific restriction system of *Streptomyces coelicolor* A3(2) and of the role played by laterally acquired nucleases. *FEMS Microbiology Letters*, *301*, 35–43.

Gottelt, M., Kol, S., Gomez-Escribano, J. P., Bibb, M., & Takano, E. (2010). Deletion of a regulatory gene within the *cpk* gene cluster reveals novel antibacterial activity in *Streptomyces coelicolor* A3(2). *Microbiology (Reading, England)*, *156*, 2343–2353.

Gregory, M. A., Till, R., & Smith, M. C. M. (2003). Integration site for *Streptomyces* phage phiBT1 and development of site-specific integrating vectors. *Journal of Bacteriology*, *185*, 5320–5323.

Gust, B. (2009). Cloning and analysis of natural product pathways. *Methods in Enzymology*, *458*, 159–180.

Gust, B., Challis, G. L., Fowler, K., Kieser, T., & Chater, K. F. (2003). PCR-targeted *Streptomyces* gene replacement identifies a protein domain needed for biosynthesis of the sesquiterpene soil odor geosmin. *Proceedings of the National Academy of Sciences of the United States of America*, *100*, 1541–1546.

Gust, B., Chandra, G., Jakimowicz, D., Yuqing, T., Bruton, C. J., & Chater, K. F. (2004). Lambda red-mediated genetic manipulation of antibiotic-producing *Streptomyces*. *Advances in Applied Microbiology*, *54*, 107–128.

Heide, L. (2009). Aminocoumarins mutasynthesis, chemoenzymatic synthesis, and metabolic engineering. *Methods in Enzymology*, *459*, 437–455.

Hojati, Z., Milne, C., Harvey, B., Gordon, L., Borg, M., Flett, F., et al. (2002). Structure, biosynthetic origin, and engineered biosynthesis of calcium-dependent antibiotics from *Streptomyces coelicolor*. *Chemistry & Biology*, *9*, 1175–1187.

Hu, H., Zhang, Q., & Ochi, K. (2002). Activation of antibiotic biosynthesis by specified mutations in the *rpoB* gene (encoding the RNA polymerase beta subunit) of *Streptomyces lividans*. *Journal of Bacteriology*, *184*, 3984–3991.

Jiang, J., He, X., & Cane, D. E. (2007). Biosynthesis of the earthy odorant geosmin by a bifunctional *Streptomyces coelicolor* enzyme. *Nature Chemical Biology, 3,* 711–715.

Kempter, C., Kaiser, D., Haag, S., Nicholson, G., Gnau, V., Walk, T., et al. (1997). CDA: Calcium-dependent peptide antibiotics from *Streptomyces coelicolor* A3(2) containing unusual residues. *Angewandte Chemie International Edition in English, 36,* 498–501.

Kieser, T., Bibb, M. J., Buttner, M. J., Chater, K. F., & Hopwood, D. A. (2000). Practical streptomyces genetics. John Innes Centre, Norwich Research Park, Colney, Norwich NR4 7UH, England.

Komatsu, M., Uchiyama, T., Omura, S., Cane, D. E., & Ikeda, H. (2010). Genome-minimized *Streptomyces* host for the heterologous expression of secondary metabolism. *Proceedings of the National Academy of Sciences of the United States of America, 107,* 2646–2651.

Laureti, L., Song, L., Huang, S., Corre, C., Leblond, P., Challis, G. L., et al. (2011). Identification of a bioactive 51-membered macrolide complex by activation of a silent polyketide synthase in *Streptomyces ambofaciens*. *Proceedings of the National Academy of Sciences of the United States of America, 108,* 6258–6263.

Lautru, S., Deeth, R. J., Bailey, L. M., & Challis, G. L. (2005). Discovery of a new peptide natural product by *Streptomyces coelicolor* genome mining. *Nature Chemical Biology, 1,* 265–269.

Lin, X., Hopson, R., & Cane, D. E. (2006). Genome mining in *Streptomyces coelicolor*: Molecular cloning and characterization of a new sesquiterpene synthase. *Journal of the American Chemical Society, 128,* 6022–6023.

Lombó, F., Velasco, A., Castro, A., de la Calle, F., Braña, A. F., Sáanchez-Puelles, J. M., et al. (2006). Deciphering the biosynthesis pathway of the antitumor thiocoraline from a marine actinomycete and its expression in two *Streptomyces* species. *ChemBioChem, 7,* 366–376.

MacNeil, D. J., Gewain, K. M., Ruby, C. L., Dezeny, G., Gibbons, P. H., & MacNeil, T. (1992). Analysis of *Streptomyces avermitilis* genes required for avermectin biosynthesis utilizing a novel integration vector. *Gene, 111,* 61–68.

Malpartida, F., & Hopwood, D. A. (1984). Molecular cloning of the whole biosynthetic pathway of a *Streptomyces* antibiotic and its expression in a heterologous host. *Nature, 309,* 462–464.

Martinez, A., Kolvek, S. J., Yip, C. L. T., Hopke, J., Brown, K. A., MacNeil, I. A., et al. (2004). Genetically modified bacterial strains and novel bacterial artificial chromosome shuttle vectors for constructing environmental libraries and detecting heterologous natural products in multiple expression hosts. *Applied and Environmental Microbiology, 70,* 2452–2463.

McDaniel, R., Ebert-Khosla, S., Hopwood, D. A., & Khosla, C. (1993). Engineered biosynthesis of novel polyketides. *Science, 262,* 1546–1550.

Mo, S., Sydor, P. K., Corre, C., Alhamadsheh, M. M., Stanley, A. E., Haynes, S. W., et al. (2008). Elucidation of the *Streptomyces coelicolor* pathway to 2-undecylpyrrole, a key intermediate in undecylprodiginine and streptorubin B biosynthesis. *Chemistry & Biology, 15,* 137–148.

Osoegawa, K., de Jong, P. J., Frengen, E., & Ioannou, P. A. (2001). Construction of bacterial artificial chromosome (BAC/PAC) libraries. *Current Protocols in Human Genetics,* chap. 5, unit 5.15.

Redenbach, M., Kieser, H. M., Denapaite, D., Eichner, A., Cullum, J., Kinashi, H., et al. (1996). A set of ordered cosmids and a detailed genetic and physical map for the 8 Mb *Streptomyces coelicolor* A3(2) chromosome. *Molecular Microbiology, 21,* 77–96.

Rodriguez, E., Menzella, H. G., & Gramajo, H. (2009). Heterologous production of polyketides in bacteria. *Methods in Enzymology, 459,* 339–365.

Shima, J., Hesketh, A., Okamoto, S., Kawamoto, S., & Ochi, K. (1996). Induction of actinorhodin production by *rpsL* (encoding ribosomal protein S12) mutations that confer streptomycin resistance in *Streptomyces lividans* and *Streptomyces coelicolor* A3(2). *Journal of Bacteriology*, *178*, 7276–7284.

Siebenberg, S., Bapat, P. M., Lantz, A. E., Gust, B., & Heide, L. (2010). Reducing the variability of antibiotic production in *Streptomyces* by cultivation in 24-square deepwell plates. *Journal of Bioscience and Bioengineering*, *109*, 230–234.

Song, L., Barona-Gomez, F., Corre, C., Xiang, L., Udwary, D. W., Austin, M. B., et al. (2006). Type III polyketide synthase beta-ketoacyl-ACP starter unit and ethylmalonyl-CoA extender unit selectivity discovered by *Streptomyces coelicolor* genome mining. *Journal of the American Chemical Society*, *128*, 14754–14755.

Sosio, M., Bossi, E., & Donadio, S. (2001). Assembly of large genomic segments in artificial chromosomes by homologous recombination in *Escherichia coli*. *Nucleic Acids Research*, *29*, E37.

Sosio, M., Giusino, F., Cappellano, C., Bossi, E., Puglia, A. M., & Donadio, S. (2000). Artificial chromosomes for antibiotic-producing actinomycetes. *Nature Biotechnology*, *18*, 343–345.

Taguchi, T., Itou, K., Ebizuka, Y., Malpartida, F., Hopwood, D. A., Surti, C. M., et al. (2000). Chemical characterisation of disruptants of the *Streptomyces coelicolor* A3(2) *actVI* genes involved in actinorhodin biosynthesis. *Journal of Antibiotics*, *53*, 144–152.

Tsao, S. W., Rudd, B. A., He, X. G., Chang, C. J., & Floss, H. G. (1985). Identification of a red pigment from *Streptomyces coelicolor* A3(2) as a mixture of prodigiosin derivatives. *Journal of Antibiotics*, *38*, 128–131.

Yanai, K., Murakami, T., & Bibb, M. J. (2006). Amplification of the entire kanamycin biosynthetic gene cluster during empirical strain improvement of *Streptomyces kanamyceticus*. *Proceedings of the National Academy of Sciences of the United States of America*, *103*, 9661–9666.

Zhao, B., Lin, X., Lei, L., Lamb, D. C., Kelly, S. L., Waterman, M. R., et al. (2008). Biosynthesis of the sesquiterpene antibiotic albaflavenone in *Streptomyces coelicolor* A3 (2). *Journal of Biological Chemistry*, *283*, 8183–8189.

Ziermann, R., & Betlach, M. C. (1999). Recombinant polyketide synthesis in *Streptomyces*: Engineering of improved host strains. *Biotechniques*, *26*, 106–110.

# Waking Up Silent Genes

CHAPTER FIFTEEN

# Toward Awakening Cryptic Secondary Metabolite Gene Clusters in Filamentous Fungi

**Fang Yun Lim\*, James F. Sanchez[†], Clay C.C. Wang[†,‡], Nancy P. Keller\*,[1]**

\*Department of Medical Microbiology and Immunology, University of Wisconsin, Madison, Wisconsin, USA
[†]Department of Pharmacology and Pharmaceutical Sciences, University of Southern California,
School of Pharmacy, Los Angeles, California, USA
[‡]Department of Chemistry, University of Southern California, College of Letters, Arts, and Sciences,
Los Angeles, California, USA
[1]Corresponding author: e-mail address: npkeller@wisc.edu

## Contents

## Abstract

Mining for novel natural compounds is of eminent importance owing to the continuous need for new pharmaceuticals. Filamentous fungi are historically known to harbor the genetic capacity for an arsenal of natural compounds, both beneficial and detrimental to humans. The majority of these metabolites are still cryptic or silent under standard laboratory culture conditions. Mining for these cryptic natural products can be an excellent source for identifying new compound classes. Capitalizing on the current knowledge on how secondary metabolite gene clusters are regulated has allowed the research community to unlock many hidden fungal treasures, as described in this chapter.

## 1. INTRODUCTION

Despite their obscure or sometimes unknown functions in the producing species, fungal natural compounds have certainly impacted humankind tremendously owing to their diverse and potent bioactivities. Compounds such as statins, cyclosporines, and penicillin are of clinical importance, whereas mycotoxins like aflatoxin and gliotoxin have adversely affected the food industry and human health. These low-molecular weight natural compounds are also termed secondary metabolites (SMs), in agreement with their "dispensable" role in the survival of an organism. Ongoing research in this field has not only increased our appreciation for SMs as an asset but has also shed light on their pivotal role in fungal biology, defense, and stress response and their potential contribution to human mycoses (Ben-Ami, Lewis, Leventakos, & Kontoyiannis, 2009; Bok et al., 2006; Losada, Ajayi, Frisvad, Yu, & Nierman, 2009; Rohlfs, Albert, Keller, & Kempken, 2007; Stanzani et al., 2005; Yin et al., 2012). As more genomes are sequenced, it is found that, as in SM producing bacteria, the genomic capacity for natural product diversity far exceeds known isolated compounds, because many of these clusters are either dormant under standard laboratory culture conditions or expressed at extremely low levels (Hertweck, 2009).

Our current understanding of fungal SM gene clusters regulation allows us to use this knowledge to artificially induce these silent clusters under normally noninducing conditions. Most of these approaches have been reviewed recently (Brakhage et al., 2008; Brakhage & Schroeckh, 2011; Chiang, Chang, Oakley, & Wang, 2011; Yin et al., 2012) and is briefly discussed in Section 2. Although various methods have been employed to mine novel compounds from cryptic clusters, including coculture with bacterial/fungal species that cohabit the same niche, heterologous expression in a different host (e.g., *Aspergillus oryzae*), and using chemical means to modify the chromatin landscape (e.g., 5-azacytidine, trichostatin A), this chapter focuses on *in silico-* and proteomic-based approaches to mine new compounds and their biosynthetic genes, molecular techniques used to force expression of cryptic SM gene clusters, and advances in isolating and identifying the natural products themselves.

## 2. ACTIVATION HANDLES OF FUNGAL SECONDARY METABOLITE GENE CLUSTERS

SM gene clusters are regulated in a multilevel fashion. The production of specific natural compounds is intimately linked to specific environmental cues. This is not surprising, given that these natural compounds have evolved to secure niches, avoid predation (Rohlfs et al., 2007), and defend against hostile environmental confrontations. Expression of these clusters is therefore cryptic under standard laboratory growth conditions. Several groups have shown that simulating these natural triggers serves to activate dormant gene clusters (Brakhage et al., 2008; Losada et al., 2009). In addition, both global and SM gene cluster-specific regulators have been used for activating many SM gene clusters (Bergmann et al., 2010; Yin et al., 2012). Altering the chromatin landscape by either genetic or chemical manipulation of chromatin modifiers has also been shown to activate many cryptic SM gene clusters (Bok et al., 2009; Cichewicz, 2010; Lee et al., 2009; Palmer & Keller, 2010; Reyes-Dominguez et al., 2010; Shwab et al., 2007; Wang et al., 2011).

### 2.1. Biosynthetic genes

The SM biosynthetic genes for any one metabolite in filamentous fungi, as in bacteria, are typically clustered together (Keller & Hohn, 1997; Walton, 2000). SM clusters usually consist of one or more backbone gene(s) such as polyketide synthases (PKSs), nonribosomal peptide synthetases (NRPSs), dimethylallyl tryptophan synthases (DMATs), and terpene cyclases (TCs), surrounded by genes for modifying enzymes including, but not limited to, oxidoreductases, oxygenases, dehydrogenases, reductases, and transferases. The backbone enzyme from a given cluster usually catalyzes the first step(s) of metabolite biosynthesis, forming a scaffold, which can then be modified, usually in multiple biosynthetic steps by cluster-specific modifying enzymes to produce the final compound. Altering the expression of these backbone genes affects the expression of other biosynthetic genes in the cluster, as seen by the deletion of the NRPS gene, *gliP*, leading to reduced expression of the *gli* genes (Cramer et al., 2006). The backbone gene can be artificially induced to high levels via promoter replacement at its native locus to yield high levels of the first intermediate (Seshime, Juvvadi, Kitamoto, Ebizuka, & Fujii, 2010).

## 2.2. Global and cluster-specific regulators

### 2.2.1 Global regulators

A hallmark discovery concerning fungal SM gene cluster regulation is the characterization of a nuclear methyltransferase–domain protein, LaeA, first characterized in *Aspergillus nidulans* (Bok & Keller, 2004). Loss of this protein decreased or eliminated production of many SMs. Since then, LaeA has been characterized in various *Aspergillus*, *Penicillium*, and *Fusarium* spp. and shown to both positively and negatively regulate production of natural compounds (Baba, Kinoshita, & Nihira, 2011; Bok et al., 2005; Kale et al., 2008; Oda, Kobayashi, Ohashi, & Sano, 2011; Perrin et al., 2007; Wiemann et al., 2010; Xing, Deng, & Hu, 2010). Altogether, these findings branded LaeA as a global regulator of secondary metabolism. Thus, manipulating the level of this protein will inevitably alter the SM profile. Many groups have since used LaeA regulation of SM gene clusters to link orphan clusters to their respective metabolites or used this protein to enhance heterologously expressed clusters (Sakai, Kinoshita, & Nihira, 2012). Other global regulators that have been identified to affect SM gene cluster expression include development-related transcription factors such as StuA (Twumasi-Boateng et al., 2009) and a bZIP transcription factor in *A. nidulans* (Yin et al., 2012).

### 2.2.2 Pathway-specific regulators

Many SM clusters contain a gene encoding a transcription factor specific for activating the enzymatic genes in the cluster. The hallmark transcription factor is AflR, encoded by a gene in the aflatoxin and sterigmatocystin (ST) clusters in various *Aspergillus* spp. Overexpression of *aflR* leads to enhanced expression of SM cluster genes and concomitant metabolite production (Flaherty & Payne, 1997). This method can be used to enhance production of SMs (reviewed in Yin & Keller, 2011). More recently, it has been shown that some SM cluster transcription factors can activate more than one SM cluster. In *A. nidulans*, induced expression of ScpR induced expression not only of *inp* cluster genes but also of the *afo* cluster genes (Bergmann et al., 2010), and *A. nidulans aflR* is involved in both ST and asperthecin biosynthesis (Yin et al., 2012).

## 2.3. Chromatin modifiers

LaeA impacts SM production through chromatin modification (Reyes-Dominguez et al., 2010). Complementary to this observation, several reports have indicated SM activation though modification of histone decorating enzymes, typically through deletion of heterochromatin enhancing enzymes usually involved in deacetylation and subsequent methylation of H3K9 residues. Histone modifiers shown to affect SM production include the histone deacetylase (HDAC), HdaA, heterochromatin protein 1 (HepA), and H3K9 methyltransferase (ClrD) (Lee et al., 2009; Palmer, Perrin, Dagenais, & Keller, 2008; Shwab et al., 2007). Epigenetic activating chemicals such as HDAC inhibitors also induce production of various new compounds. Trichostatin A, for example, induces production of penicillin and ST in *A. nidulans* (Shwab et al., 2007). Other HDAC inhibitors such as suberoylanilide hydroxamic acid (SAHA) and 5-azacytidine have been used to induce production of cladochromes and oxylipins in *Cladosporium cladosporioides* (Wang et al., 2011; Williams, Henrikson, Hoover, Lee, & Cichewicz, 2008). Several recent reviews address this link between epigenetics and fungal secondary metabolism (Cichewicz, 2010; Palmer & Keller, 2010; Strauss & Reyes-Dominguez, 2011).

## 3. MINING FOR NOVEL COMPOUNDS AND THEIR BIOSYNTHETIC GENES

### 3.1. *In silico* genomic mining

Identifying SM biosynthetic genes is a crucial part of unlocking cryptic gene clusters. It also aids in the sustainable production of these compounds via heterologous expression in a compatible host and pathway engineering to improve yield, bioactivity, and bioavailability. Backbone genes such as PKSs, NRPSs, and DMATs are multimodular enzymes made out of conserved functional domains characteristic of each type of enzyme. These can be used to search the sequenced genome for putative open reading frames for these functional domains. Recently, a number of SM cluster prediction algorithms have been developed to achieve this (Khaldi et al., 2010; Medema et al., 2011). Here we discuss two pipelines for identifying putative SM clusters in fungi, namely, *Secondary Metabolite Unique Region Finder* (SMURF) and antibiotics and *Secondary Metabolite Analysis SHell* (antiSMASH).

SMURF is an algorithm that predicts PKS-, NRPS-, and DMAT-based SM gene clusters in fungi. SMURF cluster predictions are based on two key parameters: $d$, the maximum intergenic distance (in base pairs) permitted between two adjacent genes in the same cluster, and $y$, the maximum number of SM domain-negative genes allowed within a cluster. SMURF can predict SM clusters from both nonannotated and annotated genomes. In addition, the SMURF database contains precomputed results of SM gene clusters from 27 sequenced fungal genomes that can be readily obtained and used for downstream validation and manipulation (Khaldi et al., 2010).

While SMURF is highly specific for predicting PKS-, NRPS-, and DMAT-based SM gene clusters in fungi, antiSMASH can be used to predict the whole repertoire of compound classes spanning the full tree of life. The antiSMASH pipeline is also designed to provide detailed analysis of PKS/NRPS functional domain architecture and predict core structures of their respective products as well as an analysis of orthologous gene clusters based on the identity of adjacent accessory genes (Medema et al., 2011). A comparison between antiSMASH and SMURF showed that both pipelines are of equal performance when tested against *A. fumigatus* AF293 and *Penicillium chrysogenum* Wisconsin 54-1255 (Medema et al., 2011).

## 3.2. Proteomic investigation of secondary metabolism: a proteomic approach

Although not designed for activating cryptic SM gene clusters, a recently developed technique that streamlined the workflow for new drug discovery is noteworthy. Proteomic Investigation of Secondary Metabolism (PriSM) allows investigators to identify both biosynthetic genes and their respective metabolites in tandem and bypassing the need for genome sequence information *a priori* (Bumpus, Evans, Thomas, Ntai, & Kelleher, 2009). Thus far, this alternative approach can only be applied to PKS- and NRPS-based SM clusters whose enzymes are both expressed and $>100$ kDa in size. PriSM operates on the premise that PKSs and NRPSs are large-molecular weight proteins (LMWPs; $>200$ kDa) that can be separated from the majority of cellular proteins by SDS-PAGE and subsequently selected for specific phosphopantetheinyl (Ppant)-containing peptides characteristic of such enzymes. The protocol is discussed in detail (Bumpus et al., 2009); briefly, LMWPs are isolated from total protein extracts and subjected to in-gel trypsin digestion and selective Ppant-containing peptide detection using a high-performance fourier transform mass spectrometer, followed by targeted PCR, cluster sequencing, and finally product discovery. Although in its infancy, PriSM sets forth an innovative and alternative approach of impressive potential, especially when used in combination with approaches that enable the expression of cryptic clusters.

## 4. GENETIC MANIPULATION OF BIOSYNTHETIC AND REGULATORY GENES

Many successful attempts to activate cryptic SM gene clusters involve altering the expression of genes encoding biosynthetic enzymes, developmental and pathway-specific transcription factors, and histone modifiers. Targeted gene deletions can be used to abolish expression and promoter replacements can be used to modify expression of a gene of interest (GOI). When an inducible promoter is chosen, the researcher can reversibly control the state of expression (on or off) and, in some cases, the level of expression itself. Here we discuss a rapid method for creating deletion/promoter replacement constructs, protoplast generation, and transformation in *Aspergillus* spp. This protocol is adapted from (Szewczyk et al., 2006; Yang et al., 2004; Yelton, Hamer, & Timberlake, 1984; Yu et al., 2004).

### 4.1. Design of transformation constructs

The 5′ upstream and 3′ downstream flanking regions serve as complementary sequences to allow homologous recombination at the locus of interest. These flanking regions range from 1.0 to 1.5 kb in size. For targeted gene disruption, you could replace the entire open reading frame (ORF) of the GOI with a selectable marker of choice (Fig. 15.1A); replace a small piece of the ORF internal to the GOI with a selectable marker of choice (Fig. 15.1B); or insert a selectable marker of choice within the GOI and disrupt it without removing any ORF (Fig. 15.1C). For promoter replacement, the promoter of choice has to be inserted immediately before the transcriptional start site of the GOI. The selectable marker in this case is usually fused upstream of the promoter to be inserted (Fig. 15.1D). The double-joint PCR protocol below can also be used for other applications such as fluorescent tagging and epitope tagging of genes for microscopy and purification, respectively. When this is the case, the inserted/fused fragment must be translated in-frame to the GOI. Design of in-frame cassettes can be found in (Szewczyk et al., 2006).

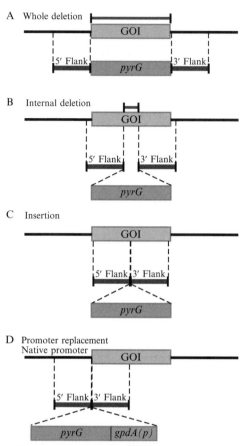

**Figure 15.1** Design of transformation cassette: (A) whole deletion of ORF from GOI, (B) internal deletion of ORF from GOI, (C) insertion of selection marker disrupting the ORF of GOI, and (D) native promoter replacement with either a constitutive or inducible promoter. (For color version of this figure, the reader is referred to the online version of this chapter.)

## 4.2. Creation of linear transformation cassettes using double-joint PCR

1. Design eight primers (Fig. 15.2):
   **1.1.** Four primers (P1–P4) for amplification of 5′ upstream and 3′ downstream flanking regions that are ∼1.0–1.5 kb in size. These will be used in PCR Reaction I.
   **1.2.** Two nested primers (P5 and P6) for selective amplification of full-length fusion product. These will be used in PCR Reaction III.
   **1.3.** Two primers (PC1 and PC2) for amplification of the selection marker. These will be used in PCR Reaction I.

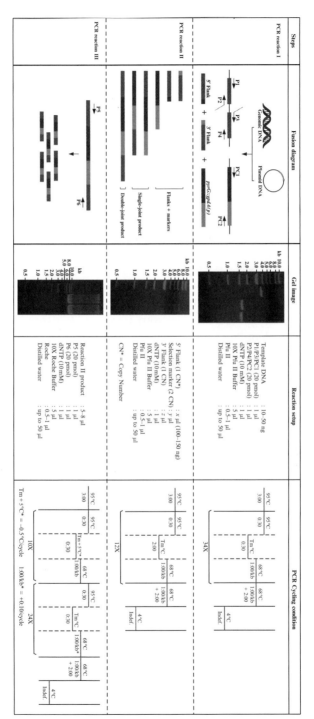

**Figure 15.2** Double-joint PCR reaction setup and cycling conditions. (See Color Insert.)

*Note*: The selection marker can either be a drug resistance gene or one that restores prototrophy to an auxotrophic mutant. In the latter case, it is important to not choose the wild type copy from the same species as it will reduce the frequency of homologous recombination at the locus of interest. For example, when transforming a *pyrG* mutant of *A. fumigatus*, it is advisable to use *Aspergillus parasiticus* (or another fungus) *pyrG*.

2. Isolate genomic DNA (gDNA) from the parental strain that will be used as template for amplifying flanking regions to the gene or locus of interest and plasmid DNA/gDNA as template to amplify the selection cassette.

3. Set up PCR Reaction I. PCR amplify both 5′ upstream and 3′ downstream flanking regions and the selectable marker using *Pfu*II Ultra (Agilent Technologies). Refer to Fig. 15.2 for reaction setup.

    *Note*: *Pfu*II Ultra is a high fidelity polymerase that minimizes PCR errors during amplification. Other high fidelity polymerases can be used instead.

4. Clean up the PCR products from Reaction I of residual template, primers, nucleotides, and in some cases, nonspecific products via gel extraction using QIAquick gel extraction kit (Qiagen).

5. Quantitate the amount of DNA for PCR Reaction II using the Nanodrop (Life Technologies) or agarose gel electrophoresis.

6. Set up PCR Reaction II. Use a fragment copy number ratio of 1:2:1 (5′ flank:marker:3′ flank). Refer to Fig. 15.2 for reaction setup.

7. Run 5–8 μl of Reaction II product on a 0.8% agarose gel. You should be able to see three to four bands corresponding to both flanking regions, the selection marker, the single-joint product, and the double-joint product (the double-joint product may or may not be visible, Fig. 15.2).

8. Set up PCR Reaction III. Selectively amplify the full-length double-joint product using nested primers P1 and P6 using Expand Long Template Polymerase (Roche). Refer to Fig. 15.2 for reaction setup.

    *Note*: This is a high fidelity polymerase suitable for amplifying large double-joint cassettes with high product yield. It has worked well in our hands, but other polymerases may be used instead. Using nested primers also decreases amplification of nonspecific products.

9. Check the quality of PCR product from Reaction III using agarose gel electrophoresis. If multiple bands are observed, the band of interest can be purified from nonspecific bands via gel extraction.

    *Note*: You can also confirm that the generated cassette is specific to the GOI by performing 1–2 restriction digests.

10. Clean up the PCR product from Reaction III by subjecting it through Sephadex$^{TM}$ G-50 DNA-grade resin (GE Healthcare).
11. Double-joint product can be stored at $-20\,^{\circ}C$ for a few months and $-80\,^{\circ}C$ indefinitely.

## 4.3. Protoplast generation

In this method, protoplasts are generated from hyphae (Szewczyk et al., 2006) rather than germlings (Yelton et al., 1984). Thus, it is crucial to allow sufficient growth of the strain. The length of incubation varies with the type of auxotrophy and strain background. For instance, tryptophan auxotrophs never recover full growth even with supplemented tryptophan, so such mutants may have to be grown longer than a pyrimidine auxotroph. Generated protoplasts are extremely sensitive to detergent and much care has to be taken to ensure that glassware, filters, and tubes are free of residual detergent.

1. Obtain spore suspension of parental strain in distilled water + 0.1% Tween-80.

   *Note*: Using parental strains deficient for NkuA or NkuB increases the frequency of transformants carrying the correct integration of the transformation cassette (da Silva Ferreira et al., 2006). In this case, you may screen a mere $10+$ transformants for gene deletion or $20+$ transformants for promoter replacement, as opposed to $60+$ transformants when using wild-type strains for NkuA and NkuB.

2. In a sterile 50-ml Erlenmeyer flask, inoculate $1 \times 10^{8}$ spores into 20 ml YG medium (5 g yeast extract, 20 g D-glucose, 400 µl trace element solution (Szewczyk et al., 2006) in 1 l distilled water).

3. Incubate for 12–14 h depending on strain background and type of auxotrophy at $30\,^{\circ}C$ and shaking at 120–150 rpm.

4. Prepare fresh $2\times$ protoplasting solution by adding 1.3 g Vinotaste Pro [Novozymes] into 10 ml for KCl, citric acid solution (8.2 g KCl, 2.1 g citric acid monohydrate, pH, to 5.8 with 1.1 $M$ KOH, in 100 ml distilled water). Filter-sterilize using 0.22-µ$M$ Millex GV filter (make sure to discard the first 0.5 ml of pass-through to remove residual detergent).

   *Note*: For *A. nidulans*, 1.3 g Vinotaste Pro is sufficient. For *A. fumigatus* and *A. flavus*, the protoplasting solution should also be supplemented with 0.1 and 0.2 g of lysing enzymes from *Trichoderma harzianum*, respectively (Sigma–Aldrich, Catalog number: L1412).

5. Harvest mycelium by gravity filtration over sterile Miracloth (Calbiochem).

6. Wash away residual detergent from mycelium with fresh YG medium.

7. With a sterile spatula, transfer mycelium from Miracloth into a fresh 125-ml Erlenmeyer flask with 16 ml protoplasting solution (1:1, v/v, YG medium: 2× protoplasting solution).

8. Incubate for 2–4 h at 30 °C and shaking at 80–100 rpm.

    *Note*: Hyphae tend to clump together in the first hour of protoplasting. Break the clump by gently pipetting up and down every now and then with a sterile serological pipette. Do not pipet after the first hour of protoplasting as now the cell walls are almost fully digested and protoplasts are sensitive to physical shearing. Monitor protoplast formation microscopically. Small or few protoplasts beyond the first hour to two of protoplasting indicate that most protoplasts are lysing.

9. In a sterile 50-ml centrifuge tube (or glass round-bottom centrifuge tube), gently overlay protoplast solution over chilled 10 ml 1.2 $M$ sucrose solution.

10. Centrifuge at $1800 \times g$ for 10 min at 4 °C in a swinging-bucket rotor. Make sure to set brakes to low or turn the brake off entirely. Hyphae will pellet and protoplasts will collect at the interphase (top of sucrose layer).

11. Collect the protoplasts with a pipette, making sure not to disturb the interphase and disperse the protoplasts. A small amount of protoplasts may still be dispersed throughout the protoplasting solution. You can collect the entire protoplasting solution if this occurs.

12. Transfer the collected protoplasts to a new sterile centrifuge tube and add at least an equal volume of 0.6 $M$ KCl. Mix thoroughly by gently inverting the centrifuge tube several times.

13. Centrifuge at $1800 \times g$ for 10 min at 4 °C to pellet the protoplasts. They will form a loose pellet, and care must be taken while decanting the supernatant post-centrifugation.

14. Resuspend protoplasts in 1 ml of 0.6 $M$ KCl and transfer to a sterile microcentrifuge tube.

15. Centrifuge at $2400 \times g$ for 2 min to pellet the protoplasts. Pipette away supernatant being careful not to dislodge the protoplast pellet.

16. Wash protoplasts twice by repeating steps 14 and 15.

17. Resuspend protoplasts in an appropriate volume of 0.6 $M$ KCl, 50 m$M$ citric acid solution (for KCl-based transformation), or STC buffer (for sorbitol-based transformation; see transformation protocol for recipe).

    *Note*: This protocol, when performed optimally, can yield up to $2 \times 10^7$ protoplasts and is sufficient for 10 transformations (Szewczyk et al., 2006). Thus, the pellet can be resuspended in up to 1 ml of the above solutions using $\sim$100 μl/transformation. Do not forget to also account for an additional 100 μl for the negative control.

## 4.4. Transformation

Both sorbitol- and KCl-based transformations have been used to generate transformants from protoplasts obtained by the above method. In our hands, the efficiency of sorbitol-based transformation is much higher than KCl-based transformation for *A. fumigatus*. Though a comparison has not been made between these two methods for *A. nidulans* and *A. flavus*, both have produced transformants from these two species. These protocols were adapted from Szewczyk et al. (2006) and Yelton et al. (1984).

### *4.4.1 Sorbitol-based transformation*

1. Dilute DNA (3–5 μg) in < 100 μl total volume of STC (1.2 *M* sorbitol, 10 m*M* CaCl$_2$, 10 m*M* Tris–HCl, pH 7.5) in a 15-ml round bottom Falcon tube.

2. Add 100 μl of protoplasts, vortex four to five times (1 s each time) at maximum speed to thoroughly mix protoplasts and DNA and incubate for 50 min on ice.

3. Add 1.25 ml of room temperature, freshly filtered 60% PEG solution (60%, w/v, PEG 3350, 50 m*M* CaCl$_2$, 50 m*M* Tris–HCl, pH 7.5). Mix gently by placing the tube on its side and rotating it multiple times. Incubate at room temperature for 20 min.

    *Note*: PEG must be stored at room temperature, as it will precipitate at 4 °C. PEG precipitate will lyse protoplasts and reduce transformation efficiency. Hence, you could filter PEG solution right before using with a 0.22-μ*M* Millex GV filter to remove precipitate.

4. Add 5 ml STC buffer and mix gently as before.

5. Plate transformation mixture on selection Sorbitol Minimal Medium (SMM) medium (10 g D-glucose, 50 ml 20× sodium nitrate salts (120 g NaNO$_3$, 10.4 g KCl, 10.4 g MgSO$_4$ 7H$_2$O, 30.4 g KH$_2$PO$_4$ in 1 l distilled water), 1 ml trace element solution (Szewczyk et al., 2006), 218.6 g sorbitol, 1.5% agar, and selection supplements) by pipetting $\sim$1 ml transformation mixture onto each plate and overlay with 5 ml of selection SMM top agar (SMM with 0.75% agar). Alternatively, you may spread the transformation mixture with a glass spreader.

    *Note*: During overlay with top agar, make sure that it is cooled to 50 °C.

6. Incubate right side up at 30 °C overnight to let dry and then transfer the plates to 37 °C to expedite growth of transformants (*A. nidulans* and *A. fumigatus*).

### 4.4.2 KCl-based transformation

1. Add DNA (10 μl of PCR Reaction III product) in < 15 μl TE buffer to 100 μl protoplasts.
2. Vortex four to five times (1 s each time) at maximum speed to thoroughly mix protoplasts and DNA.
3. Add 50 μl of room temperature, freshly filtered 25% PEG solution (25%, w/v, PEG 3350, 0.6 $M$ KCl, 50 m$M$ CaCl$_2$, 50 m$M$ Tris–HCl, pH 7.5).
4. Vortex four to five times (1 s each time) at maximum speed.
5. Incubate on ice for 25 min.
6. Add 1 ml room temperature, freshly filtered 25% PEG solution, and gently mix by turning the tube on its side and rotating it multiple times.
7. Incubate at room temperature for 25 min.
8. Plate transformation mixture on selection KCl Minimal Medium (KMM) medium (10 g D-glucose, 50 ml 20× sodium nitrate salts, 1 ml trace element solution (Szewczyk et al., 2006), 0.6 $M$ KCl, 1.5% agar, and selection supplements) by spreading transformation sample over selection plates with a glass spreader.
9. Transformants will be visible from 2 to 7 days post-transformation depending on the species and types of selection marker. Typically, transformants bearing a nutritional marker will be visible 2–3 days post-transformation while those bearing a drug resistant marker will be visible 6–7 days post-transformation.
10. The transformants can then be confirmed. Mutant screening and confirmation procedures are beyond the scope of this chapter. Briefly, we typically perform an initial PCR screen for transformants bearing the insertion at the correct locus. We then follow up with Southern analysis using two different restriction digest patterns to choose mutants with a single integration event (at the desired locus) and not ectopic integrations elsewhere in the genome.

## 5. SECONDARY METABOLITE ANALYSIS

With the ability to rapidly generate fungal mutants, a reliable protocol is needed for the purification and identification of their SMs. Here we describe methods to isolate and characterize fungal SMs using liquid chromatography/mass spectrometry (LC/MS) and NMR.

## 5.1. Small-scale cultivation for initial analysis

1. Under sterile conditions, transfer 50 ml of the liquid medium of choice into a sterile 125-ml flask. Inoculate culture to a final concentration of $5 \times 10^6$ spores/ml (or $2.5 \times 10^8$ spores in total).

2. Place the flask in a shaker at $\sim 100$ rpm. for 4 days at 37 °C (duration and temperature can vary with strain). By this time, the mycelium should have grown and appear as light-colored spheres.

    *Note*: If an inducible promoter is used, it should be added after 18 h cultivation. For *alcA* promoters, 133 μl of cyclopentanone is added to 50 ml cultivation. The volume can be scaled accordingly.

3. After cultivation, 25 ml of filtered medium is typically collected in a 50-ml Falcon tube via filtration through Miracloth (Calbiochem).

4. In the same Falcon tube, extract the medium twice with an equal amount of ethyl acetate and transfer the ethyl acetate organic layer by a glass pipette into a tared vessel.

    *Note*: When necessary, add concentrated HCl to the filtered medium until a pH of 2 is reached, typically when analyzing compounds with a carboxylic acid group. A pH paper can be used. Extract and evaporate as usual.

5. Evaporate the organic layer. We use a Turbovac LV (Caliper Lifesciences) to evaporate several vessels in parallel; alternatively, one can use a rotoevaporator to concentrate the organic layer.

6. After evaporation, the mass of the residue is determined and the residue is then dissolved in methanol at a concentration of 1 mg/ml for LC/MS analysis.

7. The solution should always be syringe-filtered to remove particulates that can clog an LC/MS column. Place $\sim 100$ μl of solution into a disposable 1-ml syringe fitted with a 13-mm filter with a PTFE membrane. With the syringe plunger, push the solvent through the filter and into an LC/MS vial. The sample is then ready to be submitted to the LC/MS. For LC/MS, we typically use two buffers, a polar "A" buffer, and a nonpolar "B" buffer. For "A" buffer, 950 ml $H_2O$, 50 ml acetonitrile, 0.5 ml formic acid, and for "B" buffer, 1 l acetonitrile, 0.5 ml formic acid.

8. A gradient with an increasing amount of the "B" buffer over time is used to separate and elute the molecules. A typical gradient is 0% B from 0 to 5 min, 0% to 100% B from 5 to 35 min, 100% B from 35 to 40 min,

100% B to 0% B from 40 to 45 min, and reequilibration with 0% B from 45 to 50 min. Set the flow rate to 125 µl/min.

## 5.2. Large-scale cultivation for isolation of metabolites

1. Once a new unknown peak of interest is identified by LC/MS, it will be necessary to grow the strain again in a larger scale to acquire enough material for NMR analysis. The amount of medium needed will depend on the titer of the metabolite. A good starting point is to use about 2 l of medium.

2. The volumes of ethyl acetate involved will mean that separatory funnels must be used in the place of Falcon tubes, and the organic material should be collected in a large round-bottom flask. Extract at least twice; four times may be optimal. Use a rotoevaporator to evaporate the organic solvent.

3. A small amount (~0.5 mg) of dried material can be dissolved in methanol and submitted to the LC/MS to determine if the scaled-up cultivation was successful. Remember to syringe filter the solution.

## 5.3. Silica gel column chromatography for semipurification

The next step is to isolate the metabolite in question. Because a crude mixture is likely to contain a number of different products, multiple purification steps may be necessary, and careful planning is needed. Typically, we use standard organic chemistry silica gel chromatography as the first round of purification in order to obtain a semipure mixture that can be further purified using other methods and to get rid of salts and lipids that could potentially complicate further purification steps.

1. Weigh out two portions of silica (particle size 0.063–0.200 mm). One portion will be two to three times the mass of the crude material. The other portion will be 30 times the crude mass.

2. Dissolve the crude material. If common organic solvents such as dichloromethane and methanol fail to dissolve the material, try adding water and then sonicate.

3. Add the smaller portion of silica into the dissolved material and evaporate the solvents completely. A substance resembling sand should be achieved. Remove any clumps with a spatula.

4. Add the larger portion of silica to a large Erlenmeyer flask. Add enough hexanes to form a slurry.

5. Pour the slurry into a glass chromatography column. Make sure to add a small amount of cotton taken from a cotton ball to the bottom of the column to prevent the silica from leaking out.

   Note: Do not use too much cotton, as it will significantly slow the flow of the solvent. The Erlenmeyer flask must be rinsed several times with hexanes to transfer all of the silica.

6. Allow solvent to drain out of the column. Using air pressure via an adapter with a regulator and release valve will greatly facilitate the running of the column, but always be careful that the valve at the bottom of the column is open before applying pressure. Remove pressure when there is no longer solvent above the silica to prevent the formation of air bubbles in the column that can jeopardize the purification process.

7. Lightly tap the edge of the column with a blunt object such as the back of a pair of scissors. This will cause the silica to settle and release air bubbles trapped in the column. Some solvent will emerge above the level of the silica. Drain until there is no solvent above the silica. Tap the edge of the column again to further pack the column. Drain.

8. Repeat tapping and draining until the silica settles no more (indicated by the lack of solvent emerging above the silica layer). The column is now packed.

9. Add more hexanes to the top of the silica so that it is ~1 cm in height. Be careful not to disrupt the silica. This can be done by slowly pipetting the solvent onto the inner side of the column in order to preserve the integrity of the packed silica.

10. Pour in the small amount of silica that was mixed with the crude fungal extract. Make sure that all of this material is off the sides of the column. If necessary, extra hexanes can be used to completely transfer the material. If so, drain the solvent until it reaches the top of the material.

11. Add standard, commercially available sand so that it is ~1 cm in height. The sand serves as a physical barrier and will help prevent disruption of the packed silica when solvent is added.

12. Elute with the solvent systems necessary to drain the metabolite(s) of interest. For every 1.5 g of crude material, use 1 l of solvent mixture. A suitable solvent system is a mixture of ethyl acetate and hexanes.

   Note: The most nonpolar metabolites will be eluted from a mixture of 10% ethyl acetate in hexanes. The more polar the metabolite, the more ethyl acetate will be needed, all the way up to 100% ethyl acetate. For especially polar metabolites, it may be necessary to use methanol in ethyl acetate. Do not exceed 10% methanol, and start gradually since each percentage of methanol added will significantly increase the polarity of the mixture. If you are unsure which solvent composition is appropriate, start with 10% ethyl acetate/hexanes, follow with a 10% increment in ethyl acetate (e.g., 20% ethyl acetate/hexanes), and so forth. Collect each fraction separately and evaporate.

13. A small amount of material can be prepared for LC/MS analysis to keep track of the status of the purification.

## 5.4. Final HPLC purification of the metabolite

Provided that the metabolite was eluted, you can proceed to HPLC purification. Dissolve the evaporated fraction(s) from silica gel chromatography completely in methanol. Try to keep solvent volumes to a minimum because it will affect the loading of the sample onto the HPLC.

1. Make sure to filter the sample prior to HPLC purification.
2. We typically use 950 ml $H_2O$, 50 ml acetonitrile, and 0.5 ml trifluoroacetic acid for buffer "A" and 1 l acetonitrile, 0.5 ml trifluoroacetic acid for buffer "B." Use a flow rate of 5.0 ml, with a UV detector at 250 nm.
3. The gradient depends on the polarity of the compound to be purified and its proximity to other metabolites. Typically, the closer the metabolites are to each other, the slower the gradient.
4. Fractions containing the product should be combined and evaporated, but first a sample should be submitted to the LC/MS to verify that it is indeed the product in question.
5. Frequently, the metabolite is now pure and, if a sufficient quantity was obtained, it is ready to be analyzed by NMR. However, sometimes the LC/MS and/or NMR reveal impurities, and further purification is necessary. Two strategies can be employed: (1) repeat the HPLC but employ a slower gradient or abolish the gradient entirely (isocratic); (2) perform a preparative TLC (See Section 5.5.).

## 5.5. Preparative TLC

Preparative TLC is appropriate for compounds that can be visualized under UV. It is also cost-effective and particularly useful as the final step of a purification scheme. However, it is not useful for purifying large amounts of compounds.

1. Dissolve the sample in a small amount of methanol. Pack a little cotton down partly through the end of a glass pipette. This "paint brush" is used

to apply the material as a line about 2 cm above the bottom of a precoated 20 × 20 cm TLC plate. No more than 10 mg of material should be applied to a single plate. Try to keep the line as thin as possible as it will affect the separation.

2. The appropriate solvent system can be determined by performing an initial analytical TLC. The right solvent composition will be similar, but not necessarily identical, to the composition used to elute the product in column chromatography.

3. Place the plate in the developing chamber. Remove the plate when the solvent is about 2 cm from the top of the plate. This will take about an hour. Let dry for another hour.

4. If the product is colorless, you will need to place the plate under a UV lamp (~250 nm) to visualize the product. With a pencil, trace the outside of the product and then remove the plate from the UV lamp.

5. Using a razor blade, scrape off the silica containing the product and collect in a flask.

6. Add 10 ml of 95:5 dichloromethane:methanol and stir for an hour. Make sure not to inhale the silica that is being scraped off.

7. Collect the product by pouring the solvent over a filter atop a funnel and into a tared flask. Evaporate the solvent.

## 5.6. NMR characterization of the metabolite

Once pure product is obtained in sufficient quantity (as much as 5 mg may be needed), you can now subject it to NMR analysis for structure elucidation. The types of NMR experiments needed depend on the nature of the product.

1. A proton NMR should be run first and preferably also a carbon NMR. Be sure that the NMR tubes are clean and dry and that deuterated solvent is used.

2. If you have an idea of the metabolite's identity you can compare the acquired spectra with published data.

3. If the metabolite's identity is still unknown, additional experiments, such as DEPT, HSQC, HMBC, COSY, and ROESY, may be needed to deduce the structure. The analysis of NMR data is beyond the scope of this chapter.

4. Often, it is not necessary to fully solve the structure of the compound of interest. Rather, you may use programs such as SciFinder Scholar to determine if there are any published compounds with similarities to the partially solved structure.

*Note*: Other software programs such as ACD/Labs can be used to either compare the acquired NMR data with those in its databases or attempt to logically solve the structure. Some computer programs also will predict the NMR spectra for a given chemical, so if you have an idea of the structure, you can see if the data match the computer prediction. A close match does not guarantee that the structure is correct, but it may help to develop an understanding of the identity of the molecule. Solving a structure through NMR can be a "back and forth" process. That is, the NMR data may provide some clues about the structure. You can then consult journal articles, use software programs, or consult with peers, which may provide inspiration for the choice of further NMR experiments to advance in solving the structure.

## ACKNOWLEDGMENTS

This research was funded by NIH 1 R01 Al065728-01 to N. P. K. and by NIH Grants PO1GM084077 from the National Institute of General Medical Sciences to N. P. K. and C. C. W.

## REFERENCES

Baba, S., Kinoshita, H., & Nihira, T. (2011). Identification and characterization of *Penicillium citrinum* VeA and LaeA as global regulators for ML-236B production. *Current Genetics*, 58, 1–11.
Ben-Ami, R., Lewis, R., Leventakos, K., & Kontoyiannis, D. (2009). *Aspergillus fumigatus* inhibits angiogenesis through the production of gliotoxin and other secondary metabolites. *Blood*, 114, 5393–5399.
Bergmann, S., Funk, A. N., Scherlach, K., Schroeckh, V., Shelest, E., Horn, U., et al. (2010). Activation of a silent fungal polyketide biosynthesis pathway through regulatory cross talk with a cryptic nonribosomal peptide synthetase gene cluster. *Applied and Environmental Microbiology*, 76, 8143–8149.
Bok, J., Balajee, S., Marr, K., Andes, D., Nielsen, K., Frisvad, J., et al. (2005). LaeA, a regulator of morphogenetic fungal virulence factors. *Eukaryotic Cell*, 4, 1574–1582.
Bok, J., Chiang, Y., Szewczyk, E., Reyes-Dominguez, Y., Davidson, A., Sanchez, J., et al. (2009). Chromatin-level regulation of biosynthetic gene clusters. *Nature Chemical Biology*, 5, 462–464.
Bok, J., Chung, D., Balajee, S., Marr, K., Andes, D., Nielsen, K., et al. (2006). GliZ, a transcriptional regulator of gliotoxin biosynthesis, contributes to *Aspergillus fumigatus* virulence. *Infection and Immunity*, 74, 6761–6768.
Bok, J., & Keller, N. (2004). LaeA, a regulator of secondary metabolism in *Aspergillus* spp. *Eukaryotic Cell*, 3, 527–535.
Brakhage, A., Schuemann, J., Bergmann, S., Scherlach, K., Schroeckh, V., & Hertweck, C. (2008). Activation of fungal silent gene clusters: A new avenue to drug discovery. *Progress in Drug Research*, 66(1), 3–12.
Brakhage, A. A., & Schroeckh, V. (2011). Fungal secondary metabolites. Strategies to activate silent gene clusters. *Fungal Genetics and Biology*, 48, 15–22.
Bumpus, S., Evans, B., Thomas, P., Ntai, I., & Kelleher, N. (2009). A proteomics approach to discovering natural products and their biosynthetic pathways. *Nature Biotechnology*, 27, 951–956.

Chiang, Y.-M., Chang, S.-L., Oakley, B. R., & Wang, C. C. C. (2011). Recent advances in awakening silent biosynthetic gene clusters and linking orphan clusters to natural products in microorganisms. *Current Opinion in Chemical Biology, 15*, 137–143.

Cichewicz, R. (2010). Epigenome manipulation as a pathway to new natural product scaffolds and their congeners. *Natural Product Reports, 27*, 11–22.

Cramer, R. A., Jr., Gamcsik, M. P., Brooking, R. M., Najvar, L. K., Kirkpatrick, W. R., Patterson, T. F., et al. (2006). Disruption of a nonribosomal peptide synthetase in *Aspergillus fumigatus* eliminates gliotoxin production. *Eukaryotic Cell, 5*, 972–980.

da Silva Ferreira, M., Kress, M., Savoldi, M., Goldman, M., Hartl, A., Heinekamp, T., et al. (2006). The *akuB*(KU80) mutant deficient for nonhomologous end joining is a powerful tool for analyzing pathogenicity in *Aspergillus fumigatus*. *Eukaryotic Cell, 5*, 207–211.

Flaherty, J. E., & Payne, G. A. (1997). Overexpression of *aflR* leads to upregulation of pathway gene transcription and increased aflatoxin production in *Aspergillus flavus*. *Applied and Environmental Microbiology, 63*, 3995–4000.

Hertweck, C. (2009). Hidden biosynthetic treasures brought to light. *Nature Chemical Biology, 5*, 450–452.

Kale, S., Milde, L., Trapp, M., Frisvad, J., Keller, N., & Bok, J. (2008). Requirement of LaeA for secondary metabolism and sclerotial production in *Aspergillus flavus*. *Fungal Genetics and Biology, 45*, 1422–1429.

Keller, N., & Hohn, T. (1997). Metabolic pathway gene clusters in filamentous fungi. *Fungal Genetics and Biology, 21*, 17–29.

Khaldi, N., Seifuddin, F. T., Turner, G., Haft, D., Nierman, W. C., Wolfe, K. H., et al. (2010). SMURF: Genomic mapping of fungal secondary metabolite clusters. *Fungal Genetics and Biology, 47*, 736–741.

Lee, I., Oh, J., Shwab, E., Dagenais, T., Andes, D., & Keller, N. (2009). HdaA, a class 2 histone deacetylase of *Aspergillus fumigatus*, affects germination and secondary metabolite production. *Fungal Genetics and Biology, 46*, 782–790.

Losada, L., Ajayi, O., Frisvad, J. C., Yu, J. J., & Nierman, W. C. (2009). Effect of competition on the production and activity of secondary metabolites in *Aspergillus* species. *Medical Mycology, 47*, S88–S96.

Medema, M. H., Blin, K., Cimermancic, P., de Jager, V., Zakrzewski, P., Fischbach, M. A., et al. (2011). antiSMASH: Rapid identification, annotation and analysis of secondary metabolite biosynthesis gene clusters in bacterial and fungal genome sequences. *Nucleic Acids Research, 39*, W339–W346.

Oda, K., Kobayashi, A., Ohashi, S., & Sano, M. (2011). *Aspergillus oryzae* LaeA regulates kojic acid synthesis genes. *Bioscience, Biotechnology, and Biochemistry, 75*, 1832–1834.

Palmer, J., Perrin, R., Dagenais, T., & Keller, N. (2008). H3K9 methylation regulates growth and development in *Aspergillus fumigatus*. *Eukaryotic Cell, 7*, 2052–2060.

Palmer, J. M., & Keller, N. P. (2010). Secondary metabolism in fungi, does location matter? *Current Opinion in Microbiology, 13*, 431–436.

Perrin, R., Fedorova, N., Bok, J., Cramer, R., Wortman, J., Kim, H., et al. (2007). Transcriptional regulation of chemical diversity in *Aspergillus fumigatus* by LaeA. *PLoS Pathogens, 3*, 50.

Reyes-Dominguez, Y., Bok, J., Berger, H., Shwab, E., Basheer, A., Gallmetzer, A., et al. (2010). Heterochromatic marks are associated with the repression of secondary metabolism clusters in *Aspergillus nidulans*. *Molecular Microbiology, 76*, 1376–1786.

Rohlfs, M., Albert, M., Keller, N., & Kempken, F. (2007). Secondary chemicals protect mould from fungivory. *Biology Letters, 3*, 523–525.

Sakai, K., Kinoshita, H., & Nihira, T. (2012). Heterologous expression system in Aspergillus oryzae for fungal biosynthetic gene clusters of secondary metabolites. *Applied Microbiology and Biotechnology, 93*, 2011–2022.

Seshime, Y., Juvvadi, P. R., Kitamoto, K., Ebizuka, Y., & Fujii, I. (2010). Identification of cypyrone B1 as the novel product of *Aspergillus oryzae* type III polyketide synthase CsyB. *Bioorganic & Medicinal Chemistry*, *18*, 4542–4546.

Shwab, E., Bok, J., Tribus, M., Galehr, J., Graessle, S., & Keller, N. (2007). Histone deacetylase activity regulates chemical diversity in *Aspergillus*. *Eukaryotic Cell*, *6*, 1656–1664.

Stanzani, M., Orciuolo, E., Lewis, R., Kontoyiannis, D., Martins, S., St John, L., et al. (2005). *Aspergillus fumigatus* suppresses the human cellular immune response via gliotoxin-mediated apoptosis of monocytes. *Blood*, *105*, 2258–2265.

Strauss, J., & Reyes-Dominguez, Y. (2011). Regulation of secondary metabolism by chromatin structure and epigenetic codes. *Fungal Genetics and Biology*, *48*, 62–69.

Szewczyk, E., Nayak, T., Oakley, C., Edgerton, H., Xiong, Y., Taheri-Talesh, N., et al. (2006). Fusion PCR and gene targeting in *Aspergillus nidulans*. *Nature Protocols*, *1*, 3111–3120.

Twumasi-Boateng, K., Yu, Y., Chen, D., Gravelat, F. N., Nierman, W. C., & Sheppard, D. C. (2009). Transcriptional profiling identifies a role for BrlA in the response to nitrogen depletion and for StuA in the regulation of secondary metabolite clusters in *Aspergillus fumigatus*. *Eukaryotic Cell*, *8*, 104–115.

Walton, J. D. (2000). Horizontal gene transfer and the evolution of secondary metabolite gene clusters in fungi: An hypothesis. *Fungal Genetics and Biology*, *30*, 167–171.

Wang, X., Sena Filho, J. G., Hoover, A. R., King, J. B., Ellis, T. K., Powell, D. R., et al. (2011). Chemical epigenetics alters the secondary metabolite composition of guttate excreted by an atlantic-forest-soil-derived *Penicillium citreonigrum*. *Journal of Natural Products*, *73*, 942–948.

Wiemann, P., Brown, D. W., Kleigrewe, K., Bok, J. W., Keller, N. P., Humpf, H.-U., et al. (2010). FfVel1 and FfLae1, components of a velvet-like complex in *Fusarium fujikuroi*, affect differentiation, secondary metabolism and virulence. *Molecular Microbiology*, *77*, 972–994.

Williams, R., Henrikson, J., Hoover, A., Lee, A., & Cichewicz, R. (2008). Epigenetic remodeling of the fungal secondary metabolome. *Organic & Biomolecular Chemistry*, *6*, 1895–1897.

Xing, W., Deng, C., & Hu, C.-H. (2010). Molecular cloning and characterization of the global regulator LaeA in *Penicillium citrinum*. *Biotechnology letters*, *32*, 1733–1737.

Yang, L., Ukil, L., Osmani, A., Nahm, F., Davies, J., De Souza, C. P., et al. (2004). Rapid production of gene replacement constructs and generation of a green fluorescent protein-tagged centromeric marker in *Aspergillus nidulans*. *Eukaryotic Cell*, *3*, 1359–1362.

Yelton, M. M., Hamer, J. E., & Timberlake, W. E. (1984). Transformation of *Aspergillus nidulans* by using a *trpC* plasmid. *Proceedings of the National Academy of Sciences of the United States of America*, *81*, 1470–1474.

Yin, W., Amaike, S., Wohlbach, D. J., Gasch, A. P., Chiang, Y.-M., Wang, C. C., et al. (2012). An *Aspergillus nidulans* bZIP response pathway hardwired for defensive secondary metabolism operates through *aflR*. *Molecular Microbiology*, *83*, 1024–1034.

Yin, W., & Keller, N. P. (2011). Transcriptional regulatory elements in fungal secondary metabolism. *Journal of Microbiology*, *49*, 329–339.

Yu, J. H., Hamari, Z., Han, K. H., Seo, J. A., Reyes-Dominguez, Y., & Scazzocchio, C. (2004). Double-joint PCR: A PCR-based molecular tool for gene manipulations in filamentous fungi. *Fungal Genetics and Biology*, *41*, 973–981.

# CHAPTER SIXTEEN

# Regulatory Cross Talk and Microbial Induction of Fungal Secondary Metabolite Gene Clusters

## Hans-Wilhelm Nützmann*,†, Volker Schroeckh*, Axel A. Brakhage*,†,1

*Department of Molecular and Applied Microbiology, Leibniz Institute for Natural Product Research and Infection Biology – Hans Knöll Institute, Jena, Germany
†Institute of Microbiology, Friedrich Schiller University, Jena, Germany
1Corresponding author: e-mail address: axel.brakhage@hki-jena.de

## Contents

## Abstract

Filamentous fungi are well-known producers of a wealth of secondary metabolites with various biological activities. Many of these compounds such as penicillin, cyclosporine, or lovastatin are of great importance for human health. Genome sequences of filamentous fungi revealed that the encoded potential to produce secondary metabolites is much higher than the actual number of compounds produced during cultivation in the laboratory. This finding encouraged research groups to develop new methods to exploit the silent reservoir of secondary metabolites. In this chapter, we present three successful strategies to induce the expression of secondary metabolite gene clusters. They are based on the manipulation of the molecular processes controlling the biosynthesis of secondary

*Methods in Enzymology*, Volume 517
ISSN 0076-6879
http://dx.doi.org/10.1016/B978-0-12-404634-4.00016-4

metabolites and the simulation of stimulating environmental conditions leading to altered metabolic profiles. The presented methods were successfully applied to identify novel metabolites. They can be also used to significantly increase product yields.

## 1. INTRODUCTION

Filamentous fungi synthesize a multitude of secondary metabolites. These molecules are thought to increase the fitness of the producing organism in its ecological niche. Their impact on humans ranges from severe pathogenic features, exemplified by the highly carcinogenic aflatoxins, to life-rescuing properties of antibiotics, exemplified by penicillins and cephalosporins. A characteristic of fungal secondary metabolism, as in bacteria, is the arrangement of the biosynthetic genes in clusters in the genome. Mining for such clusters in the rapidly increasing number of fungal genomes led to the identification of an unexpectedly high number of putative biosynthesis gene clusters. However, most of them are silent under standard laboratory conditions. Therefore, it is essential to develop methods, to induce the expression of secondary metabolite gene clusters. In previous years, cultivation-based techniques and random mutagenesis of producing strains were used to trigger the production of secondary metabolites. Fungal strains were screened under various conditions and researchers aimed at an overexpression of biosynthetic genes in native or heterologous hosts (Bode, Bethe, Hofs, & Zeeck, 2002). By contrast, recent attempts to uncover the secondary metabolite reservoir of filamentous fungi followed an indirect approach targeting whole gene clusters. The induction of gene clusters was achieved by genetic manipulation of the regulatory circuits of secondary metabolism and by simulation of environmental microbial conditions. A number of complex signaling cascades involved in the regulation of fungal secondary metabolism were identified. Their activation only occurs at certain time points of growth and in response to specific intra- and extracellular stimuli. To accomplish a tight regulation, the gene expression of secondary metabolism biosynthetic genes is controlled at different levels. The basal stage is mediated by transcription factors (Ahn & Walton, 1998; Bergmann et al., 2007; Brakhage et al., 2009; Hohn, Krishna, & Proctor, 1999; Woloshuk et al., 1994). The activity is regulated by complex signal transduction pathways with close connection to other signaling circuits, for example, development or primary metabolism (Bayram & Braus, 2012; Nahlik et al., 2010; Roze, Chanda, Wee, Awad, & Linz,

2011). Another regulation level, only recently discovered, involves chromatin-modifying enzymes, which regulate the accessibility of certain genomic regions for transcription factors (Bok et al., 2009; Nützmann et al., 2011; Reyes-Dominguez et al., 2010). Genetic engineering of fungal strains allows interference with all levels of regulation. This chapter describes three methods to activate secondary metabolite gene clusters: (i) overexpression of a pathway-specific regulatory gene, which allows the targeted activation of distinct gene clusters; (ii) methods to generate a range of genetically engineered strains with altered regulatory features of secondary metabolism; and (iii) procedures and protocols applicable for the systematic screening of fungal–bacterial interaction partners. The latter provides an efficient way to identify microbial interactions, which trigger the activation of fungal silent biosynthetic gene clusters.

## 2. PRODUCTION OF SECONDARY METABOLITES BY OVEREXPRESSION OF PATHWAY-SPECIFIC REGULATORY GENES

Pathway-specific regulatory proteins of fungal secondary metabolite gene clusters are often encoded within or directly adjacent to the target gene cluster. Typically, these proteins control the expression of the entire gene cluster and they are not expressed under cluster noninducing conditions. Thus, overexpression of their coding genes is an easy way to activate the transcription of all pathway-specific genes (Fig. 16.1). The advantage of this approach is the handling of only a single, relatively small gene and the possibility for both targeted and ectopic integration of the overexpression construct. Interestingly, recent data revealed that overexpression of a single pathway-specific regulator can also lead to the activation of additional, distantly located, secondary metabolite gene clusters (Bergmann et al., 2010). The fungal *alcA*p (promoter of alcohol dehydrogenase) and *gpdA*p (promoter of glycerinaldehyde-3-phosphate dehydrogenase) promoters were successfully applied to overexpress pathway-specific regulatory genes. As a result, novel compounds were isolated and product yields of known compounds increased (Bergmann et al., 2007; Chen et al., 2010; Chiang et al., 2009; Flaherty & Payne, 1997). A number of promoters are available for the overexpression of proteins in filamentous fungi (Table 16.1).

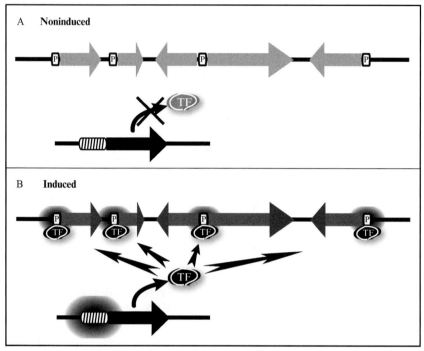

**Figure 16.1** Overexpression of a pathway-specific regulatory gene. (A) Noninducing conditions. The inducible promoter (striped) is inactive and the pathway-specific regulatory gene (black arrow) is not transcribed. The regulatory protein (TF) is not formed and the secondary metabolite gene cluster is silent. (B) Inducing conditions. The selected promoter (striped) is activated and induces the expression of the pathway-specific regulatory gene (black arrow). The regulatory protein (TF) targets specific promoter regions (P) upstream of the clustered genes and activates their transcription.

## 2.1. Overexpression of a pathway-specific regulator in *Aspergillus nidulans* (Bergmann et al., 2007)

1. Generate a gene overexpression cassette (Fig. 16.2, fusion PCR protocol 3.2).
2. Introduce the overexpression cassette by transformation into *A. nidulans* (according to Ballance & Turner, 1985).
3. Identify transformant strains containing the overexpression cassette (control PCR protocol 3.3. Southern blot).
4. Prepare stock solution of transformant strains in 0.9% (w/v) NaCl.
5. Inoculate $5 \times 10^6$ freshly harvested conidia into a 250-ml Erlenmeyer flask containing 50 ml AMM (Box 16.1) and incubate overnight (16 h) at 37 °C or at the temperature required for the appropriate strain with shaking (200 rpm).

**Table 16.1** Representative examples of promoter systems to overexpress genes in filamentous fungi

| Promoter | References[a] | Type |
|---|---|---|
| *crp*p (cyparin) | Kwon et al. (2009) | Constitutive promoter |
| *gpdA*p (glycerinaldehyde-3-phosphate dehydrogenase) | Flaherty and Payne (1997)[b] | |
| *alcA*p (alcohol dehydrogenase) | Bergmann et al. (2007)[b] | Inducible promoter |
| *acuD*p (isocitrate lyase) | Behnsen et al. (2007) | |
| *glaA*p (glucoamylase) | Santerre Henriksen et al. (1999) | |
| *thiA*p (thiamin biosynthesis) | Shoji, Maruyama, Arioka, and Kitamoto (2005) | Inducible and independent of fungal metabolism |
| hERα-ERE system | Pachlinger, Mitterbauer, Adam, and Strauss (2005) | |
| Tet-on system | Meyer et al. (2011) | |

[a]One exemplified reference was chosen.
[b]Promoters were used for gene cluster activation.

6. Prepare fresh AMM (Box 16.1) with either 2% (w/v) lactose and 10 m$M$ cyclopentanone (inducing conditions for *alcA*p) or 2% (w/v) glucose (repressing conditions).
7. Harvest fungal cell pellets using Miracloth (Calbiochem) and transfer the pellets into a new 250-ml Erlenmeyer flask containing fresh 100 ml AMM.

---

**BOX 16.1 Media**

*AMM (modified according to Pontecorvo, Roper, Hemmons, Macdonald, & Bufton, 1953)*

$NaNO_3$, 6.0 g l$^{-1}$; KCl, 0.52 g l$^{-1}$; $KH_2PO_4$, 1.52 g l$^{-1}$; $MgSO_4 \cdot 7H_2O$, 0.52 g l$^{-1}$; and 1 ml of trace element solution [$FeSO_4 \cdot 7H_2O$, 1 g l$^{-1}$; $ZnSO_4 \cdot 7H_2O$, 8.8 g l$^{-1}$; $CuSO_4 \cdot 5H_2O$, 0.4 g l$^{-1}$; $MnSO_4 \cdot 4H_2O$, 0.15 g l$^{-1}$; $Na_2B_4O_7 \cdot 10H_2O$, 0.1 g l$^{-1}$; $(NH_4)_6Mo_7O_{24} \cdot 4H_2O$, 0.05 g l$^{-1}$] per liter. The pH is adjusted to 6.5 with 10 $M$ KOH. As the carbon source, 5% (w/v) glucose is used. If required, arginine (50 μ$M$ final conc.), uracil or uridine (2.2 mg ml$^{-1}$ and 1 mg ml$^{-1}$, respectively), and p-aminobenzoic acid (3 μg ml$^{-1}$) are added.

*M79 (Prauser & Falta, 1968)*

Dextrose, 10 g l$^{-1}$; Bacto peptone, 10 g l$^{-1}$; Casamino acids, 1 g l$^{-1}$; Yeast extract, 2 g l$^{-1}$; and NaCl, 6 g l$^{-1}$. The pH was adjusted to 7.2 with 1 $M$ NaOH (resulting in pH 7.0 after autoclaving).

**8.** Incubate at 37 °C with shaking (200 rpm).

**9.** Take samples for RNA isolation after 24 h and check expression of your gene cluster of interest (see Section 4.2).

**10.** Take samples for HPLC analysis after 24–48 h (see Section 4.1).

## 3. GENETIC MANIPULATION OF SIGNAL TRANSDUCTION CASCADES

Besides regulatory genes located within gene clusters, research on fungal secondary metabolism has identified a number of additional regulatory cascades decisive for the expression of secondary metabolite gene clusters. Interfering with these cascades allows the manipulation of gene cluster expression and, thus, the secondary metabolite production level. Deletion and overexpression of genes encoding regulatory proteins and integration of genes encoding constitutively active proteins are feasible methods to alter the metabolic profile. Several groups reported the successful application of these methods in filamentous fungi leading to increased product yields as well as to the identification of novel compounds (Bok & Keller, 2004; Nahlik et al., 2010; Sakai, Kinoshita, & Nihira, 2012; Scherlach et al., 2011; Szewczyk et al., 2008). It should be noted that intervention into global signaling cascades can induce directly but also indirectly changes in the metabolic profile. Target genes for genetic manipulation can be identified either by genome mining for known regulators of secondary metabolism in the investigated or related organisms or, by a broader search for genes putatively involved in signaling processes, like protein kinases, acetyltransferases, methyltransferases, etc. When the latter method is applied, it is advantageous to create a large set of genetically engineered strains because of the risk that some manipulated genes will not affect secondary metabolism. Three genetic methods allow the fast and convenient generation of targeted gene knockout strains in the genetic model organism for filamentous fungi, *A. nidulans*.

### 3.1. Usage of Δ*ku* strains

Targeted gene deletion in filamentous fungi is carried out by homologous recombination. Thereby, a transformation cassette is integrated into the genome. The transformation construct consists of a selectable marker and regions homologous to the flanking sequence of the gene of interest. The nonhomologous end-joining recombination pathway mediates ligation of DNA constructs independent of DNA homology. Thus, it significantly decreases the efficiency of a locus-specific integration of the transformation cassette. Deletion or disruption of the genes *ku70* or *ku80*, both coding for essential proteins in nonhomologous end-joining repair, was shown to allow for more efficient integration of DNA into the genome by homologous recombination in various fungi (Hoff, Kamerewerd, Sigl, Zadra, & Kück, 2010; Nayak et al., 2006; Takahashi, Masuda, & Koyama, 2006). Therefore, *ku* mutations provide an excellent genetic background for high-throughput genetic engineering programs.

## 3.2. Generation of gene knockout cassettes by fusion PCR

Gene knockout cassettes typically consist of three DNA fragments as described above. Fusion of these fragments can be achieved by classical cloning, the GATEWAY system or fusion PCR. In contrast to the former methods, fusion PCR does not require time-consuming restriction digestion, ligation, or transformation of intermediate hosts. The protocol has been adapted to filamentous fungi (Szewczyk et al., 2006). A schematic overview of the fusion PCR protocol is given in Fig. 16.2.

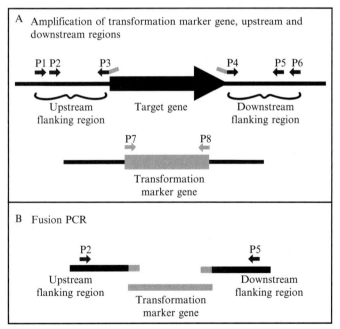

**Figure 16.2** Overview of the fusion PCR protocol. (A) The fusion PCR starts with the individual amplification of the transformation marker gene and the upstream and downstream flanking regions. The flanking regions (500–2000 bp) are amplified with primers P1 and P3, as well as P4 and P6. Primers P3 and P4 have 5′ extensions identical to the nucleotide sequence of the transformation marker gene (shaded in gray). (B) The DNA fragments amplified in A are utilized as templates for the fusion PCR. The primers P2 and P5 are used for amplification.

## 3.3. Control PCR for verification of gene deletion

After transformation of the final fusion PCR construct, it is required to verify its correct integration into the genome and the absence of wild-type contamination in the mutated strain. Two PCR reactions allow a fast and convenient verification of high numbers of transformants (Fig. 16.3). Ectopic integrations can be detected by Southern blotting.

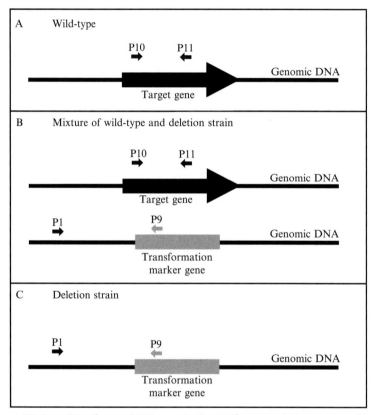

**Figure 16.3** Overview of control PCRs required to verify successful integration of a transformation construct. Two pairs of primers are necessary for the verification process. Primer P1 binds to the genome of the wild-type directly adjacent to the transformation construct. Primer P9 binds within the transformation marker gene. Primers P10 and P11 bind within the putatively deleted gene. Three scenarios are conceivable after transformation. (A) The transformation cassette did not integrate into the locus of interest. Only primers P10 and P11 will amplify a DNA product. (B) The transformation cassette was integrated into the locus of interest but the selected transformant is not free of wild-type nuclei. Both primer pairs, P10 + P11 and P1 + P9, amplify a DNA product. (C) The transformation cassette was integrated into the locus of interest and the selected transformant is free of any contamination by wild-type DNA. The primers P1 and P9 will yield a single DNA product.

## 4. MICROBIAL INDUCTION OF FUNGAL SECONDARY METABOLITE GENE CLUSTERS

Bacteria and fungi coinhabit a wide variety of environments, including soil, rhizosphere, mycorrhizae, lichens, or biofilms where interspecies interactions are most likely. Phelan, Liu, Pogliano, and Dorrestein (2012) reported that 17–42% of all predicted microbial ORFs are dedicated to microbial interactions. Such interactions involve communication via small compounds, metabolic cooperation, or competition and can be antagonistic or synergistic. "Environmental" metagenomes have been found to encode information for novel natural products (Lewis, Epstein, D'Onofrio, & Ling, 2010). The genomes of members of various genera, among them *Streptomyces* and *Aspergillus*, carry an exciting number of genes coding for the formation of secondary metabolites (Burmester et al., 2011; Sanchez, Somoza, Keller, & Wang, 2012; van Wezel & McDowall, 2011), making *Aspergillus*–streptomycete interactions especially interesting. This is further supported by the facts that both genera are represented by a high number of species and that several examples of a direct fungal–streptomycete interaction have already been published (Oh, Kauffman, Jensen, & Fenical, 2007; Pettit, 2009; Schroeckh et al., 2009; Siemieniewicz & Schrempf, 2007; Tarkka, Sarniguet, & Frey-Klett, 2009; Zuck, Shipley, & Newman, 2011). To elucidate conditions under which selected secondary metabolite gene clusters are activated, we found that transcriptome analysis is a convenient technology (Schroeckh et al., 2009). Consequently, the presented protocols describe the cocultivation of filamentous fungi and streptomycetes and the preparation of samples for transcriptome analysis using next-generation sequencing (NGS) techniques.

## 4.1. Cocultivation of streptomycetes and filamentous fungi

The described method allows identification of fungal secondary metabolites whose production is induced by streptomycetes. A main aspect which needs consideration is to guarantee that changes in the expression level of fungal genes are only due to the interaction with the streptomycete. Therefore, both organisms need a separate preculture to yield the necessary amounts of biomass followed by the transfer of these biomasses into fresh medium to start cocultivation.

1. *Fungal preculture*: Inoculate $1 \times 10^8$ spores from a 4 °C stored stock solution (in 0.9% (w/v) NaCl, 2.5% (v/v) Tween 80) into a 250-ml

Erlenmeyer flask containing 100 ml AMM (Box 16.1) and incubate overnight (16 h) at 37 °C or at the temperature required for the appropriate strain, with shaking (200 rpm).

2. *Streptomycete preculture*: Inoculate $1 \times 10^9$ spores from a liquid nitrogen-stored stock solution (in cryo-medium: PBS 1:1; cryo-medium: $K_2HPO_4$, 12.6 g l$^{-1}$; sodium citrate, 0.9 g l$^{-1}$; $MgSO_4 \cdot 7H_2O$, 0.18 g l$^{-1}$; $(NH_4)_2$ $SO_4$, 1.8 g l$^{-1}$; $KH_2PO_4$, 3.6 g l$^{-1}$; glycerol, 88 g l$^{-1}$; PBS: NaCl, 8 g l$^{-1}$; KCl, 0.2 g l$^{-1}$; $Na_2HPO_4$, 1.44 g l$^{-1}$; $KH_2PO_4$, 0.24 g l$^{-1}$, pH 7.4) into a 500-ml Erlenmeyer flask containing 100 ml M79 (Box 16.1) or other suitable medium and incubate at 28 °C for 3 days, with shaking (200 rpm).

3. Harvest fungal cell pellets using Miracloth (Calbiochem) and transfer the pellets into a new 250-ml Erlenmeyer flask containing 100 ml AMM. Add the appropriate amount of streptomycete preculture (5–20 ml) and incubate for 3–96 h at 37 °C or at the temperature required for the appropriate fungal strain with shaking (200 rpm). Take samples for transcriptome (10–20 ml, ca. 0.5 g biomass) and HPLC analysis (about 20 ml) at several time points. Remove all liquid from transcriptome analysis samples by pressing through Miracloth gauze and freeze the semi-dried biomass immediately in liquid nitrogen until isolation of RNA. Store samples for HPLC analysis (cells + medium) at 4 °C until ethyl acetate or comparable extraction.

## 4.2. Isolation of mRNA samples from mixed cultures of pro- and eukaryotic RNAs

Transcriptome analysis by RNA seq. becomes an increasingly important alternative to microarray analysis because NGS techniques not only allow high-throughput sequencing of several samples in parallel, but also circumvent hybridization primer design and the necessity to know the sequence of the given genome. When NGS samples have to be prepared from *mixed* cultures of pro- and eukaryotes, there are two specific problems: (i) how to separate the mRNA from ribosomal RNA (rRNA), and (ii) how to purify high-quality prokaryotic mRNA from contaminating eukaryotic RNA. Eukaryotic mRNA can be isolated relatively easily owing to its poly(A) tail. However, a nearly complete removal of rRNAs is required for NGS to avoid an overwhelming majority of the sequencing reads corresponding to rRNA. The protocols below describe the removal of pro- and eukaryotic rRNA by combining the RiboZero$^{TM}$ rRNA Removal Kits (Human/ Mouse/Rat) and (Gram-positive Bacteria, Epicentre Biotechnologies),

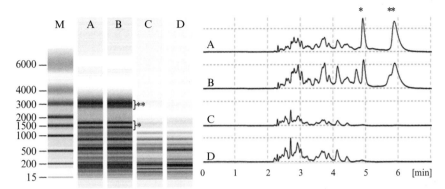

**Figure 16.4** Removal of ribosomal RNA from total RNA samples. Successful rRNA removal analyzed with the QIAxcel™ multicapillary electrophoresis system. Gel image (left) and corresponding electropherograms (right) are shown. A and B: mixed pro- and eukaryotic total RNA samples (**: 23S and 28S rRNA, respectively, *: 16S and 18S rRNA, respectively). C and D: samples A and B after rRNA removal. The electropherograms confirm the rRNA elimination.

and the preparation of prokaryotic mRNA in NGS quality from mixed pro- and eukaryotic biomasses with the MICROB*Enrich*™ and MICRO-B*Express*™ Kits (Life Technologies). Figure 16.4 shows an example of the efficiency of the used strategy. The preparation of sequencing libraries is not part of the protocols.

Harvest cocultures as described above (Section 4.1, step 3). When pro-karyotic mRNA is to be analyzed, centrifuge the Miracloth-filtered culture medium (10 min $8000 \times g$, 4 °C) to collect all prokaryotic cells. Combine the pellets. Isolate total RNA of superior quality ($OD_{260/280} = 2.10$–2.15). Perform an additional treatment of the purified RNA with a high-quality DNase. RNA can be stored at −20 °C for up to 2 months or at −80 °C for longer time. Continue with protocol 4.2.1 or 4.2.2.

### 4.2.1 Removal of pro- and eukaryotic rRNA from a total RNA mix

1. To prepare the Ribo-Zero microspheres, allow the Ribo-Zero core kit to warm to room temperature. Thaw the Ribo-Zero rRNA Removal Kit stored at −80 °C components and place them on ice.
2. Mix the room-temperature-equilibrated microspheres for 20 s to a ho-mogeneous suspension. Transfer 65 µl of microspheres into a separate 2-ml tube. Aspirate slowly the microsphere suspension to avoid air bub-bles and to ensure pipetting the full required volume. Return the unused microspheres to 4 °C.

3. Centrifuge the microspheres at $12,000 \times g$ for 3 min. Carefully discard the supernatant without disturbing the microsphere pellet. Wash the microspheres with 130 µl microsphere wash solution, vortex to resuspend the microspheres and centrifuge at $12,000 \times g$ for 3 min. Again, discard the supernatant without disturbing the microsphere pellet.

4. Add 65 µl of microsphere resuspension solution and resuspend the microspheres until a homogeneous suspension is produced. Add 1 µl of RiboGuard RNase inhibitor. Vortex briefly and store at room temperature for use in step 6.

5. Treat the total RNA with Ribo-Zero rRNA removal solution. Mix equal volumes of the rRNA removal solutions from the Human/Mouse/Rat Kit and the Gram-positive Bacteria Kit (a 50:50 mix). In a 0.2-ml RNase-free microcentrifuge tube, combine in the order given: $\times$ µl RNase-free water to obtain 40 µl final volume, 4 µl Ribo-Zero reaction buffer, 1–2.5 µg total RNA (in a maximum of 26 µl of the 50:50 rRNA removal solution). Gently mix and incubate at 68 °C for 10 min. Incubate at room temperature for a further 15 min. Return the remaining Ribo-Zero rRNA removal solution and Ribo-Zero reaction buffer to −80 °C.

6. Microsphere reaction and rRNA removal. Briefly mix the prepared microspheres (step 4) and pulse-centrifuge to collect the microsphere suspension in the bottom of the tube. Resuspend by pipetting until homogeneous. Add the (first) hybridized RNA sample from step 3, without changing the pipette tip. Immediately mix the contents of the tube by rapidly pipetting 10–15 times followed by vortexing (at medium speed) for 5 s. Place at room temperature and proceed to the next sample.

7. Incubate at room temperature for 10 min with vortex mixing (at medium speed) for 5 s every 3–4 min. Do not use a shaker platform; this does not provide sufficient mixing. At the end of the incubation, mix again and place in a water bath or heating block at 50 °C for 10 min.

8. Immediately transfer the RNA/microsphere suspension to a microsphere removal unit and centrifuge at $12,000 \times g$ for 1 min at room temperature. Save the eluate and discard the filter unit with the microspheres. The eluate contains the rRNA-depleted sample.

9. The rRNA-depleted sample should be further purified by a standard ethanol precipitation or by a column method (e.g., RNA Clean & Concentrator-5 Column, Zymo Research). The RNA can be used immediately or stored at −80 °C. Quantify the rRNA-depleted sample using RiboGreen™ RNA Reagent (Life Technologies) or a

NanoDrop$^{TM}$ spectrophotometer (Thermo Scientific). Assess quality using an Agilent 2100 Bioanalyzer (RNA6000 Pico Chip) or the QIAxcel$^{TM}$ multicapillary electrophoresis system (QIAGEN, Fig. 16.4).

### 4.2.2 Preparation of NGS-quality prokaryotic mRNA from mixed pro- and eukaryotic total RNA

1. MICROB*Enrich*$^{TM}$ removes 18S rRNA, 28S rRNA, and polyadenylated mRNAs from up to 100 μg of a purified RNA mixture (in a maximum of 30 μl RNase-free water) using magnetic beads. If the available total RNA does not fulfill the concentration or buffer requirements, perform a standard ethanol precipitation.

2. To anneal RNA and the capture oligonucleotides, add 5–100 μg RNA to 300 μl binding buffer, add 2 μl capture oligos for every 5 μg of RNA, and incubate at 70 °C for 10 min. Then, allow nucleic acids to anneal at 37 °C for 1 h. Removal of fungal rRNAs can be enhanced by adding self-designed capture oligos which contain a 3′-(A)$_{17}$-stretch and a short-linker upstream of the fungus-specific anti-18S/anti-28S sequence.

3. Prepare the "Oligo MagBeads" in the following way: for each 5 μg of RNA, transfer 25 μl "Oligo MagBeads" into a 1.5-ml tube. Capture the beads by placing the tube on a magnetic stand until all beads are arranged inside the tube near the magnet. Carefully remove the supernatant by aspiration and discard the supernatant.

4. Wash the beads with an equal volume of RNase-free water, then equilibrate the beads with an equal volume of binding buffer. Store the beads on ice until 5 min before they will be used (step 5), then warm up to room temperature.

5. To capture the eukaryotic RNA, add RNA/Capture Oligonucleotide Mix (step 2) to the prepared beads (step 4) and incubate 15 min at 37 °C. Capture the beads as described in step 3 and recover the supernatant containing the enriched bacterial RNA into a microcentrifuge tube on ice. Avoid transfer of magnetic beads.

6. Add 100 μl wash solution prewarmed to 37 °C to the captured beads to recover any remaining bacterial RNA. Remove the tube from the magnetic stand, resuspend the beads by gentle vortexing, incubate 5 min at 37 °C and recapture the beads as described. Recover the supernatant and pool it with the enriched prokaryotic RNA from step 5. Proceed immediately to the bacterial rRNA removal protocol (either next step or according to 4.2.1 steps 5 ff.)

7. The MICROB*Express* mRNA purification procedure is designed to remove 23S and 16S rRNA but not small RNAs (tRNA and 5S rRNA) from bacterial RNA. Quantify the RNA from step 6 using a NanoDrop™ spectrophotometer (Thermo Scientific). If necessary, dilute in TE (10 m$M$ Tris–HCl (pH 8.0), 1 m$M$ EDTA) to $\leq$ 10 μg μl$^{-1}$.

8. Add 4 μl Capture Oligo Mix, mix, and centrifuge briefly. Heat to 70 °C for 10 min, then incubate at 37 °C for 15 min to hybridize the capture oligonucleotides to homologous regions of the 16S and 23S rRNAs.

9. Prepare 50 μl Oligo MagBeads per sample in a 1.5-ml tube. Capture the beads and carefully remove and discard the supernatant. Wash the beads with (I) an equal volume of RNase-free water and (II) an equal volume of binding buffer. Finally, resuspend the beads in an equal volume of binding buffer and bring the slurry to 37 °C.

10. Add 50 μl prepared beads to the RNA/Capture Oligo Mix (step 8) and incubate at 37 °C for 15 min to anneal the beads to the rRNA/Capture Oligo mix. Capture the beads in a magnetic stand and carefully transfer the supernatant which contains the enriched mRNA to a microcentrifuge tube on ice. Add 100 μl wash solution prewarmed to 37 °C to the captured beads to recover any remaining mRNA as described in step 6.

11. Precipitate the mRNA by adding 1/10 vol 3 $M$ Na-acetate (pH 5.2), 4–6 μl of a DNA coprecipitant (e.g., Roti™-Clear DNA, Roth), and 2–2.5 volumes 96% (v/v) ethanol. Centrifuge 20 min at top speed (16,000 × $g$), carefully remove supernatant, wash twice with 70% (v/v) ethanol, dry the mRNA pellet 2–5 min at room temperature, and resuspend in 25 μl TE buffer. Quantify with a NanoDrop™ spectrophotometer. Enriched mRNA yields from 10 μg of high-quality total RNA is typically 1–2.5 μg.

## ACKNOWLEDGMENTS

Research in the author's laboratory was supported by the DFG excellence graduate school Jena School for Microbial Communication (JSMC) and the "Pakt für Forschung und Innovation" of the BMBF and TMBWK.

## REFERENCES

Ahn, J. H., & Walton, J. D. (1998). Regulation of cyclic peptide biosynthesis and pathogenicity in *Cochliobolus carbonum* by TOXEp, a novel protein with a bZIP basic DNA-binding motif and four ankyrin repeats. *Molecular & General Genetics, 260*, 462–469.
Ballance, D. J., & Turner, G. (1985). Development of a high-frequency transforming vector for *Aspergillus nidulans*. *Gene, 36*, 321–331.

Bayram, O., & Braus, G. H. (2012). Coordination of secondary metabolism and development in fungi: The velvet family of regulatory proteins. *FEMS Microbiology Reviews, 36,* 1–24.

Behnsen, J., Narang, P., Hasenberg, M., Gunzer, F., Bilitewski, U., Klippel, N., et al. (2007). Environmental dimensionality controls the interaction of phagocytes with the pathogenic fungi *Aspergillus fumigatus* and *Candida albicans. PLoS Pathogens, 3,* e13.

Bergmann, S., Funk, A. N., Scherlach, K., Schroeckh, V., Shelest, E., Horn, U., et al. (2010). Activation of a silent fungal polyketide biosynthesis pathway through regulatory cross talk with a cryptic nonribosomal peptide synthetase gene cluster. *Applied and Environmental Microbiology, 76,* 8143–8149.

Bergmann, S., Schümann, J., Scherlach, K., Lange, C., Brakhage, A. A., & Hertweck, C. (2007). Genomics-driven discovery of PKS–NRPS hybrid metabolites from *Aspergillus nidulans. Nature Chemical Biology, 3,* 213–217.

Bode, H. B., Bethe, B., Hofs, R., & Zeeck, A. (2002). Big effects from small changes: Possible ways to explore nature's chemical diversity. *Chembiochem, 3,* 619–627.

Bok, J. W., Chiang, Y. M., Szewczyk, E., Reyes-Dominguez, Y., Davidson, A. D., Sanchez, J. F., et al. (2009). Chromatin-level regulation of biosynthetic gene clusters. *Nature Chemical Biology, 5,* 462–464.

Bok, J. W., & Keller, N. P. (2004). LaeA, a regulator of secondary metabolism in *Aspergillus* spp. *Eukaryotic Cell, 3,* 527–535.

Brakhage, A. A., Thön, M., Spröte, P., Scharf, D. H., Al-Abdallah, Q., Wolke, S. M., et al. (2009). Aspects on evolution of fungal beta-lactam biosynthesis gene clusters and recruitment of *trans*-acting factors. *Phytochemistry, 70,* 1801–1811.

Burmester, A., Shelest, E., Glockner, G., Heddergott, C., Schindler, S., Staib, P., et al. (2011). Comparative and functional genomics provide insights into the pathogenicity of dermatophytic fungi. *Genome Biology, 12,* R7.

Chen, Y. P., Yuan, G. F., Hsieh, S. Y., Lin, Y. S., Wang, W. Y., Liaw, L. L., et al. (2010). Identification of the *mokH* gene encoding transcription factor for the upregulation of monacolin K biosynthesis in *Monascus pilosus. Journal of Agricultural and Food Chemistry, 58,* 287–293.

Chiang, Y. M., Szewczyk, E., Davidson, A. D., Keller, N., Oakley, B. R., & Wang, C. C. (2009). A gene cluster containing two fungal polyketide synthases encodes the biosynthetic pathway for a polyketide, asperfuranone, in *Aspergillus nidulans. Journal of the American Chemical Society, 131,* 2965–2970.

Flaherty, J. E., & Payne, G. A. (1997). Overexpression of *aflR* leads to upregulation of pathway gene transcription and increased aflatoxin production in *Aspergillus flavus. Applied and Environmental Microbiology, 63,* 3995–4000.

Hoff, B., Kamerewerd, J., Sigl, C., Zadra, I., & Kück, U. (2010). Homologous recombination in the antibiotic producer *Penicillium chrysogenum*: Strain DeltaPcku70 shows up-regulation of genes from the HOG pathway. *Applied Microbiology and Biotechnology, 85,* 1081–1094.

Hohn, T. M., Krishna, R., & Proctor, R. H. (1999). Characterization of a transcriptional activator controlling trichothecene toxin biosynthesis. *Fungal Genetics and Biology, 26,* 224–235.

Kwon, B. R., Kim, M. J., Park, J. A., Chung, H. J., Kim, J. M., Park, S. M., et al. (2009). Assessment of the core cryparin promoter from *Cryphonectria parasitica* for heterologous expression in filamentous fungi. *Applied Microbiology and Biotechnology, 83,* 339–348.

Lewis, K., Epstein, S., D'Onofrio, A., & Ling, L. L. (2010). Uncultured microorganisms as a source of secondary metabolites. *The Journal of Antibiotics, 63,* 468–476.

Meyer, V., Wanka, F., van Gent, J., Arentshorst, M., van den Hondel, C. A., & Ram, A. F. (2011). Fungal gene expression on demand: an inducible, tunable, and metabolism-independent expression system for *Aspergillus niger. Applied and Environmental Microbiology, 77,* 2975–2983.

Nahlik, K., Dumkow, M., Bayram, O., Helmstaedt, K., Busch, S., Valerius, O., et al. (2010). The COP9 signalosome mediates transcriptional and metabolic response to hormones, oxidative stress protection and cell wall rearrangement during fungal development. *Molecular Microbiology*, *78*, 964–979.

Nayak, T., Szewczyk, E., Oakley, C. E., Osmani, A., Ukil, L., Murray, S. L., et al. (2006). A versatile and efficient gene-targeting system for *Aspergillus nidulans*. *Genetics*, *172*, 1557–1566.

Nützmann, H. W., Reyes-Dominguez, Y., Scherlach, K., Schroeckh, V., Horn, F., Gacek, A., et al. (2011). Bacteria-induced natural product formation in the fungus *Aspergillus nidulans* requires Saga/Ada-mediated histone acetylation. *Proceedings of the National Academy of Sciences of the United States of America*, *108*, 14282–14287.

Oh, D. C., Kauffman, C. A., Jensen, P. R., & Fenical, W. (2007). Induced production of emericellamides A and B from the marine-derived fungus *Emericella* sp. in competing co-culture. *Journal of Natural Products*, *70*, 515–520.

Pachlinger, R., Mitterbauer, R., Adam, G., & Strauss, J. (2005). Metabolically independent and accurately adjustable *Aspergillus* sp. expression system. *Applied and Environmental Microbiology*, *71*, 672–678.

Pettit, R. K. (2009). Mixed fermentation for natural product drug discovery. *Applied Microbiology and Biotechnology*, *83*, 19–25.

Phelan, V. V., Liu, W. T., Pogliano, K., & Dorrestein, P. C. (2012). Microbial metabolic exchange-the chemotype-to-phenotype link. *Nature Chemical Biology*, *8*, 26–35.

Pontecorvo, G., Roper, J. A., Hemmons, L. M., Macdonald, K. D., & Bufton, A. W. (1953). The genetics of *Aspergillus nidulans*. *Advances in Genetics*, *5*, 141–238.

Prauser, H., & Falta, R. (1968). Phage sensitivity, cell wall composition and taxonomy of actinomyctes. *Zeitschrift für Allgemeine Mikrobiologie*, *8*, 39–46.

Reyes-Dominguez, Y., Bok, J. W., Berger, H., Shwab, E. K., Basheer, A., Gallmetzer, A., et al. (2010). Heterochromatic marks are associated with the repression of secondary metabolism clusters in *Aspergillus nidulans*. *Molecular Microbiology*, *76*, 1376–1386.

Roze, L. V., Chanda, A., Wee, J., Awad, D., & Linz, J. E. (2011). Stress-related transcription factor AtfB integrates secondary metabolism with oxidative stress response in aspergilli. *The Journal of Biological Chemistry*, *286*, 35137–35148.

Sakai, K., Kinoshita, H., & Nihira, T. (2012). Heterologous expression system in *Aspergillus oryzae* for fungal biosynthetic gene clusters of secondary metabolites. *Applied Microbiology and Biotechnology*, *95*, 2011–2022.

Sanchez, J. F., Somoza, A. D., Keller, N. P., & Wang, C. C. (2012). Advances in *Aspergillus* secondary metabolite research in the post-genomic era. *Natural Product Reports*, *29*, 351–371.

Santerre Henriksen, A. L., Even, S., Muller, C., Punt, P. J., van den Hondel, C. A., & Nielsen, J. (1999). Study of the glucoamylase promoter in *Aspergillus niger* using green fluorescent protein. *Microbiology*, *145*(Pt 3), 729–734.

Scherlach, K., Nützmann, H. W., Schroeckh, V., Dahse, H. M., Brakhage, A. A., & Hertweck, C. (2011). Cytotoxic pheofungins from an engineered fungus impaired in posttranslational protein modification. *Angewandte Chemie (International Ed. in English)*, *50*, 9843–9847.

Schroeckh, V., Scherlach, K., Nützmann, H. W., Shelest, E., Schmidt-Heck, W., Schuemann, J., et al. (2009). Intimate bacterial–fungal interaction triggers biosynthesis of archetypal polyketides in *Aspergillus nidulans*. *Proceedings of the National Academy of Sciences of the United States of America*, *106*, 14558–14563.

Shoji, J. Y., Maruyama, J., Arioka, M., & Kitamoto, K. (2005). Development of *Aspergillus oryzae thiA* promoter as a tool for molecular biological studies. *FEMS Microbiology Letters*, *244*, 41–46.

Siemieniewicz, K. W., & Schrempf, H. (2007). Concerted responses between the chitin-binding protein secreting *Streptomyces olivaceoviridis* and *Aspergillus proliferans*. *Microbiology*, *153*, 593–600.

Szewczyk, E., Chiang, Y. M., Oakley, C. E., Davidson, A. D., Wang, C. C., & Oakley, B. R. (2008). Identification and characterization of the asperthecin gene cluster of *Aspergillus nidulans*. *Applied and Environmental Microbiology*, *74*, 7607–7612.

Szewczyk, E., Nayak, T., Oakley, C. E., Edgerton, H., Xiong, Y., Taheri-Talesh, N., et al. (2006). Fusion PCR and gene targeting in *Aspergillus nidulans*. *Nature Protocols*, *1*, 3111–3120.

Takahashi, T., Masuda, T., & Koyama, Y. (2006). Identification and analysis of Ku70 and Ku80 homologs in the koji molds *Aspergillus sojae* and *Aspergillus oryzae*. *Bioscience, Biotechnology, and Biochemistry*, *70*, 135–143.

Tarkka, M. T., Sarniguet, A., & Frey-Klett, P. (2009). Inter-kingdom encounters: Recent advances in molecular bacterium–fungus interactions. *Current Genetics*, *55*, 233–243.

van Wezel, G. P., & McDowall, K. J. (2011). The regulation of the secondary metabolism of *Streptomyces*: New links and experimental advances. *Natural Product Reports*, *28*, 1311–1333.

Woloshuk, C. P., Foutz, K. R., Brewer, J. F., Bhatnagar, D., Cleveland, T. E., & Payne, G. A. (1994). Molecular characterization of *aflR*, a regulatory locus for aflatoxin biosynthesis. *Applied and Environmental Microbiology*, *60*, 2408–2414.

Zuck, K. M., Shipley, S., & Newman, D. J. (2011). Induced production of N-formyl alkaloids from *Aspergillus fumigatus* by co-culture with *Streptomyces peucetius*. *Journal of Natural Products*, *74*, 1653–1657.

> CHAPTER SEVENTEEN

# Waking up *Streptomyces* Secondary Metabolism by Constitutive Expression of Activators or Genetic Disruption of Repressors

## Bertrand Aigle*[,1], Christophe Corre[†,1]

*Génétique et Microbiologie, UMR UL-INRA 1128, IFR110 EFABA, Université de Lorraine, Vandœuvre-lès-Nancy, France
[†]Department of Chemistry, University of Warwick, Coventry, United Kingdom
[1]Corresponding authors: e-mail addresses: bertrand.aigle@univ-lorraine.fr; c.corre@warwick.ac.uk

## Contents

## Abstract

Streptomycete bacteria are renowned as a prolific source of natural products with diverse biological activities. Production of these metabolites is often subject to transcriptional regulation: the biosynthetic genes remain silent until the required environmental and/or physiological signals occur.

Consequently, in the laboratory environment, many gene clusters that direct the biosynthesis of natural products with clinical potential are not expressed or at very low level preventing the production/detection of the associated metabolite. Genetic

*Methods in Enzymology*, Volume 517
ISSN 0076-6879
http://dx.doi.org/10.1016/B978-0-12-404634-4.00017-6

engineering of streptomycetes can unleash the production of many new natural products. This chapter describes the overexpression of pathway-specific activators, the genetic disruption of pathway-specific repressors, and the main strategy used to identify and characterize new natural products from these engineered *Streptomyces* strains.

# 1. INTRODUCTION

Natural products, or secondary metabolites, produced by microorganisms and plants constitute the basis of a wide range of pharmacologically and agriculturally important substances. About half of the $\sim 1000$ new molecules introduced as antibacterial or antitumor drugs over the past quarter of a century are, or are derived from, natural products (Newman & Cragg, 2007). They are privileged structures in the search for biological activity owing to their intrinsic ability to interact with biological targets. Discovery of novel natural products is still of the utmost importance. Thus, for example, in human medicine, there is an urgent need for novel antibiotics to combat the threat of multidrug-resistant pathogens.

Among microorganisms, the actinomycetales, particularly the bacterial genus *Streptomyces*, are the most prolific producers of secondary metabolites (Berdy, 2005). Thus, about 65% of all antibiotics are produced by filamentous actinomycetes (Berdy, 2005; Hopwood, 2007) and 30% by fungi.

Since the sequencing of the chromosome of *Streptomyces coelicolor*, the first *Streptomyces* genome to be completely sequenced (Bentley et al., 2002), it became apparent that microbial secondary metabolic diversity had been largely underestimated. *S. coelicolor*, which had been extensively studied for nearly 50 years, was known to produce five secondary metabolites but the analysis of its genome sequence unveiled 18 additional putative secondary metabolite gene clusters (Bentley et al., 2002), most of them now characterized. Similar observations were made in other species, notably the industrial strain *Streptomyces avermitilis*, the producer of the avermectins, a group of antiparasitic agents used in human and veterinary medicine. Its genome contains 30 clusters related to secondary metabolite biosynthesis, while it was known to produce only three antibiotics (Ikeda et al., 2003). In *Streptomyces griseus*, the producer of streptomycin, only six biosynthetic pathways were previously described, while genome scanning revealed 34 clusters (Ohnishi et al., 2008).

With the increasing number of microbial genome sequencing programs (including those of fungi), candidate clusters for the production of novel natural products with potentially interesting biological properties have become ever more abundant. Nevertheless, although several strategies have been developed to gain access to these promising novel bioactive compounds, most of the gene clusters recognized by genome sequence analysis appear "silent," that is, to be not expressed, or expressed very poorly, under standard laboratory culture conditions.

Strategies have been developed to bypass this obstacle and to reveal the diversity of compounds that a strain can produce. The OSMAC (one strain/many compounds) method is based on systematic alteration of the cultivation parameters with the idea that some changes may allow expression of cryptic clusters (Bode, Bethe, Hofs, & Zeeck, 2002). This approach led, for example, to the discovery of a novel antifungal agent, the polyene ECO-02301, in *Streptomyces aizunensis* (McAlpine et al., 2005) and of three new compounds of a rare class of 22-membered macrolides in *Streptomyces* sp. strain C34 (Rateb et al., 2011). *Streptomyces* spp. are mainly found in soil and marine sediments although it has become apparent that they have also evolved to live in symbiosis with plants, fungi, and animals (Seipke, Kaltenpoth, & Hutchings, 2012). The biological function of secondary metabolites is mainly considered to be as antibiotics to fight against competitors, although several studies indicate that they can also act as signal molecules involved in cell–cell communication within microbial populations (Davies, 2006). Therefore, interactions with other microorganisms sharing the same ecological niches are expected to trigger (or to silence!) expression of so-called cryptic clusters. The first example of such a phenomenon comes from the fungus *Aspergillus nidulans* in which silent secondary metabolite gene clusters are specifically induced by interaction with bacteria, in this case, a *Streptomyces* spp. sharing the same habitat (Schroeckh et al., 2009). Similar examples have been reported for *Streptomyces*, for example, the induction of natural product biosynthesis by mycolic acid-containing bacteria, which led to the discovery of a novel antibiotic, alchivemycin A (Onaka, Mori, Igarashi, & Furumai, 2011). Heterologous expression in a dedicated host optimized for production of secondary metabolites such as the *S. coelicolor* "superhost" (Gomez-Escribano & Bibb, 2011) is another way to harness the biosynthetic potential of cryptic clusters (Corre & Challis, 2007).

An alternative way to identify products from silent metabolic gene clusters is manipulation of the transcription factor-encoding genes that govern secondary metabolite production. In *Streptomyces*, biosynthesis of secondary metabolites, which occurs in a growth phase-dependent manner, is controlled by complex regulatory mechanisms that determine the onset of production (for a recent review: van Wezel & McDowall, 2011). Two levels of control can be distinguished: pathway-specific and global controls. In both cases, they involve regulators acting either positively or negatively on natural product biosynthesis. Global regulatory genes are generally not found within secondary metabolite gene clusters and their products, also called pleiotropic regulators, affect several biosynthetic pathways. Some of them exert a unique control on natural product production, such as AfsR (positive control; Horinouchi, 2003; Horinouchi et al., 1990) or AbsA2 (negative control; Brian, Riggle, Santos, & Champness, 1996) in *S. coelicolor*. AbsA2 is one of the exceptions among global regulators since it is encoded within a biosynthetic cluster, the CDA cluster for the calcium–dependent lipopeptide antibiotic. Others also control morphological differentiation. Thus, the GntR family regulator DasR, which links nutrient stress to antibiotic production, is also essential for development (Rigali et al., 2006, 2008). Numerous transcriptional regulators with this dual function have been described, such as PhoP, ArpA, and AdpA and will not be described here. It is nevertheless interesting that the roles of these regulators may be different between different *Streptomyces* spp. For example, AfsrR in *S. griseus* also influences morphological differentiation (Umeyama, Lee, Ueda, & Horinouchi, 1999). Manipulation (deletion or overexpression) of global regulator-encoding genes can result in stimulation of secondary metabolite production, including those corresponding to cryptic clusters. For example, induction of a *S. coelicolor* cryptic type I polyketide synthase (PKS) gene cluster was observed in a mutant strain deleted for *dasR*, while in the tested conditions, no expression was observed in the wild-type strain (Rigali et al., 2008). However, it is interesting that an antibacterial activity associated with this cluster, the *cpk* cluster, was identified after deletion of a pathway-specific regulatory gene (Gottelt, Kol, Gomez-Escribano, Bibb, & Takano, 2010).

Putative pathway-specific regulators are usually located within the secondary metabolic gene clusters, together with genes for biosynthesis, secretion, and resistance (in the case of antibiotics). The rest of this chapter will focus on the pathway-specific regulatory genes. After a brief description of these regulators, we give examples of the manipulation of these genes to awaken a particular silent secondary metabolite gene cluster, a strategy that can be applicable to other *Streptomyces* spp. or more widely to other genetically manipulable actinomycetes.

## 2. AWAKENING CRYPTIC GENE CLUSTERS WITH PATHWAY-SPECIFIC ACTIVATORS

### 2.1. Pathway-specific activators in *Streptomyces*

Two main families of pathway-specific positive regulators have been described in *Streptomyces*: the SARP (*Streptomyces* antibiotic regulatory proteins) family (Wietzorrek & Bibb, 1997) and the LAL (large ATP-binding regulators of the LuxR family) family (De Schrijver & De Mot, 1999). The SARPs, which include the well-characterized proteins ActII-ORF4 and DnrI that control the production of actinorhodin in *S. coelicolor* and daunorubicin in *Streptomyces peucetius*, respectively (Gramajo, Takano, & Bibb, 1993; Stutzman-Engwall, Otten, & Hutchinson, 1992), have been found in a wide variety of secondary metabolic gene clusters, including type II and type I PKS clusters, nonribosomal peptide synthetase (NRPS) clusters, β-lactam, and lantibiotic clusters. They are characterized by a winged helix-turn-helix OmpR-like domain toward their N-terminus to bind to the promoter region of the target gene (the binding site consists of heptameric repeats) and a bacterial transcriptional activator domain at their C-terminus. With the exception of the pleiotropic regulator AfsR, which has a peculiar domain in the C-terminus, most characterized SARPs act as specific regulators at the bottom of regulatory cascades and turn on expression of the biosynthetic genes in their respective clusters. They are considered as essential transcriptional activators since their inactivation usually prevents antibiotic production.

While SARPs have only been found in actinomycetes (Bibb, 2005), LAL regulators were first identified in proteobacteria (De Schrijver & De Mot, 1999) and the prototype member of this family is MalT, an ATP-dependent transcriptional activator of the maltose regulon in *Escherichia coli* (Richet & Raibaud, 1989). They seem also to be widespread in actinomycetes and in *Streptomyces* spp. they have mostly been found not only within modular type I PKS gene clusters but also in other clusters such as hybrid PKS–NRPS-encoding clusters. They have been described as activators of polyketide biosynthesis, such as PikD in the pikromycin cluster in *Streptomyces venezuelae* (Wilson, Xue, Reynolds, & Sherman, 2001) and RapH in the rapamycin cluster in *Streptomyces hygroscopicus* (Kuscer et al., 2007). The LAL regulators contain an ATP-binding domain at the N-terminus (Walker A and B motifs) and a helix-turn-helix domain at the C-terminus, most likely responsible for DNA binding. The DNA-binding site of a LAL regulator has recently been identified in the actinomycete *Corynebacterium glutamicum* and is formed of an imperfect palindromic sequence (Zhao, Huang, Chen, Wang, & Liu, 2010).

Proteins that do not belong to large recognized families have also been found to specifically regulate the expression of biosynthetic genes. Thus PimR, which controls production of pimaricin in *Streptomyces natalensis* (Anton, Mendes, Martin, & Aparicio, 2004), represents the archetype of a new "class" of regulators, combining a SARP-like N-terminal section with a C-terminal half homologous to guanylate cyclases and LAL regulators. Among the other members of this family, SanG controls production of the peptidyl nucleoside antibiotic nikkomycin in *Streptomyces ansochromogenes* (Liu, Tian, Yang, & Tan, 2005) and PolR polyoxin biosynthesis in *Streptomyces cacaoi* subsp. *asoensis* (Li et al., 2009). Another example of these "restricted" families of proteins is typified by SrmS (Srm40), which is the ultimate regulator for expression of the spiramycin biosynthetic genes in *Streptomyces ambofaciens* (Karray, Darbon, Nguyen, Gagnat, & Pernodet, 2010). SrmS is highly similar to TylR, the essential activator of the tylosin cluster in *Streptomyces fradiae* and to AcyB2, a regulator of carbomycin biosynthesis in *Streptomyces thermotolerans* (Arisawa et al., 1993). Interestingly, SrmR (Srm22), another specific activator of the spiramycin cluster (Geistlich, Losick, Turner, & Rao, 1992), which controls expression of *srmS* (Karray et al., 2010), also does not belong to the classical families of pathway-specific activators.

The number of pathway-specific genes regulating the expression of the structural genes is variable between the gene clusters for secondary metabolites. Some clusters contain only a single regulator gene, like the actinorhodin cluster in *S. coelicolor* (*act*II-ORF4) or the streptomycin cluster in *S. griseus* (*strR*), while others code for multiple regulators (either transcriptional activators or repressors). Thus in *S. ambofaciens*, the spiramycin cluster contains four regulatory genes (Karray et al., 2010, 2007), while the *alp* cluster responsible for production of kinamycin (Bunet et al., 2011; Pang et al., 2004) contains at least five. Five regulator-encoding genes have also been identified in several other clusters, for example, in the tylosin cluster (Bate, Butler, Gandecha, & Cundliffe, 1999), in the *S. venezuelae* jadomycin cluster and in the geldanamycin gene cluster of *S. hygroscopicus* (Kim, Lee, Paik, & Hong, 2010).

All the (potential) regulatory genes identified within a cluster are not necessarily involved in regulation (either positive or negative) of the structural genes. For example, two of the spiramycin regulators appear to have no role in controlling macrolide production (Karray et al., 2010) and the SARP

TylT is not essential for tylosin production (Bate, Stratigopoulos, Bate, Stratigopoulos, & Cundliffe, 2002). In the actinorhodin cluster, ActR controls the exporters of the antibiotic but not the structural genes. However, at least one gene acts as the ultimate positive regulator: that is, it induces the expression of some or even all the structural genes at some stage of growth, and expression of this (these) final activator(s) is (are) under the control of a complex regulatory network. When several regulators are encoded within a cluster, they can be components of a complex regulatory cascade that, in response to some signal, induces the expression of the ultimate regulator, which in turn stimulates transcription of the structural genes. Such a complex cascade has been described in detail in *S. fradiae* for the regulation of tylosin production (Cundliffe, 2008) and in *S. ambofaciens* for the regulation of kinamycin production (Aigle, Pang, Decaris, & Leblond, 2005; Bunet et al., 2008, 2011). In both cases, a transcriptional repressor encoded within the cluster and belonging to the γ-butyrolactone receptor family (TylP for tylosin and AlpZ for kinamycin, see Section 3) acts at the top of the regulatory cascade and prevents indirectly (*tyl* cluster) or directly (*alp* cluster) expression of the ultimate activator (TylR and AlpV). The activity of these repressors is released by a signal molecule (still of unknown structure) which allows the onset of antibiotic production. In the case of the *alp* cluster, a second repressor, the pseudo-γ-butyrolactone receptor AlpW, acts as a late repressor to stop expression of *alpV* and consequently kinamycin biosynthesis (Bunet et al., 2011).

The pathway-specific regulatory cascades do not exclude, of course, a higher level of regulation involving global regulators.

It is interesting that the antibiotic itself, and even some biosynthetic intermediates, can play a role in the regulatory circuit. This was first demonstrated in *S. venezuelae* in which the activity of JadR1, an OmpR-type atypical response regulator indispensable for jadomycin production, is inhibited by the binding of jadomycin and some intermediates to the N-terminal receiver domain, causing JadR1 to dissociate from target promoters (Wang et al., 2009). The authors report similar regulation of RedZ, a NarL-type atypical response regulator that regulates undecylprodigiosin production in *S. coelicolor* via direct control of the expression of the ultimate regulatory gene encoding the SARP RedD.

Even if a secondary metabolic gene cluster encodes a single regulator for the control of the structural genes, transcription of this regulatory gene may be very complex. This has been well characterized for *act*II-ORF4, whose promoter is the direct target of at least three transcription factors, including the pleiotropic regulators DasR and AbsA2 (for review, van Wezel & McDowall, 2011). The expression of the ultimate regulator also depends indirectly on other global regulators.

In spite of the complex levels of regulation of pathway-specific activator genes, the deregulation of the ultimate activator by overexpression has long been known to stimulate production of the secondary metabolite under its control. Numerous examples have been described, whatever the family of positive transcriptional factors (SARP, LAL, etc.). Therefore, identifying and manipulating pathway-specific genes appears to be a suitable way to activate in a targeted fashion the expression of silent gene clusters and obtain access to novel natural products.

## 2.2. Experimental strategy to awaken a silent cluster by manipulating pathway-specific activators

In this section, we describe a strategy to wake up cryptic secondary metabolite gene clusters based on the example of the activation of a silent type I PKS gene cluster in *S. ambofaciens* ATCC23877 that, combined with comparative liquid chromatography–mass spectrometry (LC–MS) profiling, led to the identification of a novel bioactive giant macrolide complex, the stambomycins (Laureti et al., 2011).

Sequencing of the *S. ambofaciens* ATCC23877 genome, the producer of the macrolide spiramycin and the pyrrolamide congocidine, revealed more than 20-gene clusters potentially involved in the production of other secondary metabolites (Aigle et al., 2011; Choulet et al., 2006; Juguet et al., 2009; Karray et al., 2007). By a combination of biosynthetic gene deletion, manipulation of a regulatory network and an OSMAC-like approach, a genome-mined type II PKS gene cluster was shown to be responsible for the production of different forms of kinamycins, already known antibiotics of the angucyclinone family (Bunet et al., 2011; Pang et al., 2004). The products of two other gene clusters have also been reported: the known siderophores coelichelin and the desferrioxamines E/B (Barona-Gomez et al., 2006).

Among the other biosynthetic clusters whose products remained to be identified, a large type I modular PKS gene cluster in the right arm of the chromosome (Fig. 17.1), attracted attention because *in silico* analysis of the nine PKS subunits encoded by the cluster revealed that they might catalyze the synthesis of a macrolide with a novel structure (Laureti et al., 2011). RT-PCR analysis showed that several genes, including all the PKS genes, were not transcribed in different media known to elicit production of the

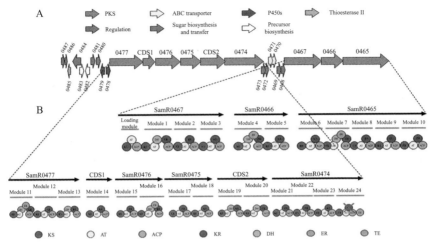

**Figure 17.1** Type I PKS gene cluster of *S. ambofaciens* ATCC23877 responsible for the production of stambomycins. (A) Genetic organization of the stambomycin biosynthetic gene cluster. The predicted functions of the genes are indicated above the map. The *samR0484* gene encodes the LAL regulator whose overexpression induces expression of the silent PKS genes. Limits of the cluster were defined based on BLAST analysis. (B) Module and domain organization of the PKS subunits encoded within the stambomycin cluster. This has been assigned using the SEARCHPKS program (Yadav, Gokhale, & Mohanty, 2003). The ketoreductase domain in module 24 is nonfunctional (two essential amino acids of the catalytic triad, a tyrosine residue, and an asparagine residue, are absent; Reid et al., 2003). KS, β-ketoacyl synthase; AT, acyltransferase; ACP, acyl carrier protein; KR, ketoreductase; DH, dehydratase; ER, enoyl reductase; TE, thioesterase. (See Color Insert.)

macrolide spiramycin, irrespective of the growth phase. Therefore, we developed a method to wake up expression of the cluster by overexpressing a pathway-specific regulatory gene. It should be noted that the size of the cluster (~150 kb) excluded heterologous expression.

1. Choice of the regulatory gene

   The first step is to identify potential regulatory genes within the cluster of interest and make, when possible, a hypothesis on their roles (activator or repressor) based on homology with other already characterized regulatory genes. This helps to define the choice/priority for the next step: either overexpression or deletion of the transcriptional regulator gene (although the two approaches are not exclusive). Two transcriptional regulators were found within the *S. ambofaciens* type I PKS gene cluster. One, *samR0484*, encodes a LAL regulator (Fig. 17.1). As described in Section 2.1, LAL regulators have been identified in several macrolide pathways and shown to act positively. The other gene, *samR0468*, codes for a response regulator and lies next to a gene encoding a sensor kinase (*samR0469*), forming a two-component signal (TCS) transduction system. TCSs have been described to act either positively or negatively on gene clusters. We chose *samR0484* as the prime candidate. Nevertheless, the same strategy was applied to the TCS but will not be described here as neither overexpression nor deletion of the TCS genes resulted in activation of the stambomycin biosynthetic genes.

**2.** Construction of a strain overexpressing an activator-encoding gene

**a.** Choice of the vector. The choice of vector for cloning the regulatory gene depends on different parameters: autonomous or site-specific integrative vector in *Streptomyces*, inducible or constitutive expression of the regulatory gene, or keeping the gene under the control of its own promoter. To avoid the need for selective pressure to maintain a plasmid in the cell and to restrict dependence on other endogenous regulators, we prefer to use an integrative vector containing a strong and constitutive promoter under which the candidate gene can be cloned. If possible, a conjugative vector is also preferred for an easy transfer of the recombinant plasmid into the *Streptomyces* host from *E. coli*. Our choice was the conjugative and integrative plasmid pIB139 (Wilkinson et al., 2002), which contains the strong and constitutive promoter *ermE*$\star$p.

**b.** PCR amplification of the regulatory gene. The open reading frame of the regulatory gene is amplified with high-fidelity polymerase using PCR primers including restriction sites at the 5′-end that are compatible with restriction sites from the multicloning site of the vector. In addition, the forward primer can contain a typical *Streptomyces* ribosomal-binding site (RBS) sequence (we chose AAAGGAGG) between the restriction site and the nucleotides matching the start codon in order to improve translation of the activator protein (Bunet et al., 2008). The RBS should be typically centered about 10 nucleotides upstream of the start codon.

   *Note*: Because the *samR0484* gene is nearly 3 kb long (2877 bp), Takara polymerase for amplification of large fragments was used with the program recommended by the supplier (2 min at 94 °C, 30 cycles of 10 s at 98 °C, and 15 min at 68 °C, followed by a final extension of 10 min at 72 °C).

**c.** Cloning. The purified PCR product (by a cleanup kit or by extracting from a gel) can be straightforwardly cloned into the expression vector after restriction with appropriate enzymes, although an intermediate cloning step into a vector dedicated for PCR product cloning (e.g., pGEMT-easy (Promega) or pJET/blunt (Fermentas) according to the property of the polymerase to add or not add a deoxyadenosine to the 3′-ends of the amplified fragment) can be envisaged. This might facilitate cloning and sequencing to check the integrity of the insert. The pIB139 derivative containing *samR0484* under the control of ermE⋆p was named pOE-0484.

**d.** Transformation of *E. coli* ET12567/pUZ8002 (Paget, Chamberlin, Atrih, Foster, & Buttner, 1999). This strain is a non–methylating strain that has to be used if the *Streptomyces* host carries a methyl-sensing restriction system. The plasmid pUZ8002 is an RP4 derivative encoding the *tra* functions required for the mobilization of the conjugative plasmid. After transformation of the thermo- or electrocompetent cells with the construct, the clones are selected with appropriate antibiotics (apramycin for pIB139 and kanamycin and chloramphenicol to maintain the selection for pUZ8002 and the *dam* mutation, respectively; Kieser, Bibb, Buttner, Chater, & Hopwood, 2000; MacNeil et al., 1992).

**e.** Conjugal transfer into *Streptomyces*. The *Streptomyces* host initially used is the wild-type strain. Nevertheless to later confirm the phenotype observed following overexpression of the transcription factor, a mutant strain deleted for one (or several) biosynthetic genes of the cluster has to be transformed with the construct. Both strains are also transformed with the vector alone and serve as negative controls. In our study, the control strain was a mutant deleted for the gene predicted to encode the first PKS subunit of the biosynthetic pathway (*samR0467*; strain ATCCΔ467; Laureti et al., 2011).

The exconjugants are analyzed by PCR and Southern blotting to confirm the presence of the plasmid integrated into the chromosome (if a site-specific integrative plasmid is used; pIB139 carries the *attP* site of the phage ΦC31) and pulsed-field gel electrophoresis to ensure that no large DNA rearrangement has occurred. If an autonomously replicating plasmid is preferred, the presence of the construct is confirmed by plasmid DNA extraction, restriction, and gel electrophoresis.

*Note*: For the generation of a mutant deleted for biosynthetic gene(s), we use the PCR-targeting technology, a gene replacement method based on Red/ET-mediated recombination (Datsenko & Wanner, 2000) and adapted for *Streptomyces* (Gust, 2009; Gust, Challis, Fowler,

Kieser, & Chater, 2003). This method is described in detail in a former issue of *Methods in Enzymology* (Gust, 2009).

3. Transcriptional analysis of the biosynthetic gene cluster by RT-PCR

The next step is to confirm that the integration of the recombinant expression plasmid into the chromosome of the host strain results in overexpression of the regulatory gene in comparison to the control strain (the strain containing the vector alone) and consequently if this over-expression leads to transcription of the biosynthetic genes.

The different strains are cultivated in liquid medium (we used a 50-ml culture in 250-ml Erlenmeyer flasks in MP5 medium, which was reported to elicit production of the macrolide spiramycin; Pernodet, Alegre, Blondelet-Rouault, & Guerineau, 1993), and 2-ml samples are harvested for RNA extraction. Total RNAs are extracted using the modified Kirby method (Kieser et al., 2000).

*Note*: If an inducible promoter (i.e., *tipA*p) is preferred, add the inducer at some point of growth (the best time of induction has to be exper-imentally determined). The RNA extraction protocol was scaled down for 2-ml samples. Lysis of the mycelium was obtained with a bath sonicator (bioruptor, Diagenode) $2 \times 10$ s, power max, in a $4\,^{\circ}\mathrm{C}$ water bath.

cDNAs are obtained after reverse transcription of 2–4 μg of DNaseI-treated total RNA with SuperScript III reverse transcriptase (Invitrogen) and high-GC-content random hexamer primers (Oligo Spiking; Eurogentec).

cDNAs are amplified using primer pairs specific to the regulatory gene used for overexpression and to the biosynthetic genes. Amplification of *hrdB* (encoding the major sigma factor) or the 16S RNA gene serves as positive control.

*Note*: Primers are designed to amplify short DNA sequence ($\sim 100$ bp) as this size is recommended if quantitative PCR is also envisaged. Control PCRs are similarly performed with RNAs untreated with reverse tran-scriptase to confirm the absence of contaminating DNA in the RNA preparations. The PCR conditions for transcriptional analysis are as fol-lows: 4 min at $94\,^{\circ}\mathrm{C}$; 28 cycles of 30 s at $94\,^{\circ}\mathrm{C}$, 30 s at $60\,^{\circ}\mathrm{C}$, and 30 s at $72\,^{\circ}\mathrm{C}$; and a final extension of 5 min at $72\,^{\circ}\mathrm{C}$.

In our study, the presence of pOE-0484 led to overexpression of the regu-latory gene and to induction of the expression of all the PKS genes and other genes of the cluster that were silent in the wild-type strain (Fig. 17.2), prov-ing the efficacy of our strategy.

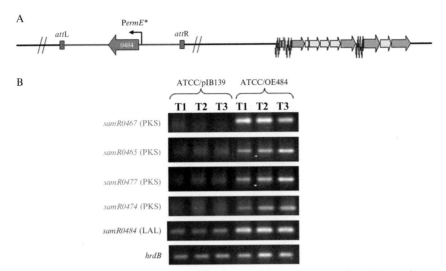

**Figure 17.2** Waking up the silent PKS genes by overexpressing the LAL regulator-encoding gene. (A) Schematic representation of the construct ATCC/pOE-0484. The figure is not to scale. (B) Transcriptional analysis by RT-PCR of the control ATCC/pIB139 and the mutant strain ATCC/OE484, grown in MP5 medium. Expression of four PKS subunit genes (*samR0467, samR0465, samR0477, samR0474*) and of the regulatory gene *samR0484* were analyzed by RT-PCR using 4 μg total RNA. The constitutively expressed *hrdB* gene coding for the major sigma factor was used as positive control. T1, exponential phase; T2, transition phase; T3, stationary phase. (For color version of this figure, the reader is referred to the online version of this chapter.)

## 3. AWAKENING CRYPTIC GENE CLUSTERS USING TRANSCRIPTIONAL REPRESSOR DISRUPTIONS

### 3.1. TetR pathway-specific repressors in *Streptomyces*

Transcriptional regulators belonging to the TetR family are bacterial regulatory proteins involved in adaptive responses (Ramos et al., 2005). Members of this family of proteins typically act as transcriptional repressors (in the absence of the cognate ligand) by binding to specific DNA sequences in operator regions. They most often repress expression of their own gene as well as that of other genes. For instance, in *E. coli*, in the absence of tetracycline, TetR represses expression not only of *tetR* but also of *tetA*, which encodes a tetracycline efflux pump. In the presence of subinhibitory concentrations of tetracycline, TetR–tetracycline complexes are formed and changes in TetR conformation result in reduction in TetR/operator binding affinity and its dissociation from the DNA. Thus, the *tetA* efflux pump gene is expressed and confers tetracycline resistance. The genome of *S. coelicolor* A3(2) alone encodes 150 TetR proteins (Willems et al., 2008).

X-ray crystal structures of a few TetR proteins have been solved and revealed an N-terminal DNA-binding domain linked to a C-terminal ligand-binding domain (Ramos et al., 2005). Interaction of TetR proteins with their cognate ligand is proposed to induce conformational changes. Among the different subfamilies of TetR proteins, ArpA-like proteins, also known as γ-butyrolactone receptors or pseudo-γ-butyrolactone receptors, are part of complex regulatory networks controlled by microbial hormones such as gamma-butyrolactone (GBL) and furan signaling molecules (2-alkyl-4-hydroxymethylfuran-3-carboxylic acid, AHFCA) (Takano, 2006). ArpA itself directly responds to the A-factor GBL (Fig. 17.3), which was one of the first signaling molecules ever discovered. ArpA represses expression of the central transcriptional activator *adpA* in the absence of A-factor. This mechanism mediates and coordinates response in *S. griseus* and results in the production of several antibiotics as well as physiological and morphological differentiation (Fig. 17.3). Since the discovery of A-factor half a century ago, GBLs are known to control antibiotic production and/or morphological differentiation in many streptomycetes. In *S. coelicolor* A3(2), SCBs (*S. coelicolor* butyrolactones) have been shown to interact with the transcriptional repressor ScbR and to control the expression of the *cpk* gene cluster

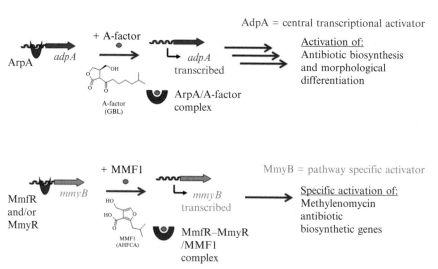

**Figure 17.3** Mechanism of transcriptional repression by ArpA and MmfR/MmyR in *S. griseus* and *S. coelicolor* A3(2), respectively. (See Color Insert.)

(Gottelt et al., 2010). In the same microorganism, a novel class of signaling molecules (AHFCAs) that specifically induce production of the methylenomycin antibiotics was recently discovered (Corre, Song, O'Rourke, Chater, & Challis, 2008; Willey & Gaskell, 2011; see also Chapter 4). These furan signaling molecules are structurally very distinct from GBLs. More recently, another novel structural class of signaling molecules, the avenolides, have been shown to control avermectin production in *S. avermitilis* by interacting with the ArpA-like transcriptional repressor AvaR1 (Kitani et al., 2011). Only one ArpA-like protein (CprB from *S. coelicolor*) has been crystallized to date but its cognate ligand remains unknown (Natsume et al., 2003).

In this section, we focus on the transcriptional regulators involved in the regulation of methylenomycin antibiotic production. Two TetR repressors (MmfR and MmyR) are encoded within the gene cluster as well as a transcriptional activator (MmyB). Molecular genetics experiments have suggested that MmfR and MmyR prevent expression of the transcriptional activator MmyB in the absence of the MMF signaling molecules (Fig. 17.3; O'Rourke et al., 2009). Interestingly, independent disruption of the two *arpA*-like genes *mmfR* and *mmyR* resulted in clearly different phenotypes: overproduction of methylenomycin antibiotic in the *mmyR* mutant but not in the *mmfR* mutant (O'Rourke et al., 2009). The specific characteristics of MmfR and MmyR are currently under investigation *in vitro*.

Disruption of *arpA*-like transcriptional repressors has been shown to enhance (or awake) production of specific antibiotics in several systems and therefore represents a great strategy for strain improvement or to discover novel natural products. In *S. coelicolor*, disruption of *scbR2* resulted in the production of metabolites with antibacterial activity linked to the *cpk* gene cluster (Gottelt et al., 2010).

## 3.2. Experimental strategy to awaken a silent cluster by genetic disruption of pathway-specific repressors

The strategy described here not only permits the awakening of silent clusters but also enhancement of the expression of poorly expressed biosynthetic machineries. It also allows one to obtain efficient antibiotic production in much simpler media than those sometimes required to obtain any expression. Following identification of putative pathway-specific repressors, genetic disruption can be carried out on a construct (cosmid or BAC) containing the entire gene cluster or directly on genomic DNA of the organism of interest.

### 3.2.1 Protocol for mmyR disruption (Fig. 17.4)

The integrative cosmid C73-787, which contains the entire methylenomycin biosynthetic gene cluster, was used to generate the *S. coelicolor* methylenomycin overproducing strain W89. The transcriptional repressor *mmyR* gene was inactivated by replacement with the apramycin resistance gene *apr* using PCR-targeting technology (Fig. 17.4; Gust et al., 2003).

1. PCR amplification of *apr* from the plasmid pCC60 (Corre et al., 2008) using long PCR primers that contain homology to the sequences upstream and downstream of *mmyR*.

2. Screening for apramycin- and kanamycin-resistant clones in *E. coli*. Check by PCR and restriction digest the engineered cosmid C73-787 *mmyR::apr*.

3. As per the protocol described in Section 2.2 (parts 2d and 2e): transformation of *E. coli* ET12567/pUZ8002 followed by conjugation with a *Streptomyces* host/superhost resulted in a new *Streptomyces* strain named W89. The cosmid C73-787 contained an integrative element which permits site-specific integration into the *Streptomyces* chromosome. Southern blot and PCR analyses are then used on the DNA isolated from the new *Streptomyces* strain to confirm integration of the mutated cosmid into the host.

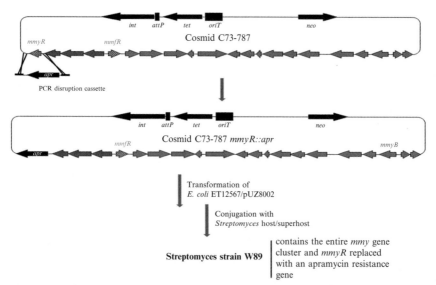

**Figure 17.4** Genetic disruption of the transcriptional repressor *mmyR* on an integrative vector containing the entire methylenomycin biosynthetic pathway and introduction into a *Streptomyces* host. (For color version of this figure, the reader is referred to the online version of this chapter.)

*Note*: When the entire gene cluster cannot be cloned into an integrative vector, transcriptional mutants have to be constructed into the *Streptomyces* strain containing the gene cluster of interest.

## 4. COMPARATIVE METABOLIC PROFILING TO IDENTIFY NOVEL METABOLITES IN THE ENGINEERED STRAINS

The following approaches are applicable to both strategies whereby a pathway-specific activator is overexpressed or a transcriptional repressor disrupted.

### 4.1. Biological assay

The first easy screen is to search for biological activity. In the case of *S. ambofaciens*, the mutant strain overexpressing the regulatory gene and the control strain containing the vector alone (ATCC/OE484 and ATCC/pIB139 in our study) are grown in liquid medium in which transcriptional analyses have been carried out or on agar plates. Antibacterial activity of the two strains is compared at different stages of growth (exponential, transitional, and stationary phases) to see if the waking up of the cryptic cluster results in a new activity compared to the control strain (or an increase of activity since the strain can produce other antibiotic compounds in the tested conditions). The supernatant of ATCC/OE484 grown in MP5 was tested for antibacterial activity but did not show any specific activity compared to the control strain (ATCC/pIB139). In the case of the methylenomycin gene cluster, a heterologous host lacking the ability to produce metabolites with significant antibacterial activity and containing the cosmid C73-787 *mmyR::apr* such as strain *Streptomyces* W89 was grown on SMMS agar plates (pH 5.0, due to the carboxylic acid group present in the methylenomycins, this pH allows a better diffusion of the antibiotic; Corre & Challis, 2005). After 48 h of incubation at 30 °C, plugs from this plate were transferred onto an agar plate (pH 5.0) freshly inoculated with a methylenomycin sensitive strain (*Bacillus subtilis* or *Micrococcus luteus*). Following incubation at 30 °C for an additional 24 h, a large zone of inhibition became apparent surrounding plugs where the strain W89 was grown. Methylenomycin was found to be overproduced in the new strain *Streptomyces* W89 compared to *S. coelicolor* A3(2). Importantly, methylenomycin production was consistent and did not rely on a complex culture medium.

*Note*: Different indicator strains should be tested: Gram-positive and Gram-negative strains as well as fungi.

## 4.2. Analytical chemistry: LC–MS analyses

To isolate the natural product of interest and/or if no specific phenotype can be linked to expression of the biosynthetic cluster (no biological activity easily detectable, etc.), comparison of the metabolic profile of mutant and wild-type strains is carried out by LC–MS analysis. In many cases, natural products are exported extracellularly by means of specific efflux pumps. These pumps can constitute a self-defense mechanism if the natural product is toxic to the producer organism. The methylenomycins as well as many small organic molecules are extracted from the acidified culture medium (pH 2) using organic solvents (ethyl acetate); extracts can be prepared at different time points (incubation at 30 °C between 2 and 7 days). The solvent is then evaporated from these organic extracts using a rotary evaporator and the extracts are resuspended in a solvent compatible with LC–MS analyses (water/methanol 1:1 mixture). Ordinarily a $C_{18}$-reverse phase column is coupled with a mass spectrometer. The MS chromatograms representing the ionized metabolites extracted from the mutant and wild-type strains can then be compared for differences. Fractions from the mutant extracts containing novel metabolites are then analyzed by accurate mass spectrometry. Thus, the molecular formula of each compound can be determined.

*Note*: Analysis of these data is facilitated if the medium in which metabolites are produced is minimal.

As some compounds accumulate intracellularly (e.g., undecylprodigiosin in *S. coelicolor*), extractions can be carried out not only on the culture medium but also on the mycelium.

*Note*: For agar culture, cellophane membranes are placed on the solid medium and spores are spread on top. After incubation at 30 °C for a few days, membranes are lifted off the plate and the mycelium is scraped off them. The mycelium is dispersed in a solvent and the mixture is sonicated for 10 min. When possible, choose the solvent for the extraction step according to the physicochemical properties of the predicted compound (hydrophilic or hydrophobic, etc.).

Confirm that the new peak(s) observed in the strain overexpressing the regulatory gene and absent from the extract of the control strain is (are) the product(s) of the gene cluster by analyzing a strain overexpressing the regulatory gene but deleted for a biosynthetic gene.

To identify the product(s) of the type I gene cluster, the supernatant and the mycelium of the two strains were extracted with methanol and extracts were analyzed by comparative metabolic profiling using LC–MS. Two major peaks were identified in the ATCC/OE484 mycelium extract but absent from the ATCC/pIB139 extract (Fig. 17.5).

Structural characterization of the new natural products is then carried out on purified material using a combination of mass spectrometry and NMR spectroscopy. Purification of these natural products most often involves HPLC purification.

## 5. CONCLUSION/PERSPECTIVES

As discussed, positive and negative transcriptional regulators are important targets not only for strain improvement but also for waking up silent biosynthetic gene clusters in streptomycetes. A better understanding of regulatory pathways that control antibiotic production in streptomycetes is currently making the approaches described here more efficient.

The stambomycins (Fig. 17.6) were the first example of *Streptomyces* natural products discovered by overexpressing a transcriptional activator and waking up a silent gene cluster. Most interestingly, LAL regulators can be identified in several cryptic clusters in genomic databases. This strategy is applicable to other *Streptomyces* spp. and is not limited to LAL regulators but other positive regulators (SARP, SrmR, etc.) can be overexpressed as described here.

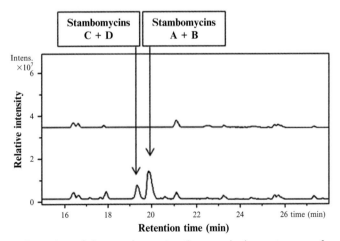

**Figure 17.5** Detection of the stambomycins. Base peak chromatograms from LC–MS analyses of methanolic mycelial extracts of ATCC/pIB139 (top) and ATCC/OE484 (bottom) strains. Peaks corresponding to stambomycins A/B and C/D in the ATCC/OE484 strain are indicated by the arrows.

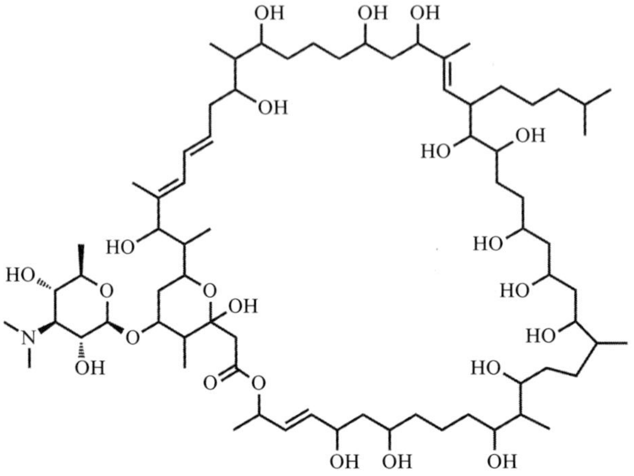

**Stambomycin B**

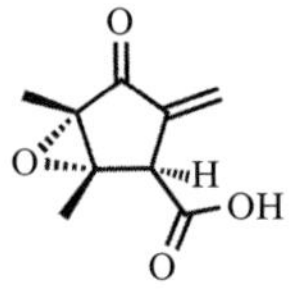

**Figure 17.6** Stambomycin B and methylenomycin A overproduced by engineering *Streptomyces* pathway-specific transcriptional regulators.

Similarly, disrupting a negative pathway–specific regulator can also result in overproduction of biologically active metabolites. TetR repressors represent a very large group of bacterial transcriptional regulators and their manipulation could be central in the exploitation of the large number of untapped biosynthetic gene clusters that have been revealed using *Streptomyces* genome mining. Many new natural products with potential clinical utility could therefore be discovered.

Importantly, the approaches described herein could also be applicable to gene clusters identified from metagenomic DNA.

## REFERENCES

Aigle, B., Bunet, R., Corre, C., Garénaux, A., Hotel, L., Huang, S., et al. (2011). Genome-guided exploration of *Streptomyces ambofaciens* secondary metabolism. In P. Dyson (Ed.), *Streptomyces* molecular biology and biotechnology (pp. 179–194). Norfolk: John Innes, Caister Academic Press.

Aigle, B., Pang, X., Decaris, B., & Leblond, P. (2005). Involvement of AlpV, a new member of the Streptomyces antibiotic regulatory protein family, in regulation of the duplicated type II polyketide synthase alp gene cluster in Streptomyces ambofaciens. *Journal of Bacteriology, 187*(7), 2491–2500.

Anton, N., Mendes, M. V., Martin, J. F., & Aparicio, J. F. (2004). Identification of PimR as a positive regulator of pimaricin biosynthesis in Streptomyces natalensis. *Journal of Bacteriology, 186*(9), 2567–2575.

Arisawa, A., Kawamura, N., Tsunekawa, H., Okamura, K., Tone, H., & Okamoto, R. (1993). Cloning and nucleotide sequences of two genes involved in the 4″-O-acylation of macrolide antibiotics from Streptomyces thermotolerans. *Bioscience, Biotechnology, and Biochemistry*, *57*(12), 2020–2025.

Barona-Gomez, F., Lautru, S., Francou, F. X., Leblond, P., Pernodet, J. L., & Challis, G. L. (2006). Multiple biosynthetic and uptake systems mediate siderophore-dependent iron acquisition in Streptomyces coelicolor A3(2) and Streptomyces ambofaciens ATCC 23877. *Microbiology*, *152*(Pt 11), 3355–3366.

Bate, N., Butler, A. R., Gandecha, A. R., & Cundliffe, E. (1999). Multiple regulatory genes in the tylosin biosynthetic cluster of Streptomyces fradiae. *Chemistry and Biology*, *6*(9), 617–624.

Bate, N., Stratigopoulos, G., Bate, N., Stratigopoulos, G., & Cundliffe, E. (2002). Differential roles of two SARP-encoding regulatory genes during tylosin biosynthesis. *Molecular Microbiology*, *43*(2), 449–458.

Bentley, S. D., Chater, K. F., Cerdeno-Tarraga, A. M., Challis, G. L., Thomson, N. R., James, K. D., et al. (2002). Complete genome sequence of the model actinomycete Streptomyces coelicolor A3(2). *Nature*, *417*(6885), 141–147.

Berdy, J. (2005). Bioactive microbial metabolites. The Journal of Antibiotics (Tokyo), *58*(1), 1–26.

Bibb, M. J. (2005). Regulation of secondary metabolism in streptomycetes. *Current Opinion in Microbiology*, *8*(2), 208–215.

Bode, H. B., Bethe, B., Hofs, R., & Zeeck, A. (2002). Big effects from small changes: Possible ways to explore nature's chemical diversity. *Chembiochem*, *3*(7), 619–627.

Brian, P., Riggle, P. J., Santos, R. A., & Champness, W. C. (1996). Global negative regulation of Streptomyces coelicolor antibiotic synthesis mediated by an absA-encoded putative signal transduction system. *Journal of Bacteriology*, *178*(11), 3221–3231.

Bunet, R., Mendes, M. V., Rouhier, N., Pang, X., Hotel, L., Leblond, P., et al. (2008). Regulation of the synthesis of the angucyclinone antibiotic alpomycin in Streptomyces ambofaciens by the autoregulator receptor AlpZ and its specific ligand. *Journal of Bacteriology*, *190*(9), 3293–3305.

Bunet, R., Song, L., Mendes, M. V., Corre, C., Hotel, L., Rouhier, N., et al. (2011). Characterization and manipulation of the pathway-specific late regulator AlpW reveals Streptomyces ambofaciens as a new producer of Kinamycins. *Journal of Bacteriology*, *193*(5), 1142–1153.

Choulet, F., Aigle, B., Gallois, A., Mangenot, S., Gerbaud, C., Truong, C., et al. (2006). Evolution of the terminal regions of the Streptomyces linear chromosome. *Molecular Biology and Evolution*, *23*(12), 2361–2369.

Corre, C., & Challis, G. L. (2005). Evidence for the unusual condensation of a diketide with a pentulose in the methylenomycin biosynthetic pathway of Streptomyces coelicolor A3(2). *Chembiochem*, *6*(12), 2166–2170.

Corre, C., & Challis, G. L. (2007). Heavy tools for genome mining. *Chemistry and Biology*, *14*(1), 7–9.

Corre, C., Song, L., O'Rourke, S., Chater, K. F., & Challis, G. L. (2008). 2-Alkyl-4-hydroxymethylfuran-3-carboxylic acids, antibiotic production inducers discovered by Streptomyces coelicolor genome mining. *Proceedings of the National Academy of Sciences of the United States of America*, *105*(45), 17510–17515.

Cundliffe, E. (2008). Control of tylosin biosynthesis in Streptomyces fradiae. *Journal of Microbiology and Biotechnology*, *18*(9), 1485–1491.

Datsenko, K. A., & Wanner, B. L. (2000). One-step inactivation of chromosomal genes in Escherichia coli K-12 using PCR products. *Proceedings of the National Academy of Sciences of the United States of America*, *97*(12), 6640–6645.

Davies, J. (2006). Are antibiotics naturally antibiotics? *Journal of Industrial Microbiology and Biotechnology*, *33*(7), 496–499.

De Schrijver, A., & De Mot, R. (1999). A subfamily of MalT-related ATP-dependent regulators in the LuxR family. *Microbiology, 145*(Pt 6), 1287–1288.

Geistlich, M., Losick, R., Turner, J. R., & Rao, R. N. (1992). Characterization of a novel regulatory gene governing the expression of a polyketide synthase gene in Streptomyces ambofaciens. *Molecular Microbiology, 6*(14), 2019–2029.

Gomez-Escribano, J. P., & Bibb, M. J. (2011). Engineering Streptomyces coelicolor for heterologous expression of secondary metabolite gene clusters. *Microbial Biotechnology, 4*(2), 207–215.

Gottelt, M., Kol, S., Gomez-Escribano, J. P., Bibb, M., & Takano, E. (2010). Deletion of a regulatory gene within the cpk gene cluster reveals novel antibacterial activity in Streptomyces coelicolor A3(2). *Microbiology, 156*(Pt 8), 2343–2353.

Gramajo, H. C., Takano, E., & Bibb, M. J. (1993). Stationary-phase production of the antibiotic actinorhodin in Streptomyces coelicolor A3(2) is transcriptionally regulated. *Molecular Microbiology, 7*(6), 837–845.

Gust, B. (2009). Chapter 7. Cloning and analysis of natural product pathways. *Methods in Enzymology, 458*, 159–180.

Gust, B., Challis, G. L., Fowler, K., Kieser, T., & Chater, K. F. (2003). PCR-targeted Streptomyces gene replacement identifies a protein domain needed for biosynthesis of the sesquiterpene soil odor geosmin. *Proceedings of the National Academy of Sciences of the United States of America, 100*(4), 1541–1546.

Hopwood, D. A. (2007). Therapeutic treasures from the deep. *Nature Chemical Biology, 3*(8), 457–458.

Horinouchi, S. (2003). AfsR as an integrator of signals that are sensed by multiple serine/threonine kinases in Streptomyces coelicolor A3(2). *Journal of Industrial Microbiology and Biotechnology, 30*(8), 462–467.

Horinouchi, S., Kito, M., Nishiyama, M., Furuya, K., Hong, S. K., Miyake, K., et al. (1990). Primary structure of AfsR, a global regulatory protein for secondary metabolite formation in Streptomyces coelicolor A3(2). *Gene, 95*(1), 49–56.

Ikeda, H., Ishikawa, J., Hanamoto, A., Shinose, M., Kikuchi, H., Shiba, T., et al. (2003). Complete genome sequence and comparative analysis of the industrial microorganism Streptomyces avermitilis. *Nature Biotechnology, 21*(5), 526–531.

Juguet, M., Lautru, S., Francou, F. X., Nezbedova, S., Leblond, P., Gondry, M., et al. (2009). An iterative nonribosomal peptide synthetase assembles the pyrrole-amide antibiotic congocidine in Streptomyces ambofaciens. *Chemistry and Biology, 16*(4), 421–431.

Karray, F., Darbon, E., Nguyen, H. C., Gagnat, J., & Pernodet, J. L. (2010). Regulation of the biosynthesis of the macrolide antibiotic spiramycin in Streptomyces ambofaciens. *Journal of Bacteriology, 192*(21), 5813–5821.

Karray, F., Darbon, E., Oestreicher, N., Dominguez, H., Tuphile, K., Gagnat, J., et al. (2007). Organization of the biosynthetic gene cluster for the macrolide antibiotic spiramycin in Streptomyces ambofaciens. *Microbiology, 153*(Pt 12), 4111–4122.

Kieser, T., Bibb, M. J., Buttner, M. J., Chater, K. F., & Hopwood, D. A. (2000). Practical Streptomyces genetics. Norfolk: John Innes.

Kim, W., Lee, J. J., Paik, S. G., & Hong, Y. S. (2010). Identification of three positive regulators in the geldanamycin PKS gene cluster of Streptomyces hygroscopicus JCM4427. *Journal of Microbiology and Biotechnology, 20*(11), 1484–1490.

Kitani, S., Miyamoto, K. T., Takamatsu, S., Herawati, E., Iguchi, H., Nishitomi, K., et al. (2011). Avenolide, a Streptomyces hormone controlling antibiotic production in Streptomyces avermitilis. *Proceedings of the National Academy of Sciences of the United States of America, 108*(39), 16410–16415.

Kuscer, E., Coates, N., Challis, I., Gregory, M., Wilkinson, B., Sheridan, R., et al. (2007). Roles of rapH and rapG in positive regulation of rapamycin biosynthesis in Streptomyces hygroscopicus. *Journal of Bacteriology, 189*(13), 4756–4763.

Laureti, L., Song, L., Huang, S., Corre, C., Leblond, P., Challis, G. L., et al. (2011). Identification of a bioactive 51-membered macrolide complex by activation of a silent polyketide synthase in Streptomyces ambofaciens. *Proceedings of the National Academy of Sciences of the United States of America, 108*(15), 6258–6263.

Li, R., Xie, Z., Tian, Y., Yang, H., Chen, W., You, D., et al. (2009). polR, a pathway-specific transcriptional regulatory gene, positively controls polyoxin biosynthesis in Streptomyces cacaoi subsp. asoensis. *Microbiology, 155*(Pt 6), 1819–1831.

Liu, G., Tian, Y., Yang, H., & Tan, H. (2005). A pathway-specific transcriptional regulatory gene for nikkomycin biosynthesis in Streptomyces ansochromogenes that also influences colony development. *Molecular Microbiology, 55*(6), 1855–1866.

MacNeil, D. J., Gewain, K. M., Ruby, C. L., Dezeny, G., Gibbons, P. H., & MacNeil, T. (1992). Analysis of Streptomyces avermitilis genes required for avermectin biosynthesis utilizing a novel integration vector. *Gene, 111*(1), 61–68.

McAlpine, J. B., Bachmann, B. O., Piraee, M., Tremblay, S., Alarco, A. M., Zazopoulos, E., et al. (2005). Microbial genomics as a guide to drug discovery and structural elucidation: ECO-02301, a novel antifungal agent, as an example. *Journal of Natural Products, 68*(4), 493–496.

Natsume, R., Takeshita, R., Sugiyama, M., Ohnishi, Y., Senda, T., & Horinouchi, S. (2003). Crystallization of CprB, an autoregulator-receptor protein from Streptomyces coelicolor A3(2). *Acta Crystallographica Section D: Biological Crystallography, 59*(Pt 12), 2313–2315.

Newman, D. J., & Cragg, G. M. (2007). Natural products as sources of new drugs over the last 25 years. *Journal of Natural Products, 70*(3), 461–477.

O'Rourke, S., Wietzorrek, A., Fowler, K., Corre, C., Challis, G. L., & Chater, K. F. (2009). Extracellular signalling, translational control, two repressors and an activator all contribute to the regulation of methylenomycin production in Streptomyces coelicolor. *Molecular Microbiology, 71*(3), 763–778.

Ohnishi, Y., Ishikawa, J., Hara, H., Suzuki, H., Ikenoya, M., Ikeda, H., et al. (2008). Genome sequence of the streptomycin-producing microorganism Streptomyces griseus IFO 13350. *Journal of Bacteriology, 190*(11), 4050–4060.

Onaka, H., Mori, Y., Igarashi, Y., & Furumai, T. (2011). Mycolic acid-containing bacteria induce natural-product biosynthesis in Streptomyces species. *Applied and Environmental Microbiology, 77*(2), 400–406.

Paget, M. S. B., Chamberlin, L., Atrih, A., Foster, S. J., & Buttner, M. J. (1999). Evidence that the extracytoplasmic function sigma factor sigma(E) is required for normal cell wall structure in Streptomyces coelicolor A3(2). *Journal of Bacteriology, 181*(1), 204–211.

Pang, X., Aigle, B., Girardet, J. M., Mangenot, S., Pernodet, J. L., Decaris, B., et al. (2004). Functional angucycline-like antibiotic gene cluster in the terminal inverted repeats of the Streptomyces ambofaciens linear chromosome. *Antimicrobial Agents and Chemotherapy, 48* (2), 575–588.

Pernodet, J. L., Alegre, M. T., Blondelet-Rouault, M. H., & Guerineau, M. (1993). Resistance to spiramycin in Streptomyces ambofaciens, the producer organism, involves at least two different mechanisms. *Journal of General Microbiology, 139*(5), 1003–1011.

Ramos, J. L., Martinez-Bueno, M., Molina-Henares, A. J., Teran, W., Watanabe, K., Zhang, X., et al. (2005). The TetR family of transcriptional repressors. *Microbiology and Molecular Biology Reviews, 69*(2), 326–356.

Rateb, M. E., Houssen, W. E., Harrison, W. T., Deng, H., Okoro, C. K., Asenjo, J. A., et al. (2011). Diverse metabolic profiles of a Streptomyces strain isolated from a hyper-arid environment. *Journal of Natural Products, 74*(9), 1965–1971.

Reid, R., Piagentini, M., Rodriguez, E., Ashley, G., Viswanathan, N., Carney, J., et al. (2003). A model of structure and catalysis for ketoreductase domains in modular polyketide synthases. *Biochemistry, 42*(1), 72–79.

Richet, E., & Raibaud, O. (1989). MalT, the regulatory protein of the Escherichia coli maltose system, is an ATP-dependent transcriptional activator. *EMBO Journal*, *8*(3), 981–987.

Rigali, S., Nothaft, H., Noens, E. E., Schlicht, M., Colson, S., Muller, M., et al. (2006). The sugar phosphotransferase system of Streptomyces coelicolor is regulated by the GntR-family regulator DasR and links N-acetylglucosamine metabolism to the control of development. *Molecular Microbiology*, *61*(5), 1237–1251.

Rigali, S., Titgemeyer, F., Barends, S., Mulder, S., Thomae, A. W., Hopwood, D. A., et al. (2008). Feast or famine: The global regulator DasR links nutrient stress to antibiotic production by Streptomyces. *EMBO Reports*, *9*(7), 670–675.

Schroeckh, V., Scherlach, K., Nutzmann, H. W., Shelest, E., Schmidt-Heck, W., Schuemann, J., et al. (2009). Intimate bacterial-fungal interaction triggers biosynthesis of archetypal polyketides in Aspergillus nidulans. *Proceedings of the National Academy of Sciences of the United States of America*, *106*(34), 14558–14563.

Seipke, R. F., Kaltenpoth, M., & Hutchings, M. I. (2012). Streptomyces as symbionts: An emerging and widespread theme? *FEMS Microbiology Reviews*, *36*(4), 862–876.

Stutzman-Engwall, K. J., Otten, S. L., & Hutchinson, C. R. (1992). Regulation of secondary metabolism in Streptomyces spp. and overproduction of daunorubicin in Streptomyces peucetius. *Journal of Bacteriology*, *174*(1), 144–154.

Takano, E. (2006). Gamma-butyrolactones: Streptomyces signalling molecules regulating antibiotic production and differentiation. *Current Opinion in Microbiology*, *9*(3), 287–294.

Umeyama, T., Lee, P. C., Ueda, K., & Horinouchi, S. (1999). An AfsK/AfsR system involved in the response of aerial mycelium formation to glucose in Streptomyces griseus. *Microbiology*, *145*(Pt 9), 2281–2292.

van Wezel, G. P., & McDowall, K. J. (2011). The regulation of the secondary metabolism of Streptomyces: New links and experimental advances. *Natural Product Reports*, *28*(7), 1311–1333.

Wang, L., Tian, X., Wang, J., Yang, H., Fan, K., Xu, G., et al. (2009). Autoregulation of antibiotic biosynthesis by binding of the end product to an atypical response regulator. *Proceedings of the National Academy of Sciences of the United States of America*, *106*(21), 8617–8622.

Wietzorrek, A., & Bibb, M. (1997). A novel family of proteins that regulates antibiotic production in streptomycetes appears to contain an OmpR-like DNA-binding fold. *Molecular Microbiology*, *25*(6), 1181–1184.

Wilkinson, C. J., Hughes-Thomas, Z. A., Martin, C. J., Bohm, I., Mironenko, T., Deacon, M., et al. (2002). Increasing the efficiency of heterologous promoters in actinomycetes. *Journal of Molecular Microbiology and Biotechnology*, *4*(4), 417–426.

Willems, A. R., Tahlan, K., Taguchi, T., Zhang, K., Lee, Z. Z., Ichinose, K., et al. (2008). Crystal structures of the Streptomyces coelicolor TetR-like protein ActR alone and in complex with actinorhodin or the actinorhodin biosynthetic precursor (S)-DNPA. *Journal of Molecular Biology*, *376*(5), 1377–1387.

Willey, J. M., & Gaskell, A. A. (2011). Morphogenetic signaling molecules of the streptomycetes. *Chemical Reviews*, *111*(1), 174–187.

Wilson, D. J., Xue, Y., Reynolds, K. A., & Sherman, D. H. (2001). Characterization and analysis of the PikD regulatory factor in the pikromycin biosynthetic pathway of Streptomyces venezuelae. *Journal of Bacteriology*, *183*(11), 3468–3475.

Yadav, G., Gokhale, R. S., & Mohanty, D. (2003). SEARCHPKS: A program for detection and analysis of polyketide synthase domains. *Nucleic Acids Research*, *31*(13), 3654–3658.

Zhao, K. X., Huang, Y., Chen, X., Wang, N. X., & Liu, S. J. (2010). PcaO positively regulates pcaHG of the beta-ketoadipate pathway in Corynebacterium glutamicum. *Journal of Bacteriology*, *192*(6), 1565–1572.

# CHAPTER EIGHTEEN

# Use and Discovery of Chemical Elicitors That Stimulate Biosynthetic Gene Clusters in *Streptomyces* Bacteria

## Jane M. Moore*, Elizabeth Bradshaw*, Ryan F. Seipke[†], Matthew I. Hutchings[†], Michael McArthur*,[1]

*Department of Molecular Microbiology, John Innes Centre, Norwich, United Kingdom
[†]School of Biological Sciences, University of East Anglia, Norwich Research Park, Norwich, United Kingdom
[1]Corresponding author: e-mail address: michael.mcarthur@jic.ac.uk

## Contents

## Abstract

Secondary metabolite production from *Streptomyces* bacteria is primarily controlled at the level of transcription. Under normal laboratory conditions, the majority of the biosynthetic pathways of *Streptomyces coelicolor* are transcriptionally silent. These are often referred to as "cryptic" pathways and it is thought that they may encode the biosynthesis of yet unseen natural products with novel structures that may be valuable leads for therapeutics and as bioactive compounds. Sequencing of microbial genomes has supported the notion that cryptic pathways are widely distributed and likely to be a

*Methods in Enzymology*, Volume 517
ISSN 0076-6879
http://dx.doi.org/10.1016/B978-0-12-404634-4.00018-8

source of new chemical diversity. Hence, techniques that can reverse the silencing will be valuable for natural product screening as well as giving access to interesting new biology.

We have focused on the identification of chemical elicitors capable of inducing expression of secondary metabolic gene clusters and to do so have drawn a parallel with fungal biology where inhibitors of histone acetylation change chromatin structure to derepress biosynthetic pathways. Similarly, we find that the same chemicals can also modify the expression of pathways in *S. coelicolor* and other *Streptomyces* spp. They variously act to increase expression from known pathways as well as inducing cryptic pathways. We hypothesize that nucleoid structure may be playing an analogous role to fungal chromatin structure in controlling transcriptional programs. Further, we speculate that microbial natural product collections could themselves be a rich source of new histone deacetylase inhibitors that have many applications in human health, such as anticancer therapeutics, beyond their traditional use as antimicrobials.

# 1. INTRODUCTION

Expression of secondary metabolic pathways in the model actinobacterium *Streptomyces coelicolor* can be controlled by both pathway-specific transcription factors and pleiotropic regulators, which in turn, are influenced by a bevy of developmental, environmental, and stress–related signaling pathways (Bibb, 2005). In addition to this complexity, the structure of the nucleoid may also have a role in global regulation of these pathways. Inside the bacterium, the genomes are compacted by association with RNAs, specialized nucleoid-associated proteins (NAPs), and differential supercoiling to form the nucleoid (Browning, Grainger, & Busby, 2010). While all genes are compacted, some are more compacted than others and this differential accessibility may be a simple mechanism for effecting transcriptional control by limiting access of the transcriptional machinery, potentially in a similar manner to how chromatin structure regulates transcription in fungi and other eukaryotes. In support of this hypothesis, we previously used *S. coelicolor* to measure DNaseI sensitivity *in vivo* and observed that transcriptionally active genes were located in regions of the chromosome with a more open nucleoid structure (McArthur & Bibb, 2006). This parallels the seminal description of changes in chromatin structure and DNaseI-sensitivity concurrent with transcriptional induction of developmentally regulated human globin genes in tissue culture cell lines (Felsenfeld, 1992).

Though it is unknown to what extent the *S. coelicolor* nucleoid contributes to determining the transcriptional program, there are clear reasons for testing the hypothesis that it does. As discussed in this chapter, genes can be induced by treatment with chemicals known to deacetylate eukaryotic histone proteins with a concomitant change in chromatin structure. Members of the histone deacetylase (HDAC) family are widespread in bacteria (Lombardi et al., 2011); treatment of *S. coelicolor* with HDAC inhibitors causes both an upregulation of certain biosynthetic pathways and an alteration in nucleoid structure as measured by *in vivo* DNaseI sensitivity (M. McArthur, unpublished data). The mechanism for how the biosynthetic pathways are induced is currently being investigated, but results to date suggest that such chemicals may be used to invigorate natural product screening efforts.

# 2. ROLE OF HDAC-LIKE ACETYLTRANSFERASES IN CONTROLLING THE EXPRESSION OF BIOSYNTHETIC PATHWAYS

## 2.1. HDACs

In eukaryotes, HDACs play an important role in controlling gene expression by antagonizing the regulatory acetylation of histone proteins, leading to alterations in chromosome structure. They act on the basic N-terminal tails of the core histone proteins that form the nucleosome, which is the simplest repeating structure of chromatin. The N-terminal tails of histones project out of the nucleosome and are the main targets for a variety of posttranslational modifications, including lysine and arginine methylation, phosphorylation, ubiquitination (Munshi, Shafi, Aliya, & Jyothy, 2009) and, of particular interest, acetylation (Sterner & Berger, 2000). Such modifications, or combinations thereof, can be interpreted by the nuclear machinery to affect transcription of the underlying genes—referred to as the "histone code" (Margueron, Trojer, & Reinberg, 2005)—usually by altering the extent of chromatin folding to make the regulatory sequences associated with the pathways more or less accessible. Histone acetylation is typically associated with transcriptional activation and deacetylation with inactivation (Luger & Richmond, 1998), but how this is achieved is unclear. There are several possibilities being considered: lysine acetylation will weaken the electrostatic interaction between the histone and DNA, leading to a loosening of the local chromatin structure and an opening of the chromatin; acetylation will affect protein–protein interactions, including those between histones, other nucleosomes and transcription factors; large chromatin remodeling complexes can recognize sites of modification and reorganize long stretches of chromatin structure (Eberharter, Ferreira, & Becker, 2005).

HDACs are members of a large family of zinc-containing enzymes distributed throughout all kingdoms (Leipe & Landsman, 1997). The mammalian enzymes are the best studied owing to their central role in aging and cancer (Rodriguez & Fraga, 2010). Eighteen HDACs have been identified in mammals and can be categorized into two distinct groups, depending on whether or not they share similarity with the yeast HDAC, Sir2. In yeast, this enzyme keeps regions of chromatin deacetylated and as a result they become tightly folded and transcriptionally inert (Fritze, Verschueren, Strich, & Easton-Esposito, 1997; Smith, Brachmann, & Boeke, 1998). Seven of the mammalian enzymes (SIRT1–7) have been found to contain a core Sir2 catalytic domain and are known as the Class III HDACs, or more commonly the sirtuins. The family can be further divided into those that catalyze the NAD-dependent deacetylation (SIRT1–3 and SIRT5) or mediate the ADP-ribosylation (SIRT4 and SIRT6) of various protein substrates (Tanner, Landry, Sternglaz, & Denu, 2000). The sirtuins have been implicated in numerous diseases and in the process of aging (Haigis & Guarente, 2006).

The remaining 11 are referred to as the classical HDACs (Marmorstein & Trievel, 2008) and can be further grouped into 3 different classes (Gregoretti, Lee, & Goodson, 2004). Of these, the Class I enzymes (HDAC1, 2, 3, and 8) share similarity with the yeast transcriptional regulator RDP3 and are primarily localized in the nucleus. They are expressed ubiquitously and have important froles in regulating cellular proliferation and survival (Minucci & Pelicci, 2006). Class II HDACs, which resemble yeast HDA1, shuttle between the cytoplasm and nucleus and they have more restricted tissue-specific expression patterns and regulatory functions. This class can be further subdivided into IIa (HDAC4, 5, 7, and 9) and IIb (HDAC6 and 10, both predominantly located in the cytoplasm) (Marks, 2010). Of these, HDAC6 is a major therapeutic target for new cancer treatments given its role in cellular proliferation, though in this case the enzyme does not directly affect patterns of gene expression, as it is primarily cytoplasmic and targets a range of nonhistone proteins such as $\alpha$-tubulin (Haggart, Koeller, Wong, Grozinger, & Schreiber, 2003) and peroxiredoxins (Parmigiani et al., 2008). The final class (IV) currently has a single member, HDAC11, which is evolutionarily conserved but the function of which is unknown (Yang & Seto, 2008).

Where do the *S. coelicolor* HDACs fit into these schemes? There are three candidates, two of which are sirtuin like (SCO0452 and SCO6464) and SCO3330 being most similar to Class I human HDAC enzymes (Fig. 18.1). SCO0452 is most similar to human SIRT4, the enzymatic

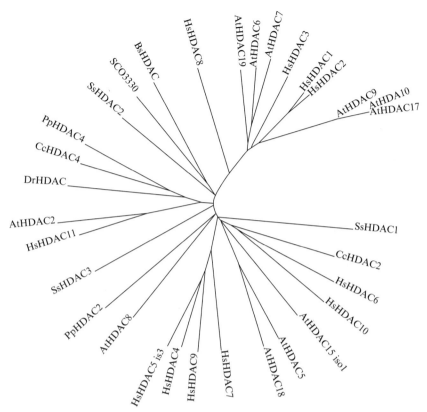

**Figure 18.1** Neighbor-joining tree of eukaryotic and prokaryotic HDAC sequences. Classes are assigned as per Gregoretti et al. (2004). Full-length sequences of the following accession numbers were used: *Arabidopsis thaliana* AtHDAC2 (gi21105771), AtHDAC5 (At9759454a), AtHDAC6 (gi15242626), AtHDAC7 (gi10176806), AtHDAC8 (gi18390898), AtHDAC9 (gi15230483), AtHDAC10 (gi175264567), AtHDAC15 (gi18401915), AtHDAC17 (gi175264341), AtHDAC18 (At9759454b), AtHDAC19 (gi2318131); *Bacillus subtilis* BsHDAC (NP_390849.1); *Caulobacter crescentus* CcHDAC2 (NP_420875.1), CcHDAC4 (NP_422442.1); *Deinococcus radiodurans* DrHDAC (NP_294557.1); *Homo sapiens* HDAC1 (gi13128860), HDAC2 (gi21411359), HDAC3 (gi13128862), HDAC4 (gi5174481), HDAC5 (gi4885531), HDAC6 (gi13128864), HDAC7 (gi13259524), HDAC8 (gi8923769), HDAC9 (gi15590680), HDAC10 (gi15213867), HDAC11 (gi13376228); *Pseudomonas putrescens* PpHDAC2 (NP_747441.1), PpHDAC4 (gi26991445); *Streptomyces coelicolor* SCO3330 (NP_627540.1); and *Sulfolobus solfataricus* SsHDAC1 (gi15896999), SsHDAC2 (gi15896979), SsHDAC3 (gi15897978).

mechanism of which is both NAD dependent and can ribosylate proteins in an ADP–dependent fashion, and SCO6464 shares significant homology with SIRT5 and its activity is NAD dependent. In comparison to the human enzymes, SCO3330 is most similar to HDAC8 (Fig. 18.1), which in humans plays a crucial role in the transcriptional regulation of developmentally important genetic pathways. It is advantageous that the homologs of the three enzymes are structurally distinct, as perhaps the *S. coelicolor* enzymes will have distinct roles and will be inhibited or stimulated by different molecules, allowing their functions to be deduced by chemical inquisition.

There has been considerable success in using recognized HDAC inhibitors to modify transcriptional regulation of fungal secondary metabolic pathways (Strauss & Reyes-Dominguez, 2010), in an exciting new area of research led by the Keller laboratory (Palmer & Keller, 2010) who identified the role of the LaeA HDAC in the expression of secondary metabolic pathways in *Aspergillus* species (Bok & Keller, 2004). On the basis of this work, we investigated whether HDAC inhibitors would have a similar role in *S. coelicolor*.

## 2.2. HDAC inhibitors

There are numerous chemical classes of HDAC inhibitors active on the classical enzymes and these include hydroxylamine acids, short-chain fatty acids, and electrophilic ketones, examples of which are given in Table 18.1. Understanding the mechanisms of their action is an important area of research as it informs improvements to the molecules that alter their pharmacokinetic properties, activity and, importantly, their specificity (Bolden, Peart, & Johnstone, 2006). In preclinical trials, HDAC inhibitors can show good efficacy in preventing tumor growth but also have poor toxicity profiles, presumably due to the lack of specificity causing inhibition of multiple HDACs controlling diverse biological functions. Many of the inhibitors of Class I and II enzymes disrupt the binding of the zinc ion. The potent inhibitor SAHA, for example, has a hydroxylamine group that binds to $Zn^{2+}$, linked by a straight alkyl chain to a hydrophobic group that interacts with the amino acids at the rim of the catalytic site to control the specificity of the inhibitor (Vannini et al., 2004). The amino acids at the rim of the catalytic site vary considerably among the different HDACs and this may provide an approach to engineer new variants with controllable specificity.

In addition to the zinc-binding site, the sirtuins also contain a cleft where the ribose and nicotinamide moieties of $NAD^+$ bind together with the targeted lysine residue (Sanders, Jackson, & Marmorstein, 2010). Given the role of these enzymes in aging, there is considerable interest in understanding how the enzymes work and small-molecule activators and inhibitors are being used to dissect the mechanism (Dittenhafer-Reed, Feldman, & Denu, 2010). Of particular interest has been the plant natural product

**Table 18.1** Properties of HDAC inhibitors used in this study

| Chemical elicitor | Class | PREDicted mode of ACTion | Source | Stock solution | Cost (£/g) |
|---|---|---|---|---|---|
| Sodium butyrate | Short-chain fatty acid | Inhibits Class I and II HDACs and general stress | Chemical synthesis | 100 mM in water | 0.17 |
| Valproic acid | Short-chain fatty acid | Inhibits Class I and II HDACs | Chemical synthesis | 10 mM in DMSO | 40.6 |
| Trichostatin A (TSA) | Hydroxyamate | Inhibits Class I and II HDACs | *Streptomyces* product | 1 mM in ethanol | 98,400 |
| SAHA | Hydroxyamate | Inhibits Class I and II HDACs | Chemical synthesis | 10 mM in DMSO | 7260 |
| Apicidin | Depsipeptides | Inhibits Class I HDACs | Fungal product | 1 mg/mL in 50% DMSO | 60,700 |
| FK-228 | Depsipeptides | Inhibits Class I HDACs | *ChromobACTerium violaceum* product | 1 mg/mL in 50% DMSO | ~10,000 |
| Trapoxin A | Depsipeptides | Inhibits Class I HDACs | *Helicoma ambiens* product | 1 mg/mL in 50% DMSO | 102,500 |
| MS-275 | Benzamides | Inhibits Class I HDACs (HDAC1 > 3) | Chemical synthesis | Not tested | 520 |

*Continued*

**Table 18.1** Properties of HDAC inhibitors used in this study—cont'd

| Chemical elicitor | Class | PREDicted mode of ACTion | Source | Stock solution | Cost (£/g) |
|---|---|---|---|---|---|
| Phenyl butyrate (PB) | Benzamides | Inhibits Class I and II HDACs | Chemical synthesis | 100 mM in PBS | 45 |
| Depudecin | Linear polyketide | Inhibits Class I and II HDACs | Fungal natural product | 1 mg/mL in 50% DMSO | 285,000 |
| Nicotinamide | Vitamin | Inhibit sirtuins (III) | Natural product | 50 mM in water | 0.95 |
| Sirtinol | – | Inhibit sirtuins (III) | Chemical synthesis | Not tested | 39,000 |
| Splitomicin | – | Inhibit sirtuins (III) | Fungal natural product | 100 mM in DMSO | 8300 |
| Quercitin | Flavonoid | ACTivates sirtuins (III) | Plant-derived flavonoid | 100 mM in 5% DMSO | 2.43 |
| Resveratrol | Phytoalexin | ACTivates sirtuins (III) | Plant phytochemical | 10 mM in 50% DMSO | 691 |
| Piceatannol | Phytoalexin | ACTivates sirtuins (III) | Plant metabolite | Not tested | 14,600 |

resveratrol that is an activator of sirtuins and seems to promote healthy aging and extend life cycles in experimental animals (Agarwal & Baur, 2011). Companies seeking to improve the properties of resveratrol have been synthesizing analogs in an attempt to improve its drug-like qualities and develop it as a treatment for type II diabetes (Milne et al., 2007).

It is of interest that both types of HDAC inhibitor have been found in chemical libraries as well as natural products from plant and bacterial sources. In the latter case, they are usually identified as antifungal agents (fungal genomes are condensed into chromatin that is partially controlled by HDACs and other modifying enzymes that are perhaps the target for these natural products). The possibility that actinobacteria have a comparably complex nucleoid structure that is affected by posttranslational modification of constituent proteins opens the possibility that bacterial natural products may prove to be a hitherto unexpectedly rich source of lead compounds for cancer therapies that target histone-modifying enzymes. Hence, another approach to developing HDAC inhibitors with improved profiles is to concentrate on finding leads from natural product libraries and such a strategy is discussed in Section 4.

## 2.3. Potential substrates of HDAC

The ubiquitous DNA-binding proteins include HU and H-NS (both from *Escherichia coli* and with homologs and analogs in most microbial genomes). HU is described as histone-like owing to its general similarity to the eukaryotic histones in as much as it is a small basic protein sufficient to bind to bacterial genomes every 200 bp or so, it exists as a dimer (paralleling the formation of the histone octamer), and it can bind to DNA in a cooperative fashion. Some controversy exists as to whether the HU is associated with the periphery of the bacterial nucleoid, coinciding with actively transcribed sequences associated with ribosomal factories in the cytoplasm, or within the nucleoid itself, with a more general distribution of binding sites throughout the bacterial genome. Large-scale proteomic studies have shown lysine acetylation to be a widespread phenomenon in bacteria, particularly on proteins involved in translation or carbohydrate metabolism. A higher level of acetylation is seen during stationary phase but the modifications are removed within two hours after transfer to fresh medium (Yu, Kim, Moon, Ryu, & Pan, 2008), suggesting that these modifications are regulatory.

Histone acetyltransferases (HATs), those that acetylate the lysines and HDACs, may influence gene expression through a number of substrates besides proteins. HATs are often able to acetylate a broad range of substrates; for example, the Gnat-family HAT AAC(6′)-ly from *Salmonella enterica* is able to acetylate histones and itself as well as a broad array of aminoglycoside antibiotics (Vetting, Magnet, Nieves, Roderick, & Blanchard, 2004). Bacterial acetylpolyamine amidohydrolases (part of the classical Class II HDAC superfamily) regulate the level of acetylation of polyamines such as spermidine or putrescine, altering their ability to bind to DNA and thereby affect gene expression (Leipe & Landsman, 1997).

Less is known about modifications on DNA-binding proteins, perhaps because these are less often detected in proteomics studies owing to their lower abundance. In *E. coli*, a HAT/sirtuin pair reversibly acetylates four transcription factors, including the helix–turn–helix domain of the global regulator RcsB, which may alter its DNA-binding ability (Thao, Chen, Zhu, & Escalante-Semerena, 2010). There is also reason to suspect that at least two of the highly abundant NAPs are regulated by modification in *Streptomyces*: Parker et al. (2010) detected phosphorylation of Lsr2 (an H-NS equivalent), and HupA (an HU homolog) appeared in two places on a 2D gel, suggesting modification (Andrew Hesketh and Mervyn Bibb, personal communication).

## 3. USE OF HDAC INHIBITORS TO STIMULATE EXPRESSION OF SECONDARY METABOLIC GENES

Initially to test whether HDAC inhibitors could affect production of secondary metabolites, we focused on the levels of production from two well-characterized gene clusters from *S. coelicolor*: those for the blue-pigmented polyketide actinorhodin (ACT) and the red nonribosomal peptide complex, the prodiginines, of which undecylprodigiosin is the major component (RED). The stage and level at which each of these is produced has been well characterized and found to be conditionally dependent on the medium or the type of agar as well as the stage of growth; most secondary metabolites are produced during stationary phase. All agars and media referred to in this chapter are described in the manual *Practical Streptomyces Genetics* (Kieser, Bibb, Buttner, Chater, & Hopwood, 2000).

### 3.1. Induction of actinorhodin production by treatment with sodium butyrate on minimal agar

Sodium butyrate is a well-characterized inhibitor of Class I and Class II HDACs (Table 18.1) used in many experiments studying eukaryotic histone modification. It is a water-soluble molecule with an unpleasant odor. Stock solutions were prepared freshly before use in the fume-hood at a concentration of 11.1 mg/mL in water (100 m$M$). Sodium butyrate was added to either an R5 medium or a minimal medium (MM; containing 1.5% Iberian agar) at concentrations up to 50 m$M$. This was to establish the minimum inhibitory concentration (MIC) of the molecule as sodium butyrate, along with some of the other HDAC inhibitors, can cause a general stress response at higher concentrations and we wished to avoid using the chemicals under these conditions. Instead, the concentration at which the molecules were used was the lowest at which a strong effect was seen on the production of the two-pigmented antibiotics. These two types of agar were chosen because of their different effects on the yield and timing of pigment production: R5 is permissive for production of the two-pigmented antibiotics, while MM supports lower yields at later time points. Hence, the system provides a convenient screen for both activators and repressors of production of well-characterized secondary metabolites. Addition of 25 m$M$ sodium butyrate-stimulated ACT production from *S. coelicolor* A3(2) strain M145 grown on MM agar and repressed production on R5 (Fig. 18.2). It is curious that sodium butyrate conditionally activates or represses production of ACT, but it is worth noting that a similar pattern of induction was observed for *N*-acetylglucosamine (Rigali et al., 2008) and this may provide important clues for our ongoing studies into the molecular mechanism(s) of how this HDAC inhibitor alters the regulation of secondary metabolism. van Wezel and colleagues have described how the DasR transcription factor acts as a nutrient sensing system, sensitive to the levels of *N*-acetylglucosamine (Rigali et al., 2008), to activate antibiotic production and development when the molecule is limiting (which is indicative of unfavorable growth

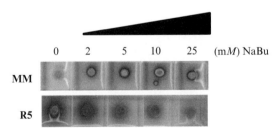

**Figure 18.2** Sodium butyrate has inhibitory and stimulatory effects on antibiotic production by *S. coelicolor* depending on which agar medium is used. *S. coelicolor* A3(2) was grown on MM agar plates (top panel) or R5 agar plates (below), supplemented with various concentrations of sodium butyrate as indicated. At high concentrations of sodium butyrate production of actinorhodin was stimulated on the usually non-permissive MM agar, while being repressed on R5 agar.

conditions or "famine") while blocking these processes under more favorable conditions. Perhaps levels of lysine acetylation somehow play a similar role to DasR in changing transcriptional programs to adapt to different environmental conditions.

## 3.2. Screening the panel of HDAC inhibitors against *S. coelicolor*

To determine the effect of each of the major classes of HDAC inhibitors and sirtuin activators on pigment production, the panel of chemicals shown in Table 18.2 was tested. The activities of these chemicals on antibiotic production of *S. coelicolor* were tested at sub-MIC concentrations on the two selected agars, the permissive R5 and the MM that supports much lower levels of pigment production. In general, the results were consistent: inhibitors of Class I and II HDACs inhibited ACT production on the rich medium while leading to overproduction on minimal plates. The short-chain fatty acids sodium butyrate and valproic acid resulted in the largest increase in production. Arguably, some of this activity could be due to a general stress response induced by these molecules but the effect was also seen with more specific inhibitors, such as SAHA. Because of the relative

**Table 18.2** Effects of various HDAC inhibitors on antibiotic production by *S. coelicolor* A3(2) strain M145

| Elicitor | Effective concentration | R5 agar | Effective concentration | MM agar |
|---|---|---|---|---|
| None | – | ACT and RED production after 3–5 days | – | Minimal production of either antibiotic after 5 days |
| Sodium butyrate | 25 m$M$ | No ACT production, RED production reduced | 25 m$M$ | Strong induction of ACT, slight induction of RED |
| Valproic acid | 0.5 m$M$ | No ACT production, RED production reduced | 1 m$M$ | Strong induction of ACT |
| SAHA | 25 μ$M$ | Downregulation of ACT and RED production | 25 μ$M$ | Slight induction of ACT |

**Table 18.2** Effects of various HDAC inhibitors on antibiotic production by *S. coelicolor* A3(2) strain M145—cont'd

| Elicitor | Effective concentration | R5 agar | Effective concentration | MM agar |
|---|---|---|---|---|
| TSA | 1.5 μ*M* | Inhibition of ACT production | 15 μ*M* | No change |
| Apicidin | 1 m*M* | Inhibition of ACT production | 1 m*M* | Strong induction of ACT |
| Phenyl butyrate | 0.1 m*M* | Inhibition of ACT production | 1 m*M* | No change |
| Nicotinamide | | No effect | | No effect |
| Splitomicin | 2.5 m*M* | No ACT production, RED production Reduced | 2.5 m*M* | No effect |
| Quercetin | 10 m*M* | Little ACT production, RED production reduced | 10 m*M* | Slight induction of ACT |
| Resveratol | 1 m*M* | ACT production inhibited | 1 m*M* | No effect |

cheapness of sodium butyrate this molecule was used in experiments in liquid media to determine whether or not it had an effect on expression from cryptic pathways of *S. coelicolor*. Using SMM medium (Kieser et al., 2000) supplemented with 25 m*M* sodium butyrate, five of the recognized cryptic pathways in *S. coelicolor* (Bentley et al., 2002; Song et al., 2006) were seen to be induced, as measured by quantitative PCR. These genes were: *SCO0381* (deoxysugar synthetase), *SCO5222* (sesquiterpene cyclase), *SCO6759* (hopanoids), *SCO7221* (germicidin), and *SCO7682* (coelibactin). Hence, treatment of *S. coelicolor* with Class I/II HDAC inhibitors can be used to activate cryptic pathways.

Chemicals active on the sirtuins produced different results to the Class I/II inhibitors. Nicotinamide was not active but this may have been due to issues with membrane permeabilization as has been observed for eukaryotes. With the activators quercetin, and to a lesser extent resveratrol, a strong inhibitory effect on both ACT and RED production was evident on R5 agar at concentrations of 25 m*M*, but there was little effect seen on MM agar.

However, surprisingly, a similar phenotype was seen for the proposed inhibitor splitomicin: inhibition of pigment production on R5 at high concentrations and only slight effects on MM. Hence, it is unclear how the sirtuins are involved in controlling secondary metabolic pathways in *S. coelicolor*, as it would be expected that activators and inhibitors would have opposite effects, unless these chemicals act in a different manner on the *Streptomyces* enzymes than they do on eukaryotic homologs. To investigate whether splitomicin and quercetin act on sirtuins, we overexpressed the *S. coelicolor* sirtuin-like proteins (Section 4).

## 3.3. Use of HDAC inhibitors in non-model actinomycetes: A proof of concept

Orthologues of HDACs are encoded by nearly every sequenced streptomycetes genome in the database and by those of other actinomycete genera known to produce antibiotics, such as *Pseudonocardia*, *Saccharopolyspora*, and *Amycolatopsis*. Thus, there is large potential to use HDAC inhibitors to increase antibiotic titers or activate gene clusters normally silenced under laboratory conditions in bacteria other than *S. coelicolor*. As a proof of concept, we performed agar-based bioassays against the human pathogen *Candida albicans* using two actinomycete strains isolated form fungus-growing ants, *Streptomyces* KY5 (R.F. Seipke and M.I. Hutchings, unpublished data) and *Pseudonocardia* P1 (Barke et al., 2010). Mannitol–soya flour (MS) agar (20 g soya flour, 20 g mannitol, 20 g agar, 1 L tap water) was prepared according to Kieser et al. (2000). MS agar was supplemented with fresh filter-sterilized sodium butyrate (Sigma-Aldrich) to a final concentration of 150 m$M$ immediately before pouring plates. Bioassays were conducted as described by Seipke et al. in Chapter 11. Briefly, the center of an MS agar plate was inoculated with 10 μL of unquantified spore stock of either P1 or KY5 and incubated at 30 °C for 10 days to allow a good-sized colony to grow. P1 and KY5 were challenged with *C. albicans* as follows: 5 mL of soft nutrient agar (SNA, 8 g nutrient broth powder, 5 g agar, 1 L deionized water) was inoculated with 200 μL of *C. albicans* culture grown overnight at 37 °C in Lennox broth (10 g Bacto-tryptone, 5 g yeast extract, 5 g NaCl, 1 L deionized water) (Kieser et al., 2000). *C. albicans*-containing SNA (5 mL) was pipetted onto the MS agar plate containing the actinomycete culture and gently swirled until the SNA was evenly distributed. (*Note*: the hydrophobic nature of actinomycete aerial hyphae prevents submersion of the colony in SNA.) After inoculation with *C. albicans*, plates were incubated at room temperature and inspected daily for ~3 days before

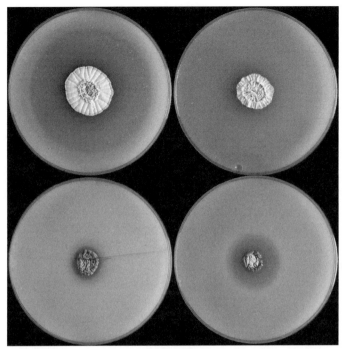

**Figure 18.3** Antifungal bioactivity of *Streptomyces* KY5 is induced by treatment with sodium butyrate. Bioassays with uninduced (plates on left) and induced *Streptomyces* KY5 (plates on right) record higher activity against *Candida albicans* (top plates) and *Pseudonocardia* P1 (bottom plates) in the presence of 150 m*M* sodium butyrate. (For color version of this figure, the reader is referred to the online version of this chapter.)

photographs were taken. Both KY5 and P1 reproducibly inhibited the growth of *C. albicans* more severely (as evident by the larger zones of inhibition) when cultured in the presence of 150 m*M* sodium butyrate compared to the 0 m*M* sodium butyrate (Fig. 18.3). This result provides exciting evidence for the use of HDAC inhibitors as tools for natural product discovery in non-model *Streptomyces* spp. and rare actinomycete genera.

## 4. SCREENING FOR HDAC INHIBITORS FROM *STREPTOMYCES* BACTERIA

### 4.1. Construction of screen

To test whether the three predicted HDACs were involved in response to the chemical elicitors, we prepared overexpression strains where each gene was cloned into a multicopy vector (pIJ86) (a gift from Prof. Mervyn Bibb) containing the strong constitutive promoter, *ermE*\*. The phenotypes of these three strains were compared to that of the parental strain, M145, harboring the empty vector. Production of ACT and RED was assessed during surface growth on MM or R5 supplemented with various HDAC inhibitors. This experimental design enabled us to both identify which HDAC protein (s) was involved in regulating antibiotic production and to determine which HDAC inhibitor interacted with a specific HDAC.

An example of the readout from such screens is given for the pIJ86-SCO6464 strain. The resultant high levels of overexpression of the sirtuin-like molecule SCO6464 creates a nonexpression phenotype for ACT and RED on R5 agar. As such this mimics the quercetin phenotype that also ablated pigment production. Addition of splitomicin to the pIJ86-SCO6464 strain restored levels of ACT and RED production to those seen for both the wild-type strain (M145) and that transformed with the empty vector (M145::pIJ86, Fig. 18.4). Hence, splitomicin seems to be acting as an inhibitor of SCO6464, consistent with its role as an inhibitor of sirtuins. Higher concentrations of splitomicin were seen to have a deleterious effect on growth as well as reducing the amount of pigment production. These results suggest that SCO6464 has a similar function to fungal sirtuins in controlling biosynthetic pathways and similar relationships were found for the other HDAC-like enzymes and the inhibitors specific for their class. Hence, we propose that such overexpression strains can be used to screen for potentially new bacterial natural products with activities, both positive and negative, against HDACs. These could have potential roles as therapeutic leads for new HDAC inhibitors for cancer therapies and the like.

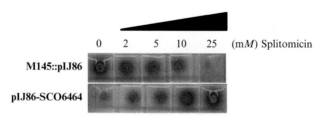

**Figure 18.4** Splitomicin inhibits the action of the sirtuin-like HDAC SCO6464 in *S. coelicolor*. Pigment production on R5 agar was compared between the parental *S. coelicolor* strain-containing empty vector (M145::pIJ86) and a strain overexpressing SCO6464 (pIJ86-SCO6464) in the presence of various concentrations of the sirtuin inhibitor splitomicin. Splitomicin was seen to reverse the overexpression phenotype at high concentrations.

# 5. CONCLUSION

Though the function of the HDACs is yet to be established in *S. coelicolor*, it is clear that they have roles in the transcriptional regulation of biosynthetic pathways. This could occur through modification of nucleoid structure with concomitant alteration to the expression of secondary metabolic gene clusters among others. If nucleoid proteins were the target of these enzymes, it would provide an interesting parallel with how such developmentally regulated genes are controlled in fungi (Bok & Keller, 2004), and this possibility is being actively researched. However, there are numerous other proposed targets for these HDAC-like enzymes, including polyamines, acetoin, and enzymes controlling primary metabolism (Leipe & Landsman, 1997). Irrespective of the mechanism by which HDAC inhibitors work, it is clear that these inexpensive chemicals can be used empirically to variously cause overexpression of some biosynthetic pathways and induction of a proportion of the cryptic pathways. That the enzymes are widely distributed in streptomycetes and other actinobacteria suggests that the use of such chemicals may induce a portion of the novel chemical diversity still untapped in these bacteria.

## REFERENCES

Agarwal, B., & Baur, J. A. (2011). Resveratol and life extension. *Annals of the New York Academy of Sciences*, *1215*, 138–143.

Barke, J., Seipke, R. F., Grüschow, S., Heavens, D., Drou, N., Bibb, M. J., et al. (2010). A mixed community of actinomycetes produce multiple antibiotics for the fungus farming ant *Acromyrmex octospinosus*. *BMC Biology*, *8*, 109.

Bentley, S. D., Chater, K. F., Cerdeño-Tárraga, A. M., Challis, G. L., Thomson, N. R., James, K. D., et al. (2002). Complete genome sequence of the model actinomycete Streptomyces coelicolor A3(2). *Nature*, *417*, 141–147.

Bibb, M. J. (2005). Regulation of secondary metabolism in streptomycetes. *Current Opinion in Microbiology*, *8*, 208–215.

Bok, J. W., & Keller, N. P. (2004). LaeA, a regulator of secondary metabolism in Aspergillus ssp. *Eukaryotic Cell*, *3*, 527–535.

Bolden, J. E., Peart, M. J., & Johnstone, R. W. (2006). Anticancer activities of histone deacetylase inhibitors. *Nature Reviews. Drug Discovery*, *5*, 769–784.

Browning, D. F., Grainger, D. C., & Busby, S. J. (2010). Effects of nucleoid-associated proteins on bacterial chromosome structure and gene expression. *Current Opinion in Microbiology*, *8*, 208–215.

Dittenhafer-Reed, K. E., Feldman, J. L., & Denu, J. M. (2010). Catalysis and mechanistic insights into Sirtuin activation. *Chembiochem*, *12*, 281–289.

Eberharter, A., Ferreira, R., & Becker, P. (2005). Dynamic chromatin: Concerted nucleosome remodelling and acetylation. *Biological Chemistry*, *386*, 745–751.

Felsenfeld, G. (1992). Chromatin as an essential part of the transcriptional mechanism. *Nature*, *355*, 219–224.

Fritze, C. E., Verschueren, K., Strich, R., & Easton-Esposito, R. (1997). Direct evidence for SIR2 modulation of chromatin structure in yeast rDNA. *The EMBO Journal, 16,* 6495–6509.

Gregoretti, I. V., Lee, Y. M., & Goodson, H. V. (2004). Molecular evolution of the histone deacetylase family: Functional implications of phylogenetic analysis. *Journal of Molecular Biology, 338,* 17–31.

Haggart, S. J., Koeller, K. M., Wong, J. C., Grozinger, C. M., & Schreiber, S. L. (2003). Domain-selective small-molecule inhibitor of histone deacetylase 6 (HDAC6)-mediated tubulin deacetylation. *Proceedings of the National Academy of Sciences of the United States of America, 100,* 4389–4394.

Haigis, M. C., & Guarente, L. P. (2006). Mammalian sirtuins—Emerging roles in physiology, aging and calorie restriction. *Genes and Development, 20,* 2913–2921.

Kieser, T., Bibb, M. J., Buttner, M. J., Chater, K. F., & Hopwood, D. A. (2000). *Practical* Streptomyces *genetics.* Norwich, UK: John Innes Foundation.

Leipe, D. D., & Landsman, D. (1997). Histone deacetylases, acetoin utilization proteins and acetylpolyamine amidohydrolases are members of an ancient protein superfamily. *Nucleic Acids Research, 25,* 3693–3697.

Lombardi, P. M., Angell, H. D., Whittington, D. A., Flynn, E. F., Rajashankar, K. R., & Christianson, D. W. (2011). Structure of prokaryotic polyamine deacetylase reveals evolutionary functional relationships with eukaryotic histone deacetylases. *Biochemistry, 50,* 1808–1817.

Luger, K., & Richmond, T. J. (1998). The histone tails of the nucleosome. *Current Opinion in Genetics and Development, 8,* 140–146.

Margueron, R., Trojer, P., & Reinberg, D. (2005). The key to development: Interpreting the histone code? *Current Opinion in Genetics and Development, 15,* 163–176.

Marks, P. A. (2010). Histone deacetylase inhibitors: A chemical genetics approach to understanding cellular functions. *Biochimica et Biophysica Acta, 1799,* 717–725.

Marmorstein, R., & Trievel, R. C. (2008). Histone modifying enzymes: Structures, mechanisms, and specificities. *Biochimica et Biophysica Acta, 1789,* 58–68.

McArthur, M., & Bibb, M. J. (2006). *In vivo* DNase I sensitivity of the *Streptomyces coelicolor* chromosome correlates with gene expression: Implications for bacterial chromosome structure. *Nucleic Acids Research, 34,* 5395–5401.

Milne, J. C., Lambert, P. D., Schenk, S., Carney, D. P., Smith, J. J., Gagne, D. J., et al. (2007). Small molecule activators of SIRT1 as therapeutics for the treatment of type 2 diabetes. *Nature, 450,* 712–716.

Minucci, S., & Pelicci, P. G. (2006). Histone deacetylase inhibitors and the promise of epigenetic (and more) treatments for cancer. *Nature Reviews. Cancer, 6,* 38–51.

Munshi, A., Shafi, G., Aliya, N., & Jyothy, A. (2009). Histone modifications dictate specific biological readouts. *Journal of Genetics and Genomics, 36,* 75–88.

Palmer, J. M., & Keller, N. P. (2010). Secondary metabolism in fungi: Does chromosomal location matter? *Current Opinion in Microbiology, 13,* 431–436.

Parker, J. L., Jones, A. M., Serazetdinova, L., Saalbach, G., Bibb, M. J., & Naldrett, M. J. (2010). Analysis of the phosphoproteome of the multicellular bacterium Streptomyces coelicolor A3(2) by protein/peptide fractionation, phosphopeptide enrichment and high-accuracy mass spectrometry. *Proteomics, 10,* 2486–2497.

Parmigiani, R. B., Xu, W. S., Venta-Perez, G., Erdjument-Bromage, H., Yaneva, M., Tempst, P., et al. (2008). HDAC6 is a specific deacetylase of peroxiredoxins and is involved in redox regulation. *Proceedings of the National Academy of Sciences of the United States of America, 105,* 9633–9638.

Rigali, S., Titgemeyer, F., Barends, S., Mulder, S., Thomae, A. W., Hopwood, D. A., et al. (2008). Feast or famine: The global regulator DasR links nutrient stress to antibiotic production by Streptomyces. *EMBO Reports, 9,* 670–675.

Rodriguez, R. M., & Fraga, M. F. (2010). Aging and cancer: Are sirtuins the link? *Future Oncology, 6*, 905–915.

Sanders, D. D., Jackson, B., & Marmorstein, R. (2010). Structural basis for sirtuin function: What we know and what we don't. *Biochimica et Biophysica Acta, 1804*, 1604–1616.

Smith, J. S., Brachmann, C. B., & Boeke, J. D. (1998). *Genetics, 149*, 1205–1219.

Song, L., Barona-Gomez, F., Corre, C., Xiang, L., Udwary, D. W., Austin, M. B., et al. (2006). Type III polyketide synthase beta-ketoacyl-ACP starter unit and methylmalonyl-CoA extender unit selectivity discovered by *Streptomyces coelicolor* genome mining. *Journal of the American Chemical Society, 128*, 14754–14755.

Sterner, D. E., & Berger, S. L. (2000). Acetylation of histones and transcription-related factors. *Microbiology and Molecular Biology Reviews, 64*, 435–459.

Strauss, J., & Reyes-Dominguez, Y. (2010). Regulation of secondary metabolism by chromatin. *Fungal Genetics and Biology, 48*, 62–69.

Tanner, K. G., Landry, J., Sternglaz, R., & Denu, J. M. (2000). Silent information of NAD-dependent histone/protein deacetylases generates a unique product, 1-O-acetyl-ADP-ribose. *Proceedings of the National Academy of Sciences of the United States of America, 97*, 14178–14182.

Thao, S., Chen, C. S., Zhu, H., & Escalante-Semerena, J. C. (2010). Nepsilon-lysine acetylation of a bacterial transcription factor inhibits its DNA-binding activity. *PLoS One, 5*, e15123.

Vannini, A., Volpari, C., Filocamo, G., Casavola, E. C., Brunetti, M., Renzoni, D., et al. (2004). Crystal structure of a eukaryotic zinc-dependent histone deacetylase, human HDAC8, complexed with a hydroxamic acid inhibitor. *Proceedings of the National Academy of Sciences of the United States of America, 101*, 15064–15069.

Vetting, M. W., Magnet, S., Nieves, E., Roderick, S. L., & Blanchard, J. S. (2004). A bacterial acetyltransferase capable of regioselective N-acetylation of antibiotics and histones. *Chemistry and Biology, 11*, 565–573.

Yang, X. J., & Seto, E. (2008). The Rpd3/Hda1 family of lysine deacetylases: From bacteria and yeast to mice and men. *Nature Reviews. Molecular Cell Biology, 9*, 206–218.

Yu, B. J., Kim, J. A., Moon, J. H., Ryu, S. E., & Pan, J. G. (2008). The diversity of lysine-acetylated proteins in *Escherichia coli*. *Journal of Microbiology and Biotechnology, 18*, 1529–1536.

CHAPTER NINETEEN

# Persister Eradication: Lessons from the World of Natural Products

**Iris Keren, Lawrence R. Mulcahy, Kim Lewis[1]**

Antimicrobial Discovery Center and Department of Biology, Northeastern University, Boston, Massachusetts, USA
[1]Corresponding author: e-mail address: k.lewis@neu.edu

## Contents

## Abstract

Persisters are specialized survivor cells that protect bacterial populations from killing by antibiotics. Persisters are dormant phenotypic variants of regular cells rather than mutants. Bactericidal antibiotics kill by corrupting their targets into producing toxic products; tolerance to antibiotics follows when targets are inactive. Transcriptome analysis of isolated persisters points to toxin/antitoxin modules as a principle component of persister formation. Mechanisms of persister formation are redundant, making it difficult to eradicate these cells. In *Escherichia coli*, toxins RelE and MazF cause dormancy by degrading mRNA; HipA inhibits translation by phosphorylating Ef-Tu; and TisB forms an anion channel in the membrane, leading to a decrease in pmf and ATP levels. Prolonged treatment of chronic infections with antibiotics selects for *hip* mutants that produce more persister cells. Eradication of tolerant persisters is a serious challenge. Some of the existing antibiotics are capable of killing persisters, pointing to ways of developing therapeutics to treat chronic infections. Mitomycin is a prodrug which is converted into a reactive compound forming adducts with DNA upon entering the cell. Prolonged treatment with aminoglycosides that cause mistranslation leading to misfolded peptides can sterilize a stationary culture of *Pseudomonas aeruginosa*, a pathogen responsible for chronic, highly tolerant infections of cystic fibrosis patients. Finally, one of the best bactericidal agents is rifampin, an inhibitor of RNA polymerase, and we suggest that it "kills" by preventing persister resuscitation.

# 1. PERSISTERS AND INFECTIOUS DISEASE

Persisters represent a small subpopulation of cells that spontaneously go into a dormant, nondividing state. When a population is treated with a bactericidal antibiotic, regular cells die, while persisters survive (Fig. 19.1). Taking samples and plating them for colony counts over time from a culture treated with antibiotic produces a biphasic pattern, with a distinct plateau of surviving persisters. In order to kill, antibiotics require active targets, which explain tolerance of dormant persisters. By contrast, resistance mechanisms prevent antibiotics from binding to their targets (Fig. 19.2).

Infectious disease is often untreatable, even when caused by a pathogen that is not resistant to antibiotics. This is the essential paradox of chronic

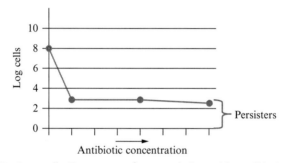

**Figure 19.1** *Persister cells.* Treatment of a population with antibiotics, resulting in biphasic killing. (For color version of this figure, the reader is referred to the online version of this chapter.)

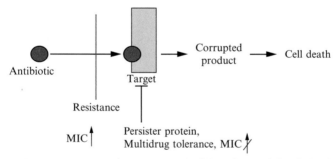

**Figure 19.2** *Resistance versus tolerance.* Bactericidal antibiotics kill cells by forcing the active target to produce corrupted products. Persister proteins act by blocking or inactivating the target, so no corrupted product can be produced. By contrast, all resistance mechanisms prevent the antibiotic from binding to the target. (For color version of this figure, the reader is referred to the online version of this chapter.)

infections. In most cases, chronic infections are accompanied by the formation of biofilms, which seem to point to the source of the problem (Costerton, Stewart, & Greenberg, 1999; Del Pozo & Patel, 2007). Biofilms have been linked to dental disease, endocarditis, cystitis, urinary tract infections, deep-seated infections, indwelling device and catheter infections, and the incurable disease of cystic fibrosis (CF). In the case of indwelling devices such as prostheses and heart valves, reoperation is the method of choice for treating the infection. Biofilms do not generally restrict penetration of antibiotics (Walters, Roe, Bugnicourt, Franklin, & Stewart, 2003) but do form a barrier for the larger components of the immune system (Jesaitis et al., 2003; Leid, Shirtliff, Costerton, & Stoodley, 2002; Vuong et al., 2004). The presence of biofilm-specific resistance mechanisms was suggested to account for the recalcitrance of infectious diseases (Stewart & Costerton, 2001). However, the bulk of cells in the biofilm is actually highly susceptible to killing by antibiotics; only a small fraction of persisters remains alive (Spoering & Lewis, 2001). Based on these findings, we proposed a simple model of a relapsing chronic infection—antibiotics kill the majority of cells, and the immune system eliminates both regular cells and persisters from the bloodstream (Lewis, 2001). The only remaining live cells are then persisters in the biofilm. Once the level of antibiotic drops, persisters repopulate the biofilm, and the infection relapses. While this is a plausible model, it is not the only one. A simpler possibility is that antibiotics fail to effectively reach at least some cells *in vivo*, resulting in a relapsing infection.

Establishing potential causality between persisters and therapy failure is not trivial, as these cells form a small subpopulation with a temporary phenotype, which precludes introducing them into an animal model of infection. We reasoned that causality could be tested based on what we know about selection for high persister (*hip*) mutants *in vitro*. Periodic application of high doses of bactericidal antibiotics leads to the selection of strains that produce increased levels of persisters (Moyed & Bertrand, 1983; Wolfson, Hooper, McHugh, Bozza, & Swartz, 1990). This is precisely what happens in the course of treating chronic infections—the patient is periodically exposed to high doses of antibiotics, which may select for *hip* mutants. But *hip* mutants would only gain advantage if the drugs effectively reach and kill the regular cells of the pathogen.

Patients with CF largely succumb to an incurable and chronic infection of the lungs with *Pseudomonas aeruginosa* (Gibson, Burns, & Ramsey, 2003). The periodic application of high doses of antibiotics provides some relief by decreasing the pathogen burden but does not clear the infection. If *hip* strains of pathogens were selected *in vivo*, they would most likely be present in a CF patient. We took advantage of a set of longitudinal *P. aeruginosa* isolates from a single patient, collected over the course of many years (Smith et al., 2006). Testing persister levels by monitoring survival after challenge with a high dose of ofloxacin showed a dramatic, 100-fold increase in surviving persister cells in the last four isolates (Mulcahy, Burns, Lory, & Lewis, 2010). Testing paired strains from additional patients showed that, in most cases, there was a considerable increase in persister levels in the late isolate from a patient. Interestingly, most of the *hip* isolates had no increase in minimum inhibitory concentration (MIC) compared to their clonal parent strain to ofloxacin, carbenicillin, and tobramycin, suggesting that classical acquired resistance plays little to no role in the recalcitrance of CF infection. These experiments directly link persisters to the clinical manifestation of the disease and suggest that persisters are responsible for the therapy failure of chronic CF infection. But why have the *hip* mutants with their striking survival phenotype evaded detection for such a long time? The main focus of research in antimicrobials has been on drug resistance, and the basic starting experiment is to test a clinical isolate for its ability to grow in the presence of elevated levels of different antibiotics, and record any increases in the MIC. This is also the standard test employed by clinical microbiology laboratories. *hip* mutants are of course missed by this test, which explains why they had remained undetected, in spite of a major effort aimed at understanding pathogen survival of antimicrobial chemotherapy. Given that *hip* mutants are the likely main culprit responsible for morbidity and mortality of the CF infection, it makes sense to test for their presence.

Is selection for *hip* mutants a general feature of chronic infections? We recently examined patients with chronic oral thrush caused by *Candida albicans* (LaFleur, Qi, & Lewis, 2010). These were cancer patients undergoing chemotherapy, and suppression of the immune system caused the fungal infection. In patients where the disease did not resolve, the *C. albicans* isolates were almost invariably *hip* mutants, as compared to patients where the disease cleared within 3 weeks of treatment with chlorhexidine. The eukaryotic *C. albicans* forms persisters (Al-Dhaheri & Douglas, 2008; Harrison, Turner, & Ceri, 2007; LaFleur, Kumamoto, & Lewis, 2006) through mechanisms that are probably analogous, rather than homologous, to that of their bacterial counterparts. Given the similar life styles of the unrelated *P. aeruginosa* and *C. albicans*, we may expect that the survival advantage of a *hip* mutation is universal. Just as multidrug resistance has become the prevalent danger in acute infections, multidrug tolerance of persisters and *hip* mutants may be the main, but largely overlooked, culprit of chronic infectious disease.

Biofilms apparently serve as a protective habitat for persisters (Harrison, Ceri, et al., 2005; Harrison, Turner, & Ceri, 2005; Harrison et al., 2009; LaFleur et al., 2006; Spoering & Lewis, 2001), allowing them to evade the immune system. However, a more general paradigm is that persisters will be critical for pathogens to survive antimicrobial chemotherapy whenever the immune response is limited. Such cases would include disseminating infections in immunocompromised patients undergoing cancer chemotherapy or infected with HIV. Persisters are also likely to play an important role in immunocompetent individuals in cases where the pathogen is located at sites poorly accessible by components of the immune system. These include the central nervous system, where pathogens cause debilitating meningitis and brain abscesses (Honda & Warren, 2009), and the gastrointestinal tract, where a hard-to-eradicate *Helicobacter pylori* causes gastroduodenal ulcers and gastric carcinoma (Peterson et al., 2000). Tuberculosis infections are chronic infection owing to the ability of the pathogen to evade the immune system. An acute infection may resolve as a result of antimicrobial therapy, but the pathogen often remains in a "latent" asymptomatic form (Barry et al., 2009). It is estimated that one in every three people carry latent *Mycobacterium tuberculosis*, and 10% of carriers develop an acute infection at some stage in their lives. Virtually nothing is known about this latent form that serves as the main reservoir of tuberculosis. One simple possibility is that persisters are equivalent to the latent form of the pathogen. The above analysis underscores the significance of drug tolerance as a barrier to effective antimicrobial chemotherapy. Given its significance—roughly half of all cases of infection—the number of studies dedicated to tolerance is miniscule as compared to publications on resistance. The formidable barriers to study persister cells account for the lack of parity between these two comparably significant fields. Hopefully, a better balance will be achieved, and the following discussion summarizes recent advances in understanding the mechanism of tolerance.

## 2. MECHANISMS OF PERSISTER FORMATION

Persisters were initially discovered in 1944 (Bigger, 1944), but the mechanism of their formation eluded us for a very long time. Only recently have the molecular mechanisms of dormancy begun to emerge. The most straightforward approach to finding an underlying mechanism of a complex function is by screening a library of transposon (Tn) insertion mutants. This produces a set of candidate genes, and subsequent analysis leads to a pathway and a mechanism. This is indeed how the basic mechanisms of sporulation, flagellation, chemotaxis, virulence, and many other functions have been established. However, screening a Tn insertion library of *Escherichia coli* for an ability to tolerate high doses of antibiotics produced no mutants completely lacking persisters (Hu & Coates, 2005; Spoering, Vulic, & Lewis, 2006). With the development of the complete, ordered *E. coli* gene knockout library by the Mori group (Baba et al., 2006; the Keio collection), it seemed reasonable to revisit the screening approach. Indeed, there always remains a possibility that Tns missed a critical gene or the library was not large enough. The use of the Keio collection largely resolves this uncertainty.

This advanced screen also failed to produce a single mutant that failed to form persisters, suggesting a high degree of redundancy (Hansen, Lewis, & Vulic, 2008). However, a number of genes did affect the ability of cells to form persisters, with some knockouts resulting in a 10-fold decrease in persister formation. The majority of hits were in global regulators, DksA, DnaKJ, HupAB, and IhfAB. This is an independent indication of redundancy—a global regulator can affect expression of several persister genes simultaneously, resulting in a phenotype. The screen also produced two interesting candidate genes that may be more directly involved in persister formation—YgfA an inhibitor of nucleotide synthesis, and YigB, which may block metabolism by depleting the pool of FMN. A similar screen of a *P. aeruginosa* mutant library was reported and confirmed that the genetic mechanisms of persister formation are redundant (De Groote et al., 2009).

The main conclusion from the screens is that persister formation does not follow the usual design theme of complex cellular functions—a single linear regulatory pathway controlling an execution mechanism. By contrast, persisters are apparently formed through a number of independent parallel mechanisms. There is a considerable adaptive advantage in this redundant design—no single compound will disable persister formation.

Screens for persister genes were useful in finding some possible candidate genes and pointing to redundancy of function. It seemed that a method better suited to uncover redundant elements would be transcriptome analysis. For this, persisters had to be isolated.

Persisters form a small and temporary population, making isolation challenging. The simplest approach is to lyse a population of growing cells with a β-lactam antibiotic and collect surviving persisters. This approach allowed us to perform the first persister transcriptome (Keren, Shah, Spoering, Kaldalu, & Lewis, 2004). A more advanced method aimed at isolating native persisters was developed, based on a hypothesis that persisters are dormant and will thus have diminished protein synthesis (Shah et al., 2006). The strains were engineered to express unstable green fluorescent protein (GFP) under the control of a ribosomal promoter. Growing cells remained GFP positive, while cells that entered dormancy rapidly dimmed owing to degradation of GFP. The difference in fluorescence allowed for the sorting of the two subpopulations. The dim cells were tolerant to ofloxacin, confirming that they are persisters.

Transcriptomes obtained by both methods pointed to downregulation of biosynthesis genes and indicated increased expression of several toxin/antitoxin (TA) modules (RelBE, MazEF, DinJYafQ, YgiU). TA modules are found on plasmids where they constitute a maintenance mechanism (Gerdes, Rasmussen, & Molin, 1986; Hayes, 2003). Typically, the toxin is a protein that inhibits an important cellular function such as translation or replication and forms an inactive complex with the antitoxin. The toxin is stable, while the antitoxin is degradable. If a daughter cell does not receive a plasmid after segregation, the antitoxin level decreases owing to proteolysis, leaving a toxin that either kills the cell or inhibits propagation. TA modules are also commonly found on bacterial chromosomes, but their role had been unknown. In *E. coli*, MazF and an unrelated toxin, RelE, induce stasis by cleaving mRNA, which of course inhibits translation, a condition that can be reversed by expression of corresponding antitoxins (Christensen & Gerdes, 2003; Pedersen et al., 2003). The ability of toxins to induce reversible stasis makes them excellent candidates for persister genes.

Ectopic expression of RelE (Keren et al., 2004) or MazF (Vázquez-Laslop, Lee, & Neyfakh, 2006) strongly increased tolerance to antibiotics. The first gene linked to persisters, *hipA* (Moyed & Bertrand, 1983), is also a toxin, and its ectopic expression causes multidrug tolerance as well (Correia et al., 2006; Falla & Chopra, 1998; Korch & Hill, 2006). Interestingly, a bioinformatics analysis indicates that HipA is a member of the Tor family of kinases, which have been extensively studied in eukaryotes (Schmelzle & Hall, 2000), but have not been previously identified in bacteria. HipA is indeed a kinase: it autophosphorylates on serine 150, and site-directed mutagenesis replacing it, or other conserved amino acids in the catalytic, and $Mg^{2+}$-binding sites abolishes its ability to stop cell growth and confer drug tolerance (Correia et al., 2006). The crystal structure of HipA in complex with its antitoxin HipB was recently resolved, and a pull-down experiment showed that the substrate of HipA is elongation factor EF-Tu (Schumacher et al., 2009). Phosphorylated EF-Tu is inactive, which leads to a block in translation and dormancy.

Deletion of potential candidates of persister genes noted above does not produce a discernible phenotype affecting persister production, possibly due to the high degree of redundancy of these elements. In *E. coli*, there are at least 15 toxin–antitoxin (TA) modules (Alix & Blanc-Potard, 2009; Pandey & Gerdes, 2005; Pedersen & Gerdes, 1999), and more than 80 in *M. tuberculosis* (Ramage, Connolly, & Cox, 2009). Recently, a strain with 10 TA mutants knocked out was constructed and it forms far fewer persisters (Maisonneuve, Shakespeare, Jorgensen, & Gerdes, 2011).

It is possible that a particular TA module is expressed under given conditions and then becomes responsible for producing the majority of persisters. Following this logic, we considered TA modules that are controlled by known transcriptional regulators.

Several TA modules contain the Lex box and are induced by the SOS response. These are *symER*, *hokE*, *yafN/yafO*, and *tisAB/istr1* (Courcelle, Khodursky, Peter, Brown, & Hanawalt, 2001; Fernandez De Henestrosa et al., 2000; Kawano, Aravind, & Storz, 2007; McKenzie, Magner, Lee, & Rosenberg, 2003; Motiejunaite, Armalyte, Markuckas, & Suziedeliene, 2007; Pedersen & Gerdes, 1999; Singletary et al., 2009; Vogel, Argaman, Wagner, & Altuvia, 2004). Fluoroquinolones induce the SOS response (Phillips, Culebras, Moreno, & Baquero, 1987), and we found that ciprofloxacin can induce persister formation (Dorr, Lewis, & Vulic, 2009). Examination of TA deletion strains showed that the level of persisters dropped dramatically, 10- to 100-fold, in a Δ*tisAB* mutant (Dorr, Vulic, & Lewis, 2010). This suggests that TisB is responsible for the formation of the majority of persisters under conditions of SOS induction. The level of persisters was unaffected in strains deleted in the other Lex box-containing TA modules. Persister levels observed in time-dependent killing experiments with ampicillin or streptomycin that do not cause DNA damage were unchanged in the Δ*tisAB* strain. TisB only had a phenotype in the presence of a functional RecA protein, confirming the dependence on the SOS pathway. Ectopic overexpression of *tisB* sharply increased the level of persisters. Drop in persisters in a deletion strain and increase upon overexpression gives reasonable confidence in the functionality of a candidate persister gene. The dependence of TisB-induced persisters on a particular regulatory pathway, the SOS response, further strengthens the case for TisB as a specialized persister protein (Fig. 19.3). Incidentally, a *tisB* mutant is not present in the otherwise fairly complete Keio knockout library, and the small ORF might have been easily missed by Tn mutagenesis as well, evading detection by the generalized screens for persister genes.

The role of TisB in persister formation is unexpected based on what we know about this type of proteins. TisB is a small, 29-amino acid hydrophobic peptide that binds to the membrane and leads to a drop in ATP levels

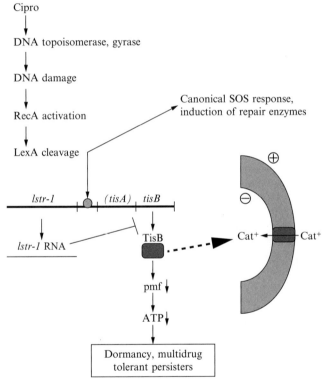

**Figure 19.3** *A model of TisB-dependent persister formation in E. coli.* A fluoroquinolone antibiotic causes DNA damage by converting the DNA gyrase and topoisomerase into endonucleases. This activates the RecA protein which in turns activates the LexA repressor, causing it to cleave. The canonical SOS response is induced, and repair enzymes that contain *lex* boxes in their promoter regions are transcribed. The Lex repressor also controls the expression of the TisB toxin, a small cationic membrane-acting agent. Decrease in the pmf and ATP shut down target functions, including DNA topoisomerase and gyrase, and a dormant persister is formed. (For color version of this figure, the reader is referred to the online version of this chapter.)

(Unoson & Wagner, 2008). Bacteria, plants, and animals all produce antimicrobial membrane-acting peptides (Garcia-Olmedo, Molina, Alamillo, & Rodriguez-Palenzuela, 1998; Sahl & Bierbaum, 1998; Zasloff, 2002). Toxins of many TA loci found on plasmids belong to this type as well. If a daughter cell does not inherit a plasmid, the concentration of a labile antitoxin decreases, and the toxin such as the membrane-acting *hok* kills the cell (Gerdes et al., 1986). High-level artificial overexpression of TisB also causes cell death by disrupting proton motive force (pmf) (Unoson & Wagner, 2008). We recently found that TisB forms an anion channel in artificial membranes, confirming its function as a typical toxic antimicrobial peptide (Gurnev, Ortenberg, Dörr, Lewis, & Bezrukov, 2012). It is remarkable from this perspective that the membrane-acting TisB under conditions of natural (mild) expression has the exact opposite effect of protecting the cell from antibiotics.

Disruption of the pmf leading to a decrease in ATP is indeed a perfect tool for rapidly creating a dormant state—antibiotic targets require energy, and a full systems shutdown will prevent them from killing persisters. The killing mechanism for the main classes of antibiotics is fairly well understood—β-lactams induce autolysins (Groicher, Firek, Fujimoto, & Bayles, 2000); aminoglycosides cause mistranslation, which produces toxic misfolded peptides (Davis, Chen, & Tai, 1986); and fluoroquinolones convert DNA gyrase/topoisomerase into an endonuclease that degrades DNA (Hooper, 2001). It was recently reported that all bactericidal antibiotics actually kill by inducing formation of reactive oxygen species (ROS) (Kohanski, Dwyer, Hayete, Lawrence, & Collins, 2007). However, it is unlikely that ROS play a significant role in killing, as we did not find a difference between the minimal bactericidal concentrations of the main classes of bactericidal antibiotics under aerobic versus anaerobic conditions (Iris Keren, unpublished data).

Fluoroquinolones such as ciprofloxacin are widely used broad spectrum antibiotics, and their ability to induce multidrug tolerant cells is unexpected and a cause of considerable concern. Induction of persister formation by fluoroquinolones may contribute to the ineffectiveness of antibiotics in eradicating infections. Indeed, preexposure with a low dose of ciprofloxacin drastically increased tolerance to subsequent exposure with a high dose, and TisB persisters are multidrug tolerant.

The finding of the role of TisB in tolerance opens an intriguing possibility of a wider link between other stress responses and persister formation. Pathogens are exposed to many stress factors in the host environment apart from DNA damaging agents—oxidants, high temperature, low pH, membrane-acting agents. It is possible that all stress responses induce the formation of surviving persisters.

While resistance and tolerance are mechanistically distinct, there is sufficient reason to believe that tolerance may be a major cause for developing resistance. Indeed, the probability of resistance development is proportional to the size of the pathogen population, and a lingering chronic infection that cannot be eradicated owing to tolerance will go on to produce resistant mutants and strains acquiring resistant determinants by transmission from other bacteria (Levin & Rozen, 2006). Combating tolerance then becomes a major component in preventing resistance.

## 3. PERSISTER ERADICATION

All antibiotics in current use were approved based on their ability to inhibit growth of rapidly dividing cells. Indeed, U.S. Food and Drug Administration does not ask for, and thus industry does not provide, data for hard-to-treat stationary, biofilm, or persister populations. Achieving the higher bar of finding compounds that kill both regular and dormant cells is problematic, given the current state of affairs in the field of antimicrobial drug discovery. Most classes of antibiotics are natural products discovered in a relatively short, 20-year period, based on screening soil actinomycetes. Developed by Selman Waksman (Waksman & Woodruff, 1940), this "Waksman platform" was also the only successful approach to antibiotic discovery we ever had. Overmining of the actinomycetes leads to the rediscovery of known compounds by the 1960s and an essential collapse of the platform. Efforts to develop a high-tech platform based on high-throughput screens and target-based rational design of synthetics has not been successful (Payne, Gwynn, Holmes, & Pompliano, 2007). Almost invariably, synthetic leads fail to penetrate well into bacterial cells. This problem is especially formidable in the case of gram-negative pathogens, equipped with powerful multidrug resistance pumps. As a result, the last several compounds to be introduced are narrow-spectrum antimicrobials that were actually discovered decades ago, and most are natural products that were initially overlooked. Given this dismal state of affairs, what are the chances of developing compounds that can sterilize an infection? One approach to the problem is to revive the Waksman platform and turn to untapped sources of secondary metabolites, such as silent operons coding for antimicrobials (Bentley et al., 2002) and uncultured bacteria (Lewis, Epstein,

D'Onofrio, & Ling, 2010), which make up the vast majority of all microbial species (Amann et al., 1990), and look for compounds that have an ability to kill nongrowing cells. Existing natural products can tell us what to look for.

There is little doubt that producers of antibiotics are confronted by the persister problem of their competitors, just as we are confronted by persisters of our pathogens. One obvious way by which bacteria can eliminate persisters is by producing generally toxic compounds, such as membrane-acting (antimicrobial peptides, detergents); redox agents; DNA intercalators. But if these compounds may kill both regular and dormant cells, why do producing organisms bother making exquisitely specific and costly, molecules such as vancomycin or rifampicin? One possible explanation is that they do so to avoid suicide. Indeed, a detergent or a DNA intercalator will be just as toxic to the producing cell, and the resistance mechanism is export. But when nutrient supply is low, export will fail, leading to intoxication. Under these conditions, making a target-specific compound that is inactive against the target in the producer seems a better option. Mitomycin is an interesting intermediate case—it forms adducts with DNA, but only in a reduced state, and its export is coupled with synthesis (He, Sheldon, & Sherman, 2001). Once a mitomycin prodrug enters the cells of a competitor, it is reduced into an active drug and covalently binds to the DNA. Interestingly, we did not observe protection from mitomycin in model persister cells overexpressing RelE (Keren et al., 2004). Mitomycin is used to treat cancer, but application as an antimicrobial is not feasible owing to its toxicity. Interestingly, an even more reactive compound than activated mitomycin is a synthetic prodrug, metronidazole. It is reduced by a microbial-specific nitrate reductase into a highly reactive compound, which indiscriminately binds covalently to proteins and DNA. Metronidazole is safe, since the toxic form is produced specifically in bacterial cells, and the highly reactive form does not escape the activating cells it targets. Applications of metronidazole are limited to pathogens living under anaerobic or microaerophilic conditions where nitrate reductase is expressed. It is possible that there are superior natural product prodrugs capable of eliminating persisters.

Another interesting opportunity in eradicating pathogens stems from observations of prolonged treatment of chronic infections with antibiotics. The acute infection may resolve spontaneously or as a result of antimicrobial therapy, but the pathogen often remains in a "latent" form (Barry et al., 2009). It is estimated that one in every three people carry latent *M. tuberculosis*, and 10% of carriers develop an acute infection at some stage in their lives. Virtually nothing is known about this latent form that serves as the main reservoir of tuberculosis. One simple possibility is that persisters are equivalent to the latent form of the pathogen.

The current "short-course" chemotherapy includes a combination of four antibiotics (isoniazid, rifampin, pyrazinamide, and ethambutol) that needs to be taken for 2 months, followed by isoniazid and rifampin for an additional 4 months. An even longer treatment course is required for curing the latent form, up to 9 months (Dye, Scheele, Dolin, Pathania, & Raviglione, 1999). Because a persistent population of pathogens remains until late in treatment, incomplete therapy is common and has resulted in the rise of multidrug-resistant and extensively drug-resistant tuberculosis (Davies, 2001; Mendoza et al., 1997). The best killing agent for *M. tuberculosis* is rifampin. Its initial bactericidal activity is low, but it can kill nonreplicating bacteria over the course of days and weeks, and at higher doses can sterilize the culture (Jayaram et al., 2003; Keren, Minami, Rubin, & Lewis, 2011; Mitchison, 2000). It has been suggested that increasing the rifampin dosage can shorten tuberculosis therapy (Jayaram et al., 2003; Ruslami et al., 2007). Unlike any of the other drugs we tested, we observed sterilization of a stationary culture of *M. tuberculosis* with rifampin after a 2–3 week incubation period (Iris Keren, unpublished data). The mechanism of killing by rifampin is unknown, and indeed puzzling. Rifampin is an inhibitor of RNA polymerase, and as such will eventually inhibit protein synthesis, and should behave similarly to chloramphenicol, a bacteriostatic inhibitor of translation. The superior killing ability of rifampin in *M. tuberculosis* may result from another anomaly in this unusual microorganism, the presence of a very large number of TA modules. *M. tuberculosis* carries around 80 TA modules which are mRNA endonucleases, belonging to the *relBE*, *higBA*, *mazEF,* and *vapBC* families (Ramage et al., 2009). The reasons for this remarkable redundancy are unknown. However, the multitude of these nucleases may act synergistically with rifampin. In the absence of new mRNA synthesis, the toxins will degrade remaining mRNA, including the messages of the TAs themselves. In the absence of protein synthesis, labile antitoxins will be hydrolyzed, and without the coding message cells will probably enter into irreversible dormancy from which they will not be able to resuscitate when rifampin levels drop.

Another case of prolonged treatment of patients with antibiotics is that of chronic *P. aeruginosa* infections of patients with CF. Application of aerosolized antibiotics results in delivery of extremely high concentrations of compound to the site of the infection, 1–2 mg/ml for tobramycin or ofloxacin.

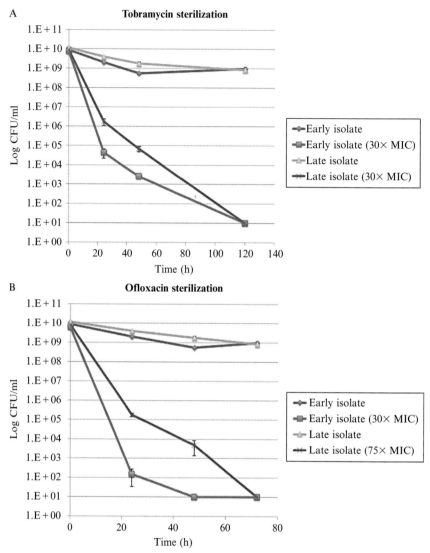

**Figure 19.4** *Sterilization of persisters with conventional antibiotics*. (A) Sterilization with tobramycin. Early isolate MIC=1. Late Isolate MIC=2. Treatment is 30× MIC with tobramycin for 120 h. Untreated long-term cultures were sampled as a control. (B) Sterilization with ofloxacin. For both the early isolate and late isolate, the MIC is 1 µg/ml. Untreated long-term cultures were sampled as a control. (For color version of this figure, the reader is referred to the online version of this chapter.)

At these enormous levels, the concentration of antibiotics is higher than the MIC of even those strains that carry resistance determinants (Geller et al., 2011). However, even this does not result in sterilization.

We were interested to test persister survival during prolonged incubation and applied ofloxacin or tobramycin at fairly high concentrations (25–75 × MIC) to a stationary culture of *P. aeruginosa* (Lawrence Mulcahy, unpublished data). Rather surprisingly, complete sterilization was observed after a 48–72 h incubation (Fig. 19.4). The longer, 72 h incubation was required for sterilization of *hip* strains. This observation opens an intriguing possibility of sterilizing a chronic infection with existing compounds, at least in the case of *P. aeruginosa*. Instead of applying extremely high levels that cannot be sustained, a more frequent introduction of lower doses sufficient for sterilization could maintain the antibiotic for the 48–72 h. Any persisters that tried to resuscitate during this time would be killed.

## REFERENCES

Al-Dhaheri, R. S., & Douglas, L. J. (2008). Absence of amphotericin B-tolerant persister cells in biofilms of some *Candida* species. *Antimicrobial Agents and Chemotherapy*, *52*(5), 1884–1887.

Alix, E., & Blanc-Potard, A. B. (2009). Hydrophobic peptides: Novel regulators within bacterial membrane. *Molecular Microbiology*, *72*(1), 5–11. http://dx.doi.org/10.1111/j.1365-2958.2009.06626.x [Research Support, Non-U.S. Gov't Review].

Amann, R. I., Binder, B. J., Olson, R. J., Chisholm, S. W., Devereux, R., & Stahl, D. A. (1990). Combination of 16S rRNA-targeted oligonucleotide probes with flow cytometry for analyzing mixed microbial populations. *Applied and Environmental Microbiology*, *56*(6), 1919–1925 [Research Support, Non-U.S. Gov't Research Support, U.S. Gov't, Non-P.H.S.].

Baba, T., Ara, T., Hasegawa, M., Takai, Y., Okumura, Y., & Baba, M. (2006). Construction of *Escherichia coli* K-12 in-frame, single-gene knockout mutants: The Keio collection. *Molecular Systems Biology*, *2*, 2006.0008.

Barry, C. E., 3rd, Boshoff, H. I., Dartois, V., Dick, T., Ehrt, S., Flynn, J., & Young, D. (2009). The spectrum of latent tuberculosis: Rethinking the biology and intervention strategies. *Nature Reviews. Microbiology*, *7*(12), 845–855. http://dx.doi.org/10.1038/nrmicro2236 [Research Support, N.I.H., Extramural Research Support, N.I.H., Intramural Research Support, Non-U.S. Gov't Review].

Bentley, S. D., Chater, K. F., Cerdeno-Tarraga, A. M., Challis, G. L., Thomson, N. R., James, K. D., & Hopwood, D. A. (2002). Complete genome sequence of the model actinomycete *Streptomyces coelicolor* A3(2). *Nature*, *417*(6885), 141–147.

Bigger, J. W. (1944). Treatment of Staphylococcal Infections with Penicillin. *The Lancet*, *244* (6320), 497–500.

Christensen, S. K., & Gerdes, K. (2003). RelE toxins from bacteria and Archaea cleave mRNAs on translating ribosomes, which are rescued by tmRNA. *Molecular Microbiology*, *48*(5), 1389–1400.

Correia, F. F., D'Onofrio, A., Rejtar, T., Li, L., Karger, B. L., Makarova, K., & Lewis, K. (2006). Kinase activity of overexpressed *HipA* is required for growth arrest and multidrug tolerance in *Escherichia coli*. *Journal of Bacteriology*, *188*(24), 8360–8367.

Costerton, J. W., Stewart, P. S., & Greenberg, E. P. (1999). Bacterial biofilms: A common cause of persistent infections. *Science, 284*(5418), 1318–1322.

Courcelle, J., Khodursky, A., Peter, B., Brown, P. O., & Hanawalt, P. C. (2001). Comparative gene expression profiles following UV exposure in wild-type and SOS-deficient *Escherichia coli*. *Genetics, 158*(1), 41–64 [Comparative Study Research Support, Non-U.S. Gov't Research Support, U.S. Gov't, P.H.S.].

Davies, P. D. (2001). Drug-resistant tuberculosis. *Journal of the Royal Society of Medicine, 94*(6), 261–263 [Editorial].

Davis, B. D., Chen, L. L., & Tai, P. C. (1986). Misread protein creates membrane channels: An essential step in the bactericidal action of aminoglycosides. *Proceedings of the National Academy of Sciences of the United States of America, 83*(16), 6164–6168.

De Groote, V., Verstraeten, N., Fauvart, M., Kint, C., Verbeeck, A., Beullens, S., & Michiels, J. (2009). Novel persistence genes in *Pseudomonas aeruginosa* identified by high-throughput screening. *FEMS Microbiology Letters, 297*(1), 73–79. http://dx.doi.org/10.1111/j.1574-6968.2009.01657.x.

Del Pozo, J., & Patel, R. (2007). The challenge of treating biofilm-associated bacterial infections. *Clinical Pharmacology and Therapeutics, 82,* 204–209.

Dorr, T., Lewis, K., & Vulic, M. (2009). SOS response induces persistence to fluoroquinolones in *Escherichia coli*. *PLoS Genetics, 5*(12), e1000760. http://dx.doi.org/10.1371/journal.pgen.1000760.

Dorr, T., Vulic, M., & Lewis, K. (2010). Ciprofloxacin causes persister formation by inducing the TisB toxin in *Escherichia coli*. *PLoS Biology, 8*(2), e1000317. http://dx.doi.org/10.1371/journal.pbio.1000317.

Dye, C., Scheele, S., Dolin, P., Pathania, V., & Raviglione, M. C. (1999). Consensus statement. Global burden of tuberculosis: estimated incidence, prevalence, and mortality by country. WHO Global Surveillance and Monitoring Project. *JAMA: The Journal of the American Medical Association, 282*(7), 677–686.

Falla, T. J., & Chopra, I. (1998). Joint tolerance to beta-lactam and fluoroquinolone antibiotics in *Escherichia coli* results from overexpression of *hipA*. *Antimicrobial Agents and Chemotherapy, 42*(12), 3282–3284.

Fernandez De Henestrosa, A. R., Ogi, T., Aoyagi, S., Chafin, D., Hayes, J. J., Ohmori, H., & Woodgate, R. (2000). Identification of additional genes belonging to the LexA regulon in *Escherichia coli*. *Molecular Microbiology, 35*(6), 1560–1572 [Research Support, Non-U.S. Gov't Research Support, U.S. Gov't, P.H.S.].

Garcia-Olmedo, F., Molina, A., Alamillo, J. M., & Rodriguez-Palenzuela, P. (1998). Plant defense peptides. *Biopolymers, 47*(6), 479–491.

Geller, D. E., Flume, P. A., Staab, D., Fischer, R., Loutit, J. S., & Conrad, D. J. (2011). Levofloxacin inhalation solution (MP-376) in patients with cystic fibrosis with *Pseudomonas aeruginosa*. *American Journal of Respiratory and Critical Care Medicine, 183*(11), 1510–1516. http://dx.doi.org/10.1164/rccm.201008-1293OC [Multicenter Study Randomized Controlled Trial Research Support, Non-U.S. Gov't].

Gerdes, K., Bech, F. W., Jorgensen, S. T., Lobner-Olesen, A., Rasmussen, P. B., Atlung, T., & von Meyenburg, K. (1986). Mechanism of postsegregational killing by the hok gene product of the parB system of plasmid R1 and its homology with the relF gene product of the *E. coli* relB operon. *The EMBO Journal, 5*(8), 2023–2029.

Gerdes, K., Rasmussen, P. B., & Molin, S. (1986). Unique type of plasmid maintenance function: Postsegregational killing of plasmid-free cells. *Proceedings of the National Academy of Sciences of the United States of America, 83*(10), 3116–3120.

Gibson, R. L., Burns, J. L., & Ramsey, B. W. (2003). Pathophysiology and management of pulmonary infections in cystic fibrosis. *American Journal of Respiratory and Critical Care Medicine, 168*(8), 918–951.

Groicher, K. H., Firek, B. A., Fujimoto, D. F., & Bayles, K. W. (2000). The *Staphylococcus aureus lrgAB* operon modulates murein hydrolase activity and penicillin tolerance. *Journal of Bacteriology, 182*(7), 1794–1801.

Gurnev, P. A., Ortenberg, R., Dörr, T., Lewis, K., & Bezrukov, S. M. (2012). Persister-promoting bacterial toxin TisB produces anion-selective pores in planar lipid bilayers. *FEBS Letters,* (submitted).

Hansen, S., Lewis, K., & Vulic, M. (2008). Role of global regulators and nucleotide metabolism in antibiotic tolerance in *Escherichia coli. Antimicrobial Agents and Chemotherapy, 52*(8), 2718–2726.

Harrison, J. J., Ceri, H., Roper, N. J., Badry, E. A., Sproule, K. M., & Turner, R. J. (2005). Persister cells mediate tolerance to metal oxyanions in *Escherichia coli. Microbiology, 151* (Pt 10), 3181–3195.

Harrison, J. J., Turner, R. J., & Ceri, H. (2005). Persister cells, the biofilm matrix and tolerance to metal cations in biofilm and planktonic *Pseudomonas aeruginosa. Environmental Microbiology, 7*(7), 981–994.

Harrison, J. J., Turner, R. J., & Ceri, H. (2007). A subpopulation of *Candida albicans* and *Candida tropicalis* biofilm cells are highly tolerant to chelating agents. *FEMS Microbiology Letters, 272*(2), 172–181.

Harrison, J. J., Wade, W. D., Akierman, S., Vacchi-Suzzi, C., Stremick, C. A., Turner, R. J., & Ceri, H. (2009). The chromosomal toxin gene yafQ is a determinant of multidrug tolerance for *Escherichia coli* growing in a biofilm. *Antimicrobial Agents and Chemotherapy, 53*(6), 2253–2258. http://dx.doi.org/10.1128/AAC.00043-09 [Research Support, Non-U.S. Gov't].

Hayes, F. (2003). Toxins-antitoxins: Plasmid maintenance, programmed cell death, and cell cycle arrest. *Science, 301*(5639), 1496–1499.

He, M., Sheldon, P. J., & Sherman, D. H. (2001). Characterization of a quinone reductase activity for the mitomycin C binding protein (MRD): Functional switching from a drug-activating enzyme to a drug-binding protein. *Proceedings of the National Academy of Sciences of the United States of America, 98*(3), 926–931. http://dx.doi.org/10.1073/pnas.031314998 [Research Support, U.S. Gov't, P.H.S.].

Honda, H., & Warren, D. K. (2009). Central nervous system infections: Meningitis and brain abscess. *Infectious Disease Clinics of North America, 23*(3), 609–623. http://dx.doi.org/10.1016/j.idc.2009.04.009 [Review].

Hooper, D. C. (2001). Mechanisms of action of antimicrobials: Focus on fluoroquinolones. *Clinical Infectious Diseases, 15*(32), S9–S15.

Hu, Y., & Coates, A. R. (2005). Transposon mutagenesis identifies genes which control antimicrobial drug tolerance in stationary-phase *Escherichia coli. FEMS Microbiology Letters, 243*(1), 117–124.

Jayaram, R., Gaonkar, S., Kaur, P., Suresh, B. L., Mahesh, B. N., Jayashree, R., & Balasubramanian, V. (2003). Pharmacokinetics-pharmacodynamics of rifampin in an aerosol infection model of tuberculosis. *Antimicrobial Agents and Chemotherapy, 47*(7), 2118–2124.

Jesaitis, A. J., Franklin, M. J., Berglund, D., Sasaki, M., Lord, C. I., Bleazard, J. B., & Lewandowski, Z. (2003). Compromised host defense on *Pseudomonas aeruginosa* biofilms: Characterization of neutrophil and biofilm interactions. *The Journal of Immunology, 171* (8), 4329–4339.

Kawano, M., Aravind, L., & Storz, G. (2007). An antisense RNA controls synthesis of an SOS-induced toxin evolved from an antitoxin. *Molecular Microbiology, 64*(3), 738–754.

Keren, I., Minami, S., Rubin, E., & Lewis, K. (2011). Characterization and transcriptome analysis of *Mycobacterium tuberculosis* persisters. *mBio, 2*(3), e00100–e00111. http://dx. doi.org/10.1128/mBio.00100-11 e00100-11 [pii] mBio.00100-11 [pii].

Keren, I., Shah, D., Spoering, A., Kaldalu, N., & Lewis, K. (2004). Specialized persister cells and the mechanism of multidrug tolerance in *Escherichia coli*. *Journal of Bacteriology, 186* (24), 8172–8180.

Kohanski, M. A., Dwyer, D. J., Hayete, B., Lawrence, C. A., & Collins, J. J. (2007). A common mechanism of cellular death induced by bactericidal antibiotics. *Cell, 130*(5), 797–810. http://dx.doi.org/10.1016/j.cell.2007.06.049 [Research Support, U.S. Gov't, Non-P.H.S.].

Korch, S. B., & Hill, T. M. (2006). Ectopic overexpression of wild-type and mutant *hipA* genes in *Escherichia coli*: Effects on macromolecular synthesis and persister formation. *Journal of Bacteriology, 188*(11), 3826–3836.

LaFleur, M. D., Kumamoto, C. A., & Lewis, K. (2006). *Candida albicans* biofilms produce antifungal-tolerant persister cells. *Antimicrobial Agents and Chemotherapy, 50*(11), 3839–3846.

LaFleur, M. D., Qi, Q., & Lewis, K. (2010). Patients with long-term oral carriage harbor high-persister mutants of *Candida albicans*. *Antimicrobial Agents and Chemotherapy, 54* (1), 39–44.

Leid, J. G., Shirtliff, M. E., Costerton, J. W., & Stoodley, A. P. (2002). Human leukocytes adhere to, penetrate, and respond to *Staphylococcus aureus* biofilms. *Infection and Immunity, 70*(11), 6339–6345.

Levin, B. R., & Rozen, D. E. (2006). Non-inherited antibiotic resistance. *Nature Reviews. Microbiology, 4*(7), 556–562.

Lewis, K. (2001). Riddle of biofilm resistance. *Antimicrobial Agents and Chemotherapy, 45*, 999–1007.

Lewis, K., Epstein, S., D'Onofrio, A., & Ling, L. L. (2010). Uncultured microorganisms as a source of secondary metabolites. *The Journal of Antibiotics, 63*(8), 468–476. http://dx.doi.org/10.1038/ja.2010.87 [Review].

Maisonneuve, E., Shakespeare, L. J., Jorgensen, M. G., & Gerdes, K. (2011). Bacterial persistence by RNA endonucleases. *Proceedings of the National Academy of Sciences of the United States of America, 108*(32), 13206–13211. http://dx.doi.org/10.1073/pnas.1100186108 [Research Support, Non-U.S. Gov't].

McKenzie, G. J., Magner, D. B., Lee, P. L., & Rosenberg, S. M. (2003). The dinB operon and spontaneous mutation in *Escherichia coli*. *Journal of Bacteriology, 185*(13), 3972–3977 [Research Support, Non-U.S. Gov't Research Support, U.S. Gov't, Non-P.H.S. Research Support, U.S. Gov't, P.H.S.].

Mendoza, M. T., Gonzaga, A. J., Roa, C., Velmonte, M. A., Jorge, M., Montoya, J. C., & Ang, C. F. (1997). Nature of drug resistance and predictors of multidrug-resistant tuberculosis among patients seen at the Philippine General Hospital, Manila, Philippines. *The International Journal of Tuberculosis and Lung Disease: The Official Journal of the International Union Against Tuberculosis and Lung Disease, 1*(1), 59–63 [Clinical Trial Research Support, Non-U.S. Gov't].

Mitchison, D. A. (2000). Role of individual drugs in the chemotherapy of tuberculosis. *The International Journal of Tuberculosis and Lung Disease, 4*(9), 796–806.

Motiejunaite, R., Armalyte, J., Markuckas, A., & Suziedeliene, E. (2007). Escherichia coli dinJ-yafQ genes act as a toxin-antitoxin module. *FEMS Microbiology Letters, 268*(1), 112–119.

Moyed, H. S., & Bertrand, K. P. (1983). *hipA*, a newly recognized gene of *Escherichia coli* K-12 that affects frequency of persistence after inhibition of murein synthesis. *Journal of Bacteriology, 155*(2), 768–775.

Mulcahy, L. R., Burns, J. L., Lory, S., & Lewis, K. (2010). Emergence of *Pseudomonas aeruginosa* strains producing high levels of persister cells in patients with cystic fibrosis. *Journal of Bacteriology, 192*(23), 6191–6199. http://dx.doi.org/10.1128/JB.01651-09 JB.01651-09 [pii].

Pandey, D. P., & Gerdes, K. (2005). Toxin-antitoxin loci are highly abundant in free-living but lost from host-associated prokaryotes. *Nucleic Acids Research, 33*(3), 966–976.

Payne, D. J., Gwynn, M. N., Holmes, D. J., & Pompliano, D. L. (2007). Drugs for bad bugs: Confronting the challenges of antibacterial discovery. *Nature Reviews. Drug Discovery, 6* (1), 29–40. http://dx.doi.org/10.1038/nrd2201 [Review].

Pedersen, K., & Gerdes, K. (1999). Multiple *hok* genes on the chromosome of *Escherichia coli. Molecular Microbiology, 32*(5), 1090–1102.

Pedersen, K., Zavialov, A. V., Pavlov, M. Y., Elf, J., Gerdes, K., & Ehrenberg, M. (2003). The bacterial toxin RelE displays codon-specific cleavage of mRNAs in the ribosomal A site. *Cell, 112*(1), 131–140.

Peterson, W. L., Fendrick, A. M., Cave, D. R., Peura, D. A., Garabedian-Ruffalo, S. M., & Laine, L. (2000). Helicobacter pylori-related disease: Guidelines for testing and treatment. *Archives of Internal Medicine, 160*(9), 1285–1291 [Research Support, Non-U.S. Gov't].

Phillips, I., Culebras, E., Moreno, F., & Baquero, F. (1987). Induction of the SOS response by new 4-quinolones. *The Journal of Antimicrobial Chemotherapy, 20*(5), 631–638.

Ramage, H. R., Connolly, L. E., & Cox, J. S. (2009). Comprehensive functional analysis of *Mycobacterium tuberculosis* toxin-antitoxin systems: Implications for pathogenesis, stress responses, and evolution. *PLoS Genetics, 5*(12), e1000767. http://dx.doi.org/10.1371/journal.pgen.1000767 [Research Support, N.I.H., Extramural Research Support, Non-U.S. Gov't].

Ruslami, R., Nijland, H. M., Alisjahbana, B., Parwati, I., van Crevel, R., & Aarnoutse, R. E. (2007). Pharmacokinetics and tolerability of a higher rifampin dose versus the standard dose in pulmonary tuberculosis patients. *Antimicrobial Agents and Chemotherapy, 51*(7), 2546–2551. http://dx.doi.org/10.1128/AAC.01550-06 AAC.01550-06 [pii].

Sahl, H. G., & Bierbaum, G. (1998). Lantibiotics: Biosynthesis and biological activities of uniquely modified peptides from gram-positive bacteria. *Annual Review of Microbiology, 52*, 41–79. http://dx.doi.org/10.1146/annurev.micro.52.1.41 [Research Support, Non-U.S. Gov't Review].

Schmelzle, T., & Hall, M. N. (2000). TOR, a central controller of cell growth. *Cell, 103*(2), 253–262.

Schumacher, M. A., Piro, K. M., Xu, W., Hansen, S., Lewis, K., & Brennan, R. G. (2009). Molecular mechanisms of *HipA*-mediated multidrug tolerance and its neutralization by *HipB. Science, 323*(5912), 396–401.

Shah, D., Zhang, Z., Khodursky, A., Kaldalu, N., Kurg, K., & Lewis, K. (2006). Persisters: A distinct physiological state of *E. coli. BMC Microbiology, 6*(1), 53–61.

Singletary, L. A., Gibson, J. L., Tanner, E. J., McKenzie, G. J., Lee, P. L., Gonzalez, C., & Rosenberg, S. M. (2009). An SOS-regulated type 2 toxin-antitoxin system. *Journal of Bacteriology, 191*(24), 7456–7465. http://dx.doi.org/10.1128/JB.00963-09 [Research Support, N.I.H., Extramural].

Smith, E. E., Buckley, D. G., Wu, Z., Saenphimmachak, C., Hoffman, L. R., D'Argenio, D. A., & Olson, M. V. (2006). Genetic adaptation by *Pseudomonas aeruginosa* to the airways of cystic fibrosis patients. *Proceedings of the National Academy of Sciences of the United States of America, 103*(22), 8487–8492.

Spoering, A. L., & Lewis, K. (2001). Biofilms and planktonic cells of *Pseudomonas aeruginosa* have similar resistance to killing by antimicrobials. *Journal of Bacteriology, 183*(23), 6746–6751.

Spoering, A. L., Vulic, M., & Lewis, K. (2006). GlpD and PlsB participate in persister cell formation in *Escherichia coli. Journal of Bacteriology, 188*(14), 5136–5144.

Stewart, P. S., & Costerton, J. W. (2001). Antibiotic resistance of bacteria in biofilms. *The Lancet, 358*(9276), 135–138.

Unoson, C., & Wagner, E. G. (2008). A small SOS-induced toxin is targeted against the inner membrane in *Escherichia coli*. *Molecular Microbiology, 70*(1), 258–270.

Vázquez-Laslop, N., Lee, H., & Neyfakh, A. A. (2006). Increased persistence in *Escherichia coli* caused by controlled expression of toxins or other unrelated proteins. *Journal of Bacteriology, 188*(10), 3494–3497. http://dx.doi.org/10.1128/JB.188.10.3494-3497.2006.

Vogel, J., Argaman, L., Wagner, E. G., & Altuvia, S. (2004). The small RNA IstR inhibits synthesis of an SOS-induced toxic peptide. *Current Biology: CB, 14*(24), 2271–2276. http://dx.doi.org/10.1016/j.cub.2004.12.003 [Comparative Study Research Support, Non-U.S. Gov't Research Support, U.S. Gov't, Non-P.H.S.].

Vuong, C., Voyich, J. M., Fischer, E. R., Braughton, K. R., Whitney, A. R., DeLeo, F. R., & Otto, M. (2004). Polysaccharide intercellular adhesin (PIA) protects *Staphylococcus epidermidis* against major components of the human innate immune system. *Cellular Microbiology, 6*(3), 269–275.

Waksman, S. A., & Woodruff, H. B. (1940). The soil as a source of microorganisms antagonistic to disease-producing bacteria. *Journal of Bacteriology, 40*(4), 581–600.

Walters, M. C., 3rd, Roe, F., Bugnicourt, A., Franklin, M. J., & Stewart, P. S. (2003). Contributions of antibiotic penetration, oxygen limitation, and low metabolic activity to tolerance of Pseudomonas aeruginosa biofilms to ciprofloxacin and tobramycin. *Antimicrobial Agents and Chemotherapy, 47*(1), 317–323.

Wolfson, J. S., Hooper, D. C., McHugh, G. L., Bozza, M. A., & Swartz, M. N. (1990). Mutants of *Escherichia coli* K-12 exhibiting reduced killing by both quinolone and beta-lactam antimicrobial agents. *Antimicrobial Agents and Chemotherapy, 34*(10), 1938–1943.

Zasloff, M. (2002). Antimicrobial peptides in health and disease. *The New England Journal of Medicine, 347*(15), 1199–1200. http://dx.doi.org/10.1056/NEJMe020106 [Comment Editorial].

# AUTHOR INDEX

Note: Page numbers followed by "*f*" indicate figures, and "*t*" indicate tables.

# SUBJECT INDEX

Note: Page numbers followed by "*f*" indicate figures, and "*t*" indicate tables.

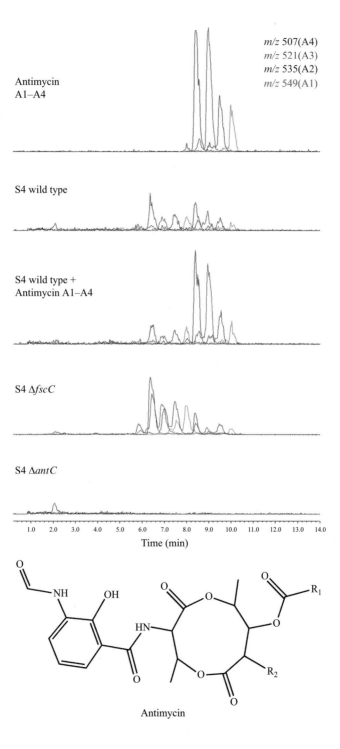

Antimycin
A1–A4

*m/z* 507(A4)
*m/z* 521(A3)
*m/z* 535(A2)
*m/z* 549(A1)

S4 wild type

S4 wild type +
Antimycin A1–A4

S4 Δ*fscC*

S4 Δ*antC*

1.0  2.0  3.0  4.0  5.0  6.0  7.0  8.0  9.0  10.0  11.0  12.0  13.0  14.0

Time (min)

Antimycin

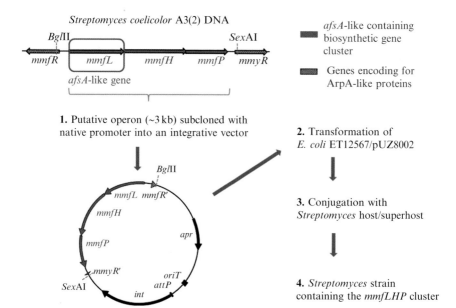

**John D. Sidda and Christophe Corre, Figure 4.3** Heterologous expression of the minicluster of genes *mmfL, H, P* that direct the biosynthesis of the methylenomycin furans (MMFs) in *Streptomyces*.

**Ryan F. Seipke *et al.*, Figure 3.5** LC–MS analysis of *Streptomyces albus* S4 wild-type and mutant strains compared to antimycin standards. The extracted ion chromatograms for antimycins A1–A4 are shown. Eight compounds consistent with the mass of antimycin A1–A4 were produced by S4 wild type and S4 Δ*fscC*, but not by the Δ*antC* mutant. Coinjection of antimycin A1–A4 with the S4 wild-type extract demonstrated that antimycin A1–A4 have the same retention time as four of the eight compounds produced by S4 wild type. Antimycin A1: $R_1 = CH(CH_3)CH_2CH_3$, $R_2 = (CH_2)_5CH_3$. Antimycin A2: $R_1 = CH(CH_3)_2$, $R_2 = (CH_2)_5CH_3$. Antimycin A3: $R_1 = CH(CH_3)CH_2CH_3$, $R_2 = (CH_2)_3CH_3$. Antimycin A4: $R1 = CH(CH_3)_2$ $R_2 = (CH_2)_3CH_3$. Reproduced from Seipke, Barke, Brearley, et al. (2011); the UV–Visible spectra and positive mode electrospray ionization mass spectra for antimycin A1–A4 and the eight antimycin compounds produced by S4 wild type are presented there.

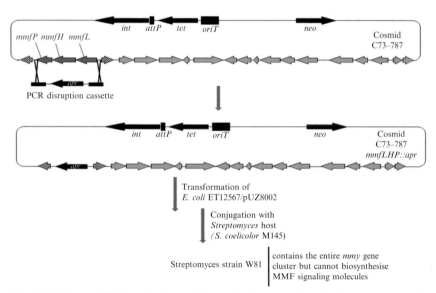

**John D. Sidda and Christophe Corre, Figure 4.4** Construction of a *Streptomyces* strain in which the methylenomycin antibiotic pathway has been silenced by disrupting the machinery that direct the biosynthesis of methylenomycin furan signaling molecules (MMFs).

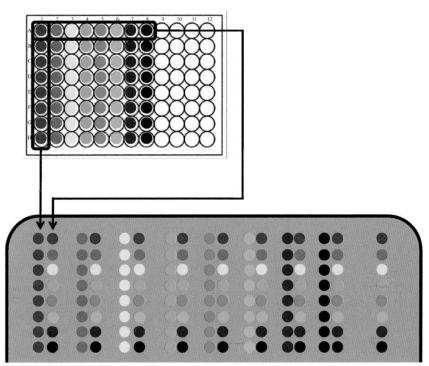

**Mohammad R. Seyedsayamdost** *et al.*, **Figure 5.1** Cartoon outline for generating screens within a group of eight bacterial strains. The strains of interest are arrayed in a 96-well plate as shown in the top panel. These are then transferred to an agar plate as shown in the bottom panel yielding an array that includes each of the 28 possible interactions twice and each strain versus itself once. A column to the right includes each strain by itself for comparison.

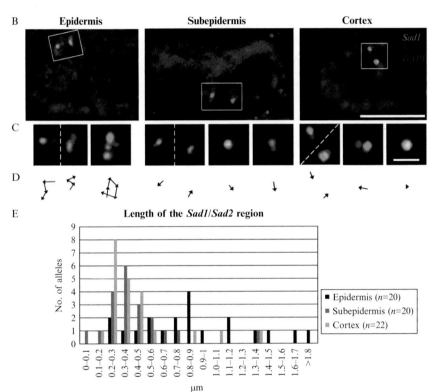

**Anne Osbourn *et al.*, Figure 6.5** Chromatin decondensation in the *Sad1/Sad2* region. (A) Diagram showing the *Sad1* (green) and *Sad2* (red) probes used for DNA *in situ* hybridization (coding regions in lighter shades of green and red). (B) The *Sad1/Sad2* region is decondensed in the nuclei of epidermal cells (left) compared with those of the subepidermis (center) and the cortex (right). In each panel, a single nucleus is shown. Bar = 5 μm. (C) Detailed views of individual gene regions. Dashed lines separate loci on two adjacent chromosomes. Bar = 1 μm. The images shown in (B) and (C) are overlays of several optical sections, section spacing 0.2 μm. Blue, chromatin stained with DAPI. (D) Line drawings showing the path length from the center of each fluorescence focus in (C) to the center of its nearest neighbor, starting from *Sad1*. (E) Length distributions of the *Sad1/Sad2* region in nuclei of epidermal, subepidermal, and cortical cells (reproduced with permission from Wegel et al., 2009).

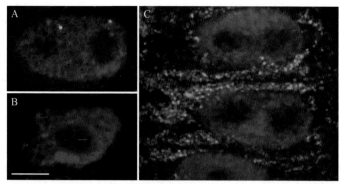

**Anne Osbourn *et al.*, Figure 6.6** Visualization of *Sad1* and *Sad2* transcripts in nuclei of oat root-tip cells. Nascent *Sad1* (red) and *Sad2* (green) transcripts; blue, chromatin stained with 4′,6-diamidino-2-phenylindole (DAPI). (A) Detection of nascent *Sad1* and *Sad2* transcripts in the nucleus of an epidermal root-tip cell. (B) In contrast to the epidermis, nuclei of the subepidermis, like the one shown, only express *Sad1*. (C) Distinctive localization patterns of *Sad1* and *Sad2* transcripts in the cytoplasm of root epidermal cells. All images are overlays of several optical sections with section spacing of 0.2 mm. Bar = 5 mm. *Adapted from Wegel et al. (2009).*

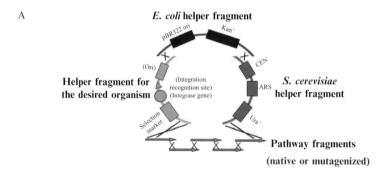

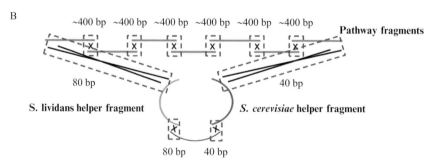

**Zengyi Shao and Huimin Zhao, Figure 10.1** (A) The DNA assembler-based strategy for efficient manipulation of natural product biosynthetic pathways. Various genetic modifications are introduced in the pathway fragments to be assembled. (B) The overlap lengths between adjacent fragments in assembly. *Reproduced by permission of The Royal Society of Chemistry.*

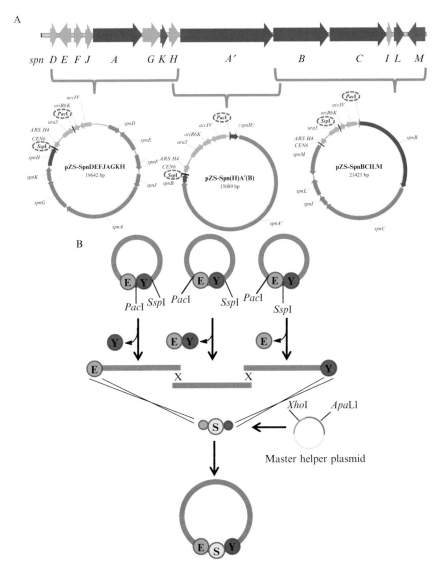

**Zengyi Shao and Huimin Zhao, Figure 10.5** The two-step strategy for assembling the spectinabilin biosynthetic pathway. (A) The first step involves construction of three intermediate plasmids, separating the four PKS genes into three plasmids. Two restriction sites, *SspI* and *PacI*, are engineered at the appropriate positions of the intermediate plasmids. (B) In the second step, restriction digestion by *SspI* and *PacI* generates three intermediate fragments, which are cotransformed with a fragment obtained from restriction digestion of the master helper plasmid by *ApaLI* and *XhoI*. *Reproduced by permission of The Royal Society of Chemistry.*

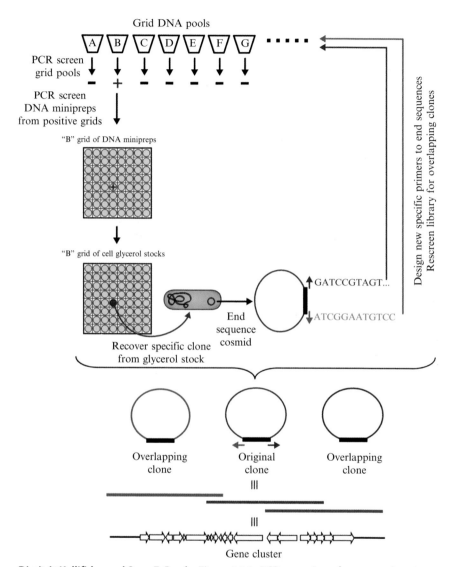

**Dimitris Kallifidas and Sean F. Brady, Figure 11.1** PCR screening of an arrayed environmental DNA library. Once an initial clone of interest is recovered, new primers are designed based on the end sequencing of this clone. These new primers are then used to rescreen the library for overlapping clones. The screening cycle is repeated until the full biosynthetic pathway is recovered.

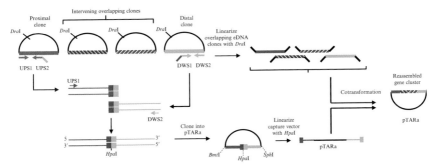

**Dimitris Kallifidas and Sean F. Brady, Figure 11.2** TAR reassembly of gene clusters capture on overlapping cosmid clones. The proximal and the distal ends of the outermost clones are amplified using upstream and downstream sets of primers (UPS1/UPS2, DWS1/DWS2), respectively. Primers, UPS2 and DWS1, are reverse complement sequences and therefore these two amplicons can be linked with a second round of PCR. The resulting amplicons, consisting of clone-specific homology arms, are ligated into pTARa capture vector. Linearized capture vector (*Hpa*I) and linearized overlapping clones (*Dra*I) are cotransformed in yeast for the recombination reaction.

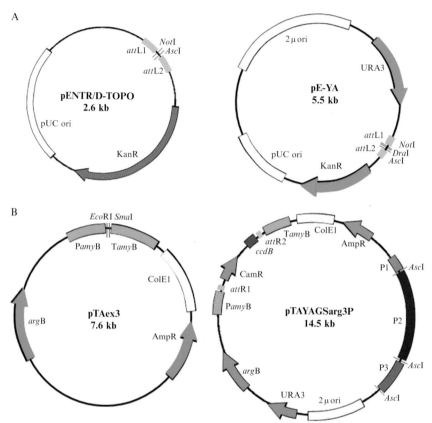

**Khomaizon A. K. Pahirulzaman *et al.*, Figure 12.1** Structures and origins of the toolbox plasmids. (A) pE-YA (right) was developed from pENTR/D-TOPO (left) as a basic gene assembly/modification vector. Coding regions constructed by homologous recombination in yeast can be transferred to expression cassettes by Gateway®-mediated site-specific recombination *in vitro*. See text for construction details. (B) pTAYAGSarg3P (right) was developed from pTAex3 (left) as a multigene expression vector. Coding regions can be placed under the control of all four promoters by homologous recombination in yeast, but the *amyB* expression cassette was designed to be filled by Gateway®-mediated site-specific recombination *in vitro*. Note that in alternative versions of this plasmid, pTAYAGSbar3P and pTAYAGSble3P, the *argB* selectable marker is replaced by P*trpC::bar* and P*trpC::ble*, respectively. See text for construction details.

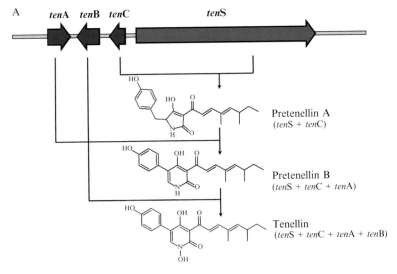

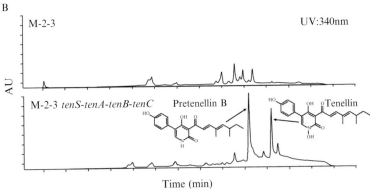

**Khomaizon A. K. Pahirulzaman *et al.*, Figure 12.2** Tenellin biosynthesis. (A) The tenellin biosynthetic pathway. Combined activities of the tenellin synthase (PKS–NRPS encoded by *ten*S) and a *trans*-acting enoyl reductase (encoded by *ten*C) produce the first pathway intermediate, pretenellin A. Ring expansion by a first cytochrome P450 (encoded by *ten*A) converts pretenellin A to pretenellin B, which is *N*-hydroxylated by a second cytochrome P450 (encoded by *ten*B) to tenellin. (B) Tenellin production in *A. oryzae* mediated by one-step transformation. The four chimaeric genes in the multigene expression plasmid pTAYAargTenellin are P*amy*B::*ten*S, P*adh*::*ten*A, P*gpd*A::*ten*C, and P*eno*::*ten*B.

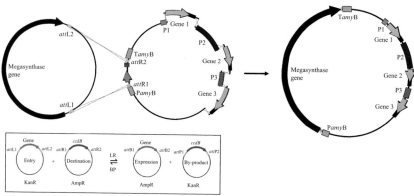

**Khomaizon A. K. Pahirulzaman *et al.*, Figure 12.4** Reconstruction of a four-gene biosynthetic pathway in a multigene expression vector. The vector (middle) is shown digested with *Asc*I to produce three fragments and ready to accept three coding regions (+UTRs) by homologous recombination in yeast. Site-specific recombination between the vector *att*R sites and the *att*L sites of an entry vector (left) is used to introduce a fourth coding region (typically of a megasynthase) into the *amy*B expression cassette. The two recombination reactions can be performed in either order to produce the fully constructed plasmid (right). The inset sketches the Gateway® reactions.

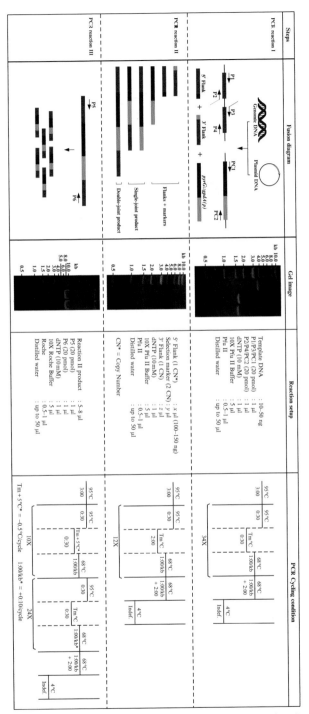

**Fang Yun Lim *et al.*, Figure 15.2** Double-joint PCR reaction setup and cycling conditions.

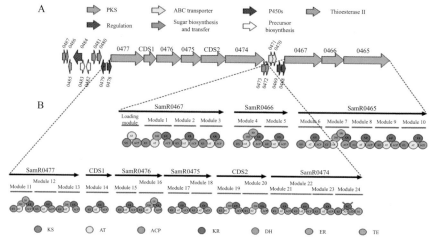

**Bertrand Aigle and Christophe Corre, Figure 17.1** Type I PKS gene cluster of *S. ambofaciens* ATCC23877 responsible for the production of stambomycins. (A) Genetic organization of the stambomycin biosynthetic gene cluster. The predicted functions of the genes are indicated above the map. The *samR0484* gene encodes the LAL regulator whose overexpression induces expression of the silent PKS genes. Limits of the cluster were defined based on BLAST analysis. (B) Module and domain organization of the PKS subunits encoded within the stambomycin cluster. This has been assigned using the SEARCHPKS program (Yadav, Gokhale, & Mohanty, 2003). The ketoreductase domain in module 24 is nonfunctional (two essential amino acids of the catalytic triad, a tyrosine residue, and an asparagine residue, are absent; Reid et al., 2003). KS, β-ketoacyl synthase; AT, acyltransferase; ACP, acyl carrier protein; KR, ketoreductase; DH, dehydratase; ER, enoyl reductase; TE, thioesterase.

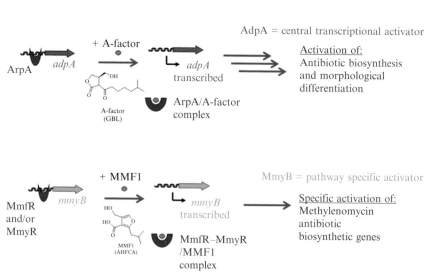

**Bertrand Aigle and Christophe Corre, Figure 17.3** Mechanism of transcriptional repression by ArpA and MmfR/MmyR in *S. griseus* and *S. coelicolor* A3(2), respectively.